Natural Fibres and their Composites

Natural Fibres and their Composites

Editor

Vincenzo Fiore

MDPI • Basel • Beijing • Wuhan • Barcelona • Belgrade • Manchester • Tokyo • Cluj • Tianjin

Editor
Vincenzo Fiore
University of Palermo
Italy

Editorial Office
MDPI
St. Alban-Anlage 66
4052 Basel, Switzerland

This is a reprint of articles from the Special Issue published online in the open access journal *Polymers* (ISSN 2073-4360) (available at: https://www.mdpi.com/journal/polymers/special_issues/Nat_Fib_Compos).

For citation purposes, cite each article independently as indicated on the article page online and as indicated below:

LastName, A.A.; LastName, B.B.; LastName, C.C. Article Title. *Journal Name* **Year**, *Volume Number*, Page Range.

ISBN 978-3-0365-0164-2 (Hbk)
ISBN 978-3-0365-0165-9 (PDF)

Cover image courtesy of Vincenzo Fiore.

© 2021 by the authors. Articles in this book are Open Access and distributed under the Creative Commons Attribution (CC BY) license, which allows users to download, copy and build upon published articles, as long as the author and publisher are properly credited, which ensures maximum dissemination and a wider impact of our publications.
The book as a whole is distributed by MDPI under the terms and conditions of the Creative Commons license CC BY-NC-ND.

Contents

About the Editor vii

Preface to "Natural Fibres and their Composites" ix

Vincenzo Fiore
Natural Fibres and Their Composites
Reprinted from: *Polymers* **2020**, *12*, 2380, doi:10.3390/polym12102380 1

Silvio Pompei, Jacopo Tirillò, Fabrizio Sarasini and Carlo Santulli
Development of Thermoplastic Starch (TPS) Including Leather Waste Fragments
Reprinted from: *Polymers* **2020**, *12*, 1811, doi:10.3390/polym12081811 5

José Herminsul Mina Hernandez, Edward Fernando Toro Perea, Katherine Caicedo Mejía and Claudia Alejandra Meneses Jacobo
Effect of Fique Fibers in the Behavior of a New Biobased Composite from Renewable Mopa-Mopa Resin
Reprinted from: *Polymers* **2020**, *12*, 1573, doi:10.3390/polym12071573 17

Rafiqah S. Ayu, Abdan Khalina, Ahmad Saffian Harmaen, Khairul Zaman, Tawakkal Isma, Qiuyun Liu, R. A. Ilyas and Ching Hao Lee
Characterization Study of Empty Fruit Bunch (EFB) Fibers Reinforcement in Poly(Butylene) Succinate (PBS)/Starch/Glycerol Composite Sheet
Reprinted from: *Polymers* **2020**, *12*, 1571, doi:10.3390/polym12071571 33

Mariana D. Stanciu, Horatiu Teodorescu Draghicescu, Florin Tamas and Ovidiu Mihai Terciu
Mechanical and Rheological Behaviour of Composites Reinforced with Natural Fibres
Reprinted from: *Polymers* **2020**, *12*, 1402, doi:10.3390/polym12061402 47

Laura Aliotta, Alessandro Vannozzi, Luca Panariello, Vito Gigante, Maria-Beatrice Coltelli and Andrea Lazzeri
Sustainable Micro and Nano Additives for Controlling the Migration of a Biobased Plasticizer from PLA-Based Flexible Films
Reprinted from: *Polymers* **2020**, *12*, 1366, doi:10.3390/polym12061366 69

Francisco Claudivan da Silva, Helena P. Felgueiras, Rasiah Ladchumananandasivam, José Ubiragi L. Mendes, Késia Karina de O. Souto Silva and Andrea Zille
Dog Wool Microparticles/Polyurethane Composite for Thermal Insulation
Reprinted from: *Polymers* **2020**, *12*, 1098, doi:10.3390/polym12051098 95

Pietro Russo, Ilaria Papa, Vito Pagliarulo and Valentina Lopresto
Polypropylene/Basalt Fabric Laminates: Flexural Properties and Impact Damage Behavior
Reprinted from: *Polymers* **2020**, *12*, 1079, doi:10.3390/polym12051079 111

Hom Nath Dhakal, Elwan Le Méner, Marc Feldner, Chulin Jiang and Zhongyi Zhang
Falling Weight Impact Damage Characterisation of Flax and Flax Basalt Vinyl Ester Hybrid Composites
Reprinted from: *Polymers* **2020**, *12*, 806, doi:10.3390/polym12040806 123

Pietro Russo, Libera Vitiello, Francesca Sbardella, Jose I. Santos, Jacopo Tirillò, Maria Paola Bracciale, Iván Rivilla and Fabrizio Sarasini
Effect of Carbon Nanostructures and Fatty Acid Treatment on the Mechanical and Thermal Performances of Flax/Polypropylene Composites
Reprinted from: *Polymers* **2020**, *12*, 438, doi:10.3390/polym12020438 141

Albert Serra, Quim Tarrés, Miquel-Àngel Chamorro, Jordi Soler, Pere Mutjé, Francesc X. Espinach and Fabiola Vilaseca
Modeling the Stiffness of Coupled and Uncoupled Recycled Cotton Fibers Reinforced Polypropylene Composites
Reprinted from: *Polymers* **2019**, *11*, 1725, doi:10.3390/polym11101725 163

Ulisses Oliveira Costa, Lucio Fabio Cassiano Nascimento, Julianna Magalhães Garcia, Sergio Neves Monteiro, Fernanda Santos da Luz, Wagner Anacleto Pinheiro and Fabio da Costa Garcia Filho
Effect of Graphene Oxide Coating on Natural Fiber Composite for Multilayered Ballistic Armor
Reprinted from: *Polymers* **2019**, *11*, 1356, doi:10.3390/polym11081356 177

Menglong Wang, Tao Hai, Zhangbin Feng, Deng-Guang Yu, Yaoyao Yang and SW Annie Bligh
The Relationships between the Working Fluids, Process Characteristics and Products from the Modified Coaxial Electrospinning of Zein
Reprinted from: *Polymers* **2019**, *11*, 1287, doi:10.3390/polym11081287 195

Anni Wang, Guijun Xian and Hui Li
Effects of Fiber Surface Grafting with Nano-Clay on the Hydrothermal Ageing Behaviors of Flax Fiber/Epoxy Composite Plates
Reprinted from: *Polymers* **2019**, *11*, 1278, doi:10.3390/polym11081278 207

Bo Wang, Erik Valentine Bachtiar, Libo Yan, Bohumil Kasal and Vincenzo Fiore
Flax, Basalt, E-Glass FRP and Their Hybrid FRP Strengthened Wood Beams: An Experimental Study
Reprinted from: *Polymers* **2019**, *11*, 1255, doi:10.3390/polym11081255 225

About the Editor

Vincenzo Fiore is an Associate Professor in Technology and Material Science at the Engineering Department of the University of Palermo. He was a Research Fellow at the University of Palermo from April 2012 to August 2020.

He graduated with honors in "Material Engineering" from the University of Messina in July 2004, and he then earned his PhD in "Economic analysis, technological innovation and management for territorial development policies" from the University of Palermo in April 2008, presenting a thesis on "Innovative technologies for the manufacturing of composite structures in the nautical field".

His research interest is focused on fiber-reinforced composite materials, with an emphasis on the following topics:

—Manufacturing and testing of fiber-reinforced composite materials;
—Extraction and characterization of new lignocellulosic fibers to be used for the reinforcement of polymeric matrices;
—Manufacturing and testing of adhesive, mechanical and mixed joints between similar and dissimilar materials (i.e., metal to composite, glass to metal, glass to composite);
—Manufacturing and testing of new eco-friendly materials with enhanced insulating properties;
—Evaluation of the aging resistance of traditional and innovative materials in hostile environments;
—Analysis of the viscoelastic behavior of metal, glass, composite structures and natural materials

He is the author/co-author of more than 50 publications in peer-reviewed journals, 2 patents, 4 book chapters and more than 30 conference presentations, seminars and invited lectures. He has supervised/co-supervised more than 60 master's theses and has more than 10 years of teaching experience.

Preface to "Natural Fibres and their Composites"

Natural fibers have several promising characteristics, such as low density and other specific properties, low price, easy processing, health advantages, renewability and recyclability. As a result, natural fibers have received increasing attention over the last several years as alternatives to their synthetic counterparts for the reinforcement of polymer-based composites.

However, it is well known that the hydrophilic nature of natural fibers renders them highly susceptible to moisture absorption and low resistance to humid and wet environments. Moreover, these fibers exhibit limited and highly variable mechanical properties, as well as weak adhesion to hydrophobic polymers. For these reasons, the use of natural fibers in industrial applications, such as those in the automotive, marine and infrastructure industries, are often limited to non-structural or semi-structural interior components.

To overcome these drawbacks, natural fibers can be pre-treated or used in combination with specific additive and/or synthetic fibers. These approaches have been widely exploited in the literature, and the resulting composites have shown a suitable balance of mechanical properties, thermal stability, aging tolerance in humid or aggressive environments, cost and environmental friendliness.

In this context, the present Special Issue comprises 14 peer-reviewed original research articles about polymer composite materials reinforced with natural fibers.

In particular, the main topics include

—Investigation of the effect of novel natural reinforcements and nano and/or micro additives on the performance of thermoplastic or thermoset-based composites;
—Use of natural fibers in hybrid composites for different applications;
—Evaluation of basalt fiber as an eco-friendly alternative to glass fiber;
—Analysis of the mechanical and rheological responses of composites reinforced with different natural fibers;
—Use of recycled natural fibers for the reinforcement of polymeric composites;
—Theoretical modeling of the mechanical behavior of natural fiber-reinforced composites.

All of the papers included in this Special Issue provide some very valuable insights into the use of natural fibers for the reinforcement of composite materials. Thus, this volume will be useful for students, designers and engineers who aim to develop a deeper understanding on this important and emerging research subject.

Vincenzo Fiore
Editor

Editorial

Natural Fibres and Their Composites

Vincenzo Fiore

Department of Engineering, University of Palermo, Viale delle Scienze, Edificio 6, 90128 Palermo, Italy; vincenzo.fiore@unipa.it

Received: 28 September 2020; Accepted: 10 October 2020; Published: 15 October 2020

Due to several promising properties, such as their low density and specific properties, low price, easy processing, health advantages, renewability and recyclability, increasing attention was paid in the last years to natural fibres as alternatives to synthetic counterparts for the reinforcement of polymeric based composites.

On the other hand, it is well known that the hydrophilic nature of natural fibres leads to high susceptibility to moisture absorption and low resistance to humid and wet environmental conditions. Moreover, these fibres evidence limited and highly variable mechanical properties as well as weak adhesion with hydrophobic polymers. For these reasons, the use of natural fibres in industrial applications such as automotive, marine and infrastructure, are often limited to non-structural or semi-structural interior components.

To overcome these drawbacks, natural fibres can be pre-tread or used together with specific additive and/or synthetic fibres. These approaches have been widely exploited in literature, and the resulting composites have shown a suitable balance of mechanical properties, thermal stability, ageing tolerance against humid or aggressive environments, cost and environment care.

In this context, the present Special Issue comprises fourteen peer-reviewed original research articles about polymer composite materials reinforced with natural fibres.

The main topic includes the investigation of the effect of novel lignocellulosic reinforcements and nano or micro additives on performances of thermoplastic or thermoset based composites.

The use of novel lignocellulosic fibres is studied in several contributions to this Special Issue. The team of Andrea Zille [1] analyzes the effect of adding dog wool fibres on the properties of polyurethane (PU)-based eco-composite foam. They show that tensile and compression strengths, hydration capacity and thermal capacity are improved whereas the foam dilatation with heating decreased with increasing the amount of dog wool microparticles, thus demonstrating the potential of this animal-derived waste for insulation applications, with a low cost and minimal environmental impact. Mina Hernandez et al. [2] develop a fully bio-based composite using a natural resin obtained from Mopa-Mopa (Elaeagia Pastoensis Mora) plant as matrix and fique fibres as reinforcement. Thanks to easy processing and good physicochemical and mechanical characteristics, it is shown that the bio-based composite can be used as wood–plastic for the replacement of plastic and/or natural wood products widely used today in several applications. Similarly, Pompei et al. [3] show that the introduction of leather fragments in a self-produced thermoplastic starch (TPS) based on starch plasticized with glycerol and cross-linked using citric acid proved to be promising. The composite biodegradability allows its possible application in products where contact with soil and progressive non-toxic degradation is required, such as it is the case for the on-site production of mulching films.

Wang et al. [4] studied the relationships between the working fluids, process characteristics and products from the modified coaxial electrospinning of zein focusing their attention on the control of the processing process during the manufacturing process as well as on the prediction and maintenance of the nanofibre quality. In particular, using an electrospinnable zein solution as the core fluid and LiCl solutions as the sheath working liquids, a series of modified coaxial electrospinning processes are performed, thus successfully preparing a number of zein nanoribbons. In a further paper [5] concerning the use of novel lignocellulosic fibres, a mixture of thermoplastic polybutylene succinate (PBS), tapioca starch, glycerol and empty fruit bunch fibre was prepared by a melt compounding method and characterized by means of mechanical, thermal and immersion tests.

The effect of the addition of nano or micro additive as filler in natural fibre reinforced composite is investigated in several papers [6–9].

In this context, the team of Andrea Lazzeri [6] prepared plasticized poly(lactic acid) (PLA)/poly(butylene succinate) (PBS) blend-based films containing chitin nanofibrils and calcium carbonate by extrusion and compression moulding methods. The diffusion coefficient experimental data shows that the adding of chitin nanofibrils can slow the plasticizer migration. However, the best result was achieved with micrometric calcium carbonate while nanometric calcium carbonate results less effective due to bio polyesters' chain scission. On the other hand, the use of both micrometric calcium carbonate and chitin nanofibrils was counterproductive due to the agglomeration phenomena that were observed. Russo et al. [7] evaluate the effects of two carbon nanostructures (graphene nanoplatelets (GNPs) and carbon nanotubes (CNTs)), of a chemical modification with a fatty acid and of maleated polypropylene, with the aim of mitigating the highly hydrophilic nature of flax fibres thus increasing their compatibility with apolar polypropylene. On the bases of the experimental data, the authors state that these simple treatments, potentially prone to further optimization, can represent a step toward producing natural fibre composites with mechanical profiles compatible with semi-structural applications. Similarly, Wang et al. [8] improved the hygrothermal resistance of flax fibre reinforced epoxy composites through grafting flax fabric with nanoclay. In more detail, the introduction of nano-clay reduces both moisture uptake and the coefficient of diffusion of composites that show better dimensional stability than the untreated ones.

In the paper by Costa et al. [9], the performance of graphene oxide coated curaua fibre reinforced epoxy composites in a multilayered armour system intended for ballistic protection is evaluated showing that this innovative composite attends the standard ballistic requirement remaining intact, differently from the non-coated curaua fibre similar composite.

Another topic addressed in this special issue deals with the use of natural fibres in hybrid composites useful for different applications. Dhakal et al. [10] present an interesting study on the evaluation of the low-velocity falling weight impact behaviour of flax-basalt vinyl ester (VE) hybrid composites, showing that the hybrid system possesses higher impact energy and peak load than flax fibre reinforced composite. Hence, the experimental results indicate that the hybridization is a promising strategy for enhancing the toughness properties of natural fibre composites. Flax, Basalt, E-Glass FRP and their hybrid combinations are used to strengthen wood beams in the paper by Wang et al. [11]. The bending tests performed on strengthened wood beams shows that all hybrid FRPs exhibit no significant enhancement in load carrying capacity but larger maximum deflection compared to the single type of FRP composite.

The last three papers belonging to this special issue deals with the investigation of flexural properties and impact damage behaviour of basalt fibre reinforced polypropylene composites [12], the mechanical and rheological behaviour of composites reinforced with different natural fibres [13] and the theoretical modelling of the stiffness of recycled cotton fibres reinforced polypropylene composites [14]. In particular, the paper by Serra et al. [14] highlight the opportunity of recovering textile cotton fibres, useless for the textile industry, to obtain composite materials with promising performances. Moreover, the use of two different micromechanics models allowed evaluating the impact of the morphology of the fibres on Young's modulus of a composite.

In summary, the papers contributed to this Special Issue give some very nice insights on the use of natural fibres as reinforcement of composite materials thus making this volume useful for students as well as for designers and engineers that would like to develop a deeper understanding on the use of this important and emerging research subject.

Funding: This research received no external funding.

Conflicts of Interest: The authors declare no conflict of interest.

References

1. Da Silva, F.C.; Felgueiras, H.P.; Ladchumananandasivam, R.; Mendes, J.U.L.; de Silva, K.K.O.S.; Zille, A. Dog Wool Microparticles/Polyurethane Composite for Thermal Insulation. *Polymers* **2020**, *12*, 1098. [CrossRef] [PubMed]
2. Hernandez, J.H.M.; Perea, E.F.T.; Mejía, K.C.; Jacobo, C.A.M. Effect of Fique Fibres in the Behavior of a New Biobased Composite from Renewable Mopa-Mopa Resin. *Polymers* **2020**, *12*, 1573. [CrossRef] [PubMed]
3. Pompei, S.; Tirillò, J.; Sarasini, F.; Santulli, C. Development of Thermoplastic Starch (TPS) Including Leather Waste Fragments. *Polymers* **2020**, *12*, 1811. [CrossRef] [PubMed]
4. Wang, M.; Hai, T.; Feng, Z.; Yu, D.-G.; Yang, Y.; Bligh, S.W.A. The Relationships between the Working Fluids, Process Characteristics and Products from the Modified Coaxial Electrospinning of Zein. *Polymers* **2019**, *11*, 1287. [CrossRef] [PubMed]
5. Ayu, R.S.; Khalina, A.; Harmaen, A.S.; Zaman, K.; Isma, T.; Liu, Q.; Ilyas, R.A.; Lee, C.H. Characterization Study of Empty Fruit Bunch (EFB) Fibres Reinforcement in Poly(Butylene) Succinate (PBS)/Starch/Glycerol Composite Sheet. *Polymers* **2020**, *12*, 1571. [CrossRef] [PubMed]
6. Aliotta, L.; Vannozzi, A.; Panariello, L.; Gigante, V.; Coltelli, M.-B.; Lazzeri, A. Sustainable Micro and Nano Additives for Controlling the Migration of a Biobased Plasticizer from PLA-Based Flexible Films. *Polymers* **2020**, *12*, 1366. [CrossRef] [PubMed]
7. Russo, P.; Vitiello, L.; Sbardella, F.; Santos, J.I.; Tirillò, J.; Bracciale, M.P.; Rivilla, I.; Sarasini, F. Effect of Carbon Nanostructures and Fatty Acid Treatment on the Mechanical and Thermal Performances of Flax/Polypropylene Composites. *Polymers* **2020**, *12*, 438. [CrossRef] [PubMed]
8. Wang, A.; Xian, G.; Li, H. Effects of Fiber Surface Grafting with Nano-Clay on the Hydrothermal Ageing Behaviors of Flax Fiber/Epoxy Composite Plates. *Polymers* **2019**, *11*, 1278. [CrossRef] [PubMed]
9. Costa, U.O.; Nascimento, L.F.C.; Garcia, J.M.; Monteiro, S.N.; Luz, F.S.; Pinheiro, W.A.; Filho, F.C.G. Effect of Graphene Oxide Coating on Natural Fibre Composite for Multilayered Ballistic Armor. *Polymers* **2019**, *11*, 1356. [CrossRef] [PubMed]
10. Dhakal, H.; Le Méner, E.; Feldner, M.; Jiang, C.; Zhang, Z. Falling Weight Impact Damage Characterisation of Flax and Flax Basalt Vinyl Ester Hybrid Composites. *Polymers* **2020**, *12*, 806. [CrossRef] [PubMed]
11. Wang, B.; Bachtiar, E.V.; Yan, L.; Kasal, B.; Fiore, V. Flax, Basalt, E-Glass FRP and Their Hybrid FRP Strengthened Wood Beams: An Experimental Study. *Polymers* **2019**, *11*, 1255. [CrossRef] [PubMed]
12. Russo, P.; Papa, I.; Pagliarulo, V.; LoPresto, V. Polypropylene/Basalt Fabric Laminates: Flexural Properties and Impact Damage Behavior. *Polymers* **2020**, *12*, 1079. [CrossRef] [PubMed]
13. Stanciu, M.D.; Teodorescu, H.D.; Tamas, F.; Terciu, O.M. Mechanical and Rheological Behaviour of Composites Reinforced with Natural Fibres. *Polymers* **2020**, *12*, 1402. [CrossRef] [PubMed]
14. Serra, A.; Tarrés, Q.; Chamorro, M.À.; Soler, J.; Mutjé, P.; Espinach, F.X.; Vilaseca, F. Modeling the Stiffness of Coupled and Uncoupled Recycled Cotton Fibers Reinforced Polypropylene Composites. *Polymers* **2019**, *11*, 1725. [CrossRef] [PubMed]

Publisher's Note: MDPI stays neutral with regard to jurisdictional claims in published maps and institutional affiliations.

© 2020 by the author. Licensee MDPI, Basel, Switzerland. This article is an open access article distributed under the terms and conditions of the Creative Commons Attribution (CC BY) license (http://creativecommons.org/licenses/by/4.0/).

Article

Development of Thermoplastic Starch (TPS) Including Leather Waste Fragments

Silvio Pompei [1], Jacopo Tirillò [2], Fabrizio Sarasini [2] and Carlo Santulli [3,*]

[1] School of Architecture and Design, Università di Camerino, viale della Rimembranza, 63100 Ascoli Piceno, Italy; silvio.pompei@studenti.unicam.it
[2] Department of Chemical Engineering Materials Environment and UdR INSTM, Sapienza Università di Roma, Via Eudossiana 18, 00184 Roma, Italy; jacopo.tirillo@uniroma1.it (J.T.); fabrizio.sarasini@uniroma1.it (F.S.)
[3] School of Sciences and Technologies, Università di Camerino, via Gentile III da Varano 7, 62032 Camerino, Italy
* Correspondence: carlo.santulli@unicam.it; Tel.: +39-380-652-2232

Received: 6 May 2020; Accepted: 11 August 2020; Published: 12 August 2020

Abstract: A thermoplastic starch (TPS) material is developed, based on corn starch plasticized with glycerol and citric acid in a 9:3:1 ratio and further bonded with isinglass and mono- and diglycerides of fatty acids (E471). In TPS, leather fragments, in the amount of 7.5 15 or 22.5 g/100 g of dry matter, were also introduced. The mixture was heated at a maximum temperature of 80 °C, then cast in an open mold to obtain films with thickness in the range 300 ± 50 microns. The leather fragments used were based on collagen obtained from production waste from shoemaking and tanned with tannins obtained from smoketree (*Rhus cotinus*), therefore free from chromium. Thermogravimetric (TGA) tests suggested that material degradation started at a temperature around 285 °C, revealing that the presence of leather fragments did not influence the occurrence of this process in TPS. Tensile tests indicated an increase in tensile properties (strength and Young's modulus) with increasing leather content, albeit coupled, especially at 22.5 wt%, with a more pronounced brittle behavior. Leather waste provided a sound interface with the bulk of the composite, as observed under scanning electron microscopy. The production process indicated a very limited degradation of the material after exposure to UV radiation for eight days, as demonstrated by the slight attenuation of amide I (collagen) and polysaccharide FTIR peaks. Reheating at 80 °C resulted in a weight loss not exceeding 3%.

Keywords: leather waste; thermoplastic starch; mechanical characterization; thermal characterization

1. Introduction

Leather is obtained from animal hides, which are based on collagen, a protein molecule constituted by sequential chains of amino acids, twisted and bound in a strong molecular structure made of fibers. The process of leather fabrication does require its conservation for manipulation through tanning, which involves the fixation of tannins to the collagen matrix. The effect of tannins is enabling preservation, making leather imputrescible, therefore treatable with chemicals, while improving its hardness and strength. If trivalent chromium salts are used for this purpose, leather waste is classified as a special refuse, which requires specific methods for handling to prevent chromium from penetrating into the soil or underground water [1]. To improve leather sustainability and ease its disposal, vegetable tannins can be employed, such as those obtained from smoketree (*Rhus cotinus*). This was a traditional use in some regions, such as Marche and Tuscany, of Central Italy, and has currently a reprise, other than on sustainability grounds, also in connection with the revived use of natural colors, of some of which tannin is an essential element [2]. In this way, the structural and functional advantages of chromium tanning are preserved, while, on the other side, leather waste can be disposed of e.g., in other biodegradable materials [3]. The waste from leather treated with vegetable tanning proved to be microporous and functional for sustainable use, as it was proposed after chemical activation with alkali, for the removal of volatile organic compounds (VOCs), such as toluene [4].

Starting from these considerations, the waste from vegetable-tanned leather could represent a candidate for possible inclusion into biopolymers such as thermoplastic starches (TPS) based on starch-glycerol. This procedure demonstrated to be suitable for the introduction of fillers in powder form, such as clay, even with limited control of their dimensions [5], or garden waste, such as the one from Opuntia, in fibrous form [6]. In general terms, this waste could be introduced in the matrix without any further treatment, although obviously with quite limited performance. As a consequence, these were able to effectively include waste, mainly from the agricultural food and non-food sector [7,8]. The aforementioned limited performance, in the case of film fabrication, can derive also from the difficult control of flowability: in this sense, citric acid, a cheap and easily available chemical, can provide some better compatibility with starch, therefore easing film processing [9]. The limitation of TPS is given by their glass transition temperature T_g, which is usually around 80 °C, yet strongly dependent on the amount of water added. Above T_g, thermoplastic starch loses its mechanical properties, and considerably swells in an irreversible way [10]. This phenomenon requires putting some attention into the creation of a suitable mixture for inclusion of other materials, especially when, as it is the case with films, the dimensional tolerance is very limited. The main waste constituting leather fragment is collagen, which has been added with starch in applications linked to food preservation (e.g., sausage casing): the addition of not exceedingly large amounts of starch (less than 50 wt%) to collagen was demonstrated as being effective for enhancing film strength and improving durability [11].

This study aims particularly to disclose the possibility of re-using environmentally friendly waste material, such as vegetable-tanned leather, which is currently disposed in general waste, because the limited and local characteristics of the productive system do not suggest transportation for recycling as an economically viable option. The application proposed involves its inclusion in self-developed thermoplastic starch (TPS), with the idea to optimize the composition of TPS, in view of its compatibility with waste, to be used as filler. The biodegradability of the composite obtained allows its possible application in products where contact with soil and progressive non-toxic degradation is required, such as it is the case for the on-site production of mulching films.

2. Materials and Methods

2.1. Material Development

2.1.1. Raw Materials Used

The mixture used for the production of the film included commercial corn starch for alimentary use, glycerol of 99.5% purity (E422), citric acid from lemon juice and isinglass to ease bonding, in the respective proportions 9:3:1:0.7. Moreover, 0.15 parts of mono- and diglycerides of fatty acids (E471) were added to prevent the formation of mold and to ease the manufacturing of a flat film of material during laying up. In this way, a thermoplastic starch (TPS) has been obtained. Smoketree-tanned leather in irregular fragments having their largest dimension between 10 and 500 microns approximately, was added to the material in the proportions of 7.5, 15 and 22.5 wt% over the dry weight of the mixture. The three composites obtained were defined ex-post as TPS_ leather 1/2, TPS_leather 1 and TPS_leather 3/2, considering 1 as the ideal proportion for the production of the final material, which requires some deformation to be retained in the flat sheet obtained. These proportions were set after a large number of experiments carried out starting from October 2015 in the Course of "Sperimentazione di Materiali Innovativi per il Design" (Experimentation on Innovative Materials for Design) in University of Camerino and completed during 2018 with the experimental aims that are illustrated in Figure 1: the last image of Figure 1 shows the material developed, which is here reported as "TPS_Leather 1", further developed in sheets with dimensions 300 × 200 mm. The contents of the different raw materials in percentage over dry weight of the film are reported in Table 1. The three composites are compared between them, and to the pure TPS developed with the aforementioned proportions.

Figure 1. Experimental process leading to the development of the film.

Table 1. Different components of the film.

Component	Content over Dry Weight (%)
Corn starch	55.2
Glycerol	18.5
Citric acid	6.2
Isinglass	4.3
E471	0.8
Leather fragments	7.5 (leather 1/2), 15 (leather 1) or 22.5 (leather 3/2)

2.1.2. Film Production

Film production was obtained by adding the mixture with some amount of water to allow obtaining a sufficient fluidity, then heating it at a temperature not exceeding 80 °C, reached by its continuous mixing in an uncovered container in 10 min. The mixture was then cast on a silicon plate, 1 mm thick, with the assistance of a steel lamina, again 1 mm thick, to level it inside a steel frame with a thickness of 1.5 mm. The frame is removed just after the cast process and the film is left cooling naturally on the silicon sheet. The phases of the production process are reported in Figure 2.

Figure 2. The different phases of the film casting.

The natural process of film drying, which typically lasted 5 days, made it possible to finally obtain a film of rather constant thickness, in the order of 300 ± 50 microns. From Figure 3, it is possible to observe the interfacial adhesion of leather fragments to the substrate, despite the irregularity of the filler. The films, cut into rectangular strips with maximum dimensions around 150 × 100 mm, are supposed to be coupled in the way reported in Figure 4, therefore partially superimposed to each other. They would serve as the support for texturized lawn modules, able to support small seeds for the growth of plants, therefore, the requirement is more on the effective integration of leather filler than in the creation of larger pieces.

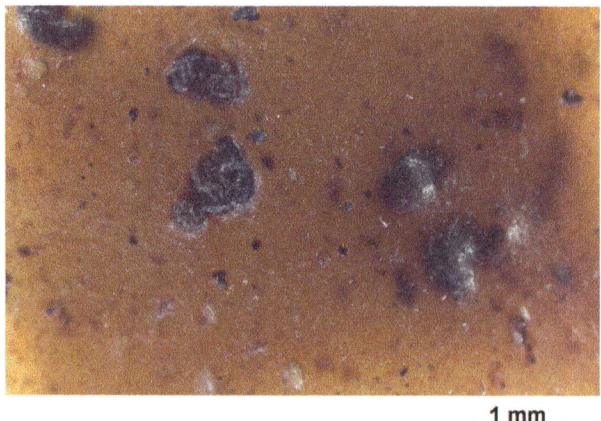

Figure 3. Macrograph of the material with fragments of leather waste.

Figure 4. Coupling of the different films strips with lawn seeds prepared for application.

2.2. Experimental Methods

2.2.1. Ageing Tests

Ageing tests were carried out by exposing materials to a gallium-doped Helios (Helios Quartz Group SA, Novazzano, Switzerland) medium pressure mercury lamp, model HMPL, which has emission peaks in the region between 400 and 430 nm, therefore in the IR range, yet it also has considerable emissions in the UV range between 200 and 400 nm. The light exposition was continuative for a period of 192 h. FTIR analysis was carried out using a Perkin Elmer (Milan, Italy) Spectrometer 100 in attenuated total reflection (ATR) mode. A spectral resolution of 3 cm^{-1} in the range (4000–600 cm^{-1}) with 512 scans was adopted to record the spectra. Two spectra were carried out, to compare those obtained from new and aged materials.

Tests with Radwag (Radom, Poland) MA 110.R thermobalance involved a program of heating with 5 min at 40 °C, 5 min at 60 °C and 8 min at 80 °C, therefore for a total duration of 18 min.

Heating was applied on square samples of 20 mm side, removed by cutting from the rectangles of materials, on which mass was measured every 10 s with an accuracy of ±1 mg.

2.2.2. Mechanical Characterization of Films

Specimens for the mechanical characterization were cut from the films in accordance with UNI EN ISO 527-2 (Type 1BA samples with a gauge length of 30 mm). Tensile tests were performed at room temperature in displacement control with a crosshead speed of 2.5 mm/min by using a Zwick/Roell Z010 (Ulm, Germany) universal testing machine. The results are the average of five tests.

2.2.3. Thermal Characterization of Films

The thermal stability of the films was investigated by thermogravimetric analysis (TGA). To this purpose, a SETSYS Evolution system by Setaram (Caluire, France) was used, heating the samples from 25 °C to 800 °C, with a heating rate of 10 °C/min in a nitrogen atmosphere.

2.2.4. Morphological Characterization by SEM

The fracture morphology of samples failed in tension was investigated by field emission scanning electron microscopy (Mira3 by Tescan, Brno, Czech Republic). Specimens were sputter coated with gold prior to analysis.

3. Results and Discussion

The tensile test results, reported in Table 2, indicated in general a limited variation in terms of stress, strain and stiffness, with respect to other typical materials including organic waste. A comparison was also possible with similar thermoplastic starches with no fillers e.g., those examined in [12]. This reference suggests that the plasticization effect is in the present study much more contained, as observed by the elongation at break for the pure TPS, which just exceeds 50%, making the material quite controllable and with predictable properties during use. In particular, a considerably lower deformation was measured with respect to what was observed in other studies, such as [13], on glycerol and citric acid on corn starch tensile properties.

Table 2. Summary of tensile test results.

	σ (MPa)	E (MPa)	ε (%)
TPS_neat	2.30 ± 0.02	52.17 ± 7.54	58.60 ± 2.70
TPS_Leather 1/2	2.65 ± 0.09	71.24 ± 6.55	45.30 ± 4.30
TPS_Leather 1	4.75 ± 0.21	200.18 ± 8.04	10.05 ± 2.26
TPS_Leather 3/2	7.27 ± 0.28	577.40 ± 18.79	2.50 ± 0.50

The mechanical properties of the developed films compare quite favorably with those reported in other studies, especially for what concerns the Young's modulus. In particular, in [14], for materials based on cornstarch and glycerol/water as plasticizer reinforced with bacterial and vegetable cellulose (1% and 5% w/w), a strength in the range 0.5–3.5 MPa and a Young's modulus in the range 1–20 MPa was reported, while in [15,16], for thermoplastic starch/clay nanocomposites, strength spanned the range from 2 to 28.1 MPa and the modulus from 7.5 to 196 MPa. As far as the introduction of leather fragments is concerned, the effect of 7.5 wt% over 100 parts of TPS (TPS_Leather 1/2) appears limited on the modification of TPS properties, whereas the material results considerably stronger and stiffer when higher amounts of leather fragments are used. However, as inferred from the typical tensile curves in Figure 5, the introduction of 22.5 wt% of leather fragments resulted in a material failing basically with no plastic behavior, which limits its applications, therefore the intermediate solution of 15 wt% of leather fragments appeared the most suitable for the use proposed.

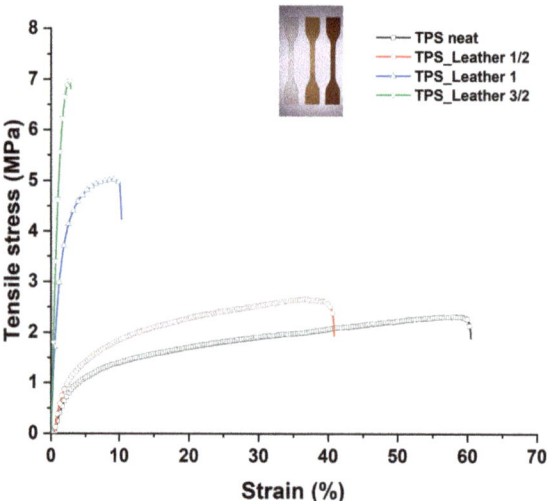

Figure 5. Typical tensile stress-strain curves.

The onset of material degradation started around 285 °C, as determined from the graphical method suggested in [17], for corn starch-glycerol materials, and indicates a typical value for thermoplastic starches. This can suggest that the effect of leather fragments on material degradation is likely to be quite limited. However, the higher amounts introduced offered some reduction of the degradation rate in the region around 250–300 °C, as observable specifically from the inset in Figure 6b. Studies on collagen indicate water loss taking place up to around 125 °C, while the degradation onset of leather is around 300–320 °C [18]. In this case, it is suggested that water loss occurs seamless with desorption of water as the effect of softening of thermoplastic starch: as a matter of fact, the trend appears to be linear up to over slightly 200 °C. Regarding the residual mass at the end of thermal degradation process, a previous study on acrylonitrile-butadiene-styrene (ABS) resin—leather waste composites suggested that leather powder alone was leaving just below 20% of the initial mass at 800 °C [19]. The data found here are basically in line with this indication, suggesting that the presence of leather increases the amount of material normally left after the degradation of thermoplastic starches obtained using corn starch with similar amounts of glycerol [20].

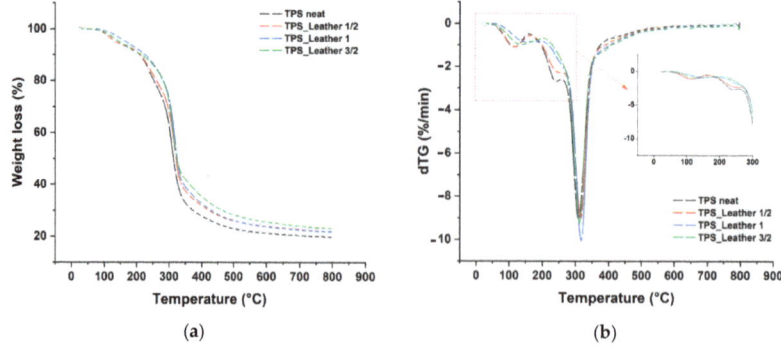

Figure 6. Thermogravimetric curve (TGA) (**a**) and first derivative of the thermogravimetric curve (DTG) (**b**) of thermoplastic starch (TPS) and the different composite materials.

The microscopic characterization of the fracture surfaces indicates for pure TPS a variable fragmentation with loss of material and creation of voids, which is attributable to plasticization effect (Figure 7). In contrast, in the case where leather fragments are introduced, the occurrence of some step-like regions is detected (Figure 8a), which is often encountered in the presence of protein-based structures and likely to be related to the internal layered structure of leather particles [21]. Some pulled-out particles were also observed, as shown in Figure 8b, which suggests the need to improve the interfacial adhesion, because the subsequent detachment of few leather fragments from the matrix would result in the coalescence of voids and sudden failure. On the other side, in most cases the interface proves effective, such as in Figure 9, though its performance might be variable, due to the random distribution of waste particles: in practice, with a lower amount of leather fragments, their boundaries are more recognizable in the TPS matrix (Figure 9a,b), whereas, with their increasing amount, the deformation of the matrix due to the insertion of the filler and the subsequent loading would rather conceal them (Figure 9c). Looking in more detail at the structure of the TPS matrix, some typical situations are recognized during loading, in particular differential deformation, leading in the most critical cases to the widespread formation of cracks (Figure 10). However, it is promising that, in most cases, leather fragments have been shown to provide a strong adhesion to the matrix, although the margins for improvement are surely to be recognized.

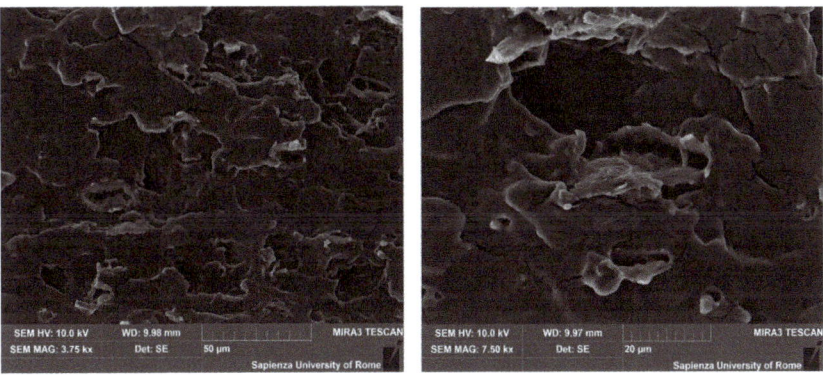

Figure 7. SEM images of the pure TPS fracture surface.

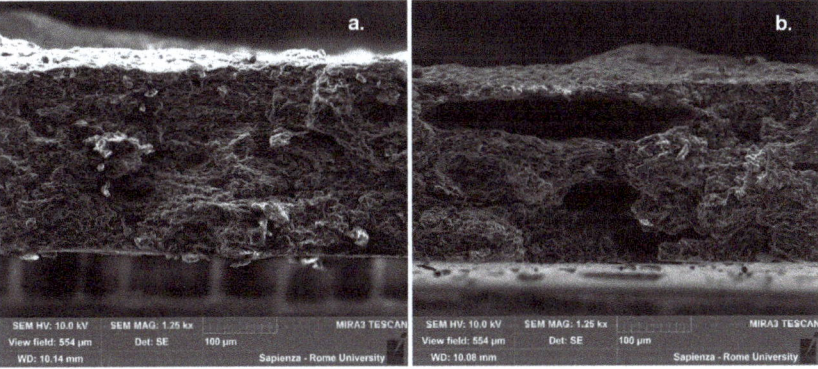

Figure 8. Fracture surface of the composite, without (**a**) and with (**b**) pulled-out leather particles at failure.

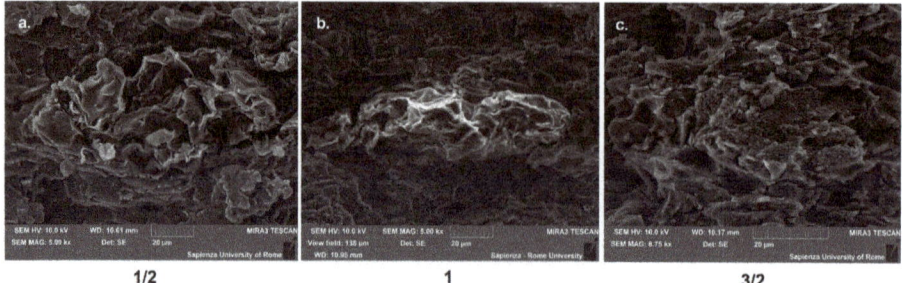

Figure 9. Leather particles embedded in the TPS matrix, 1/2 (**a**), 1 (**b**) 3/2 (**c**).

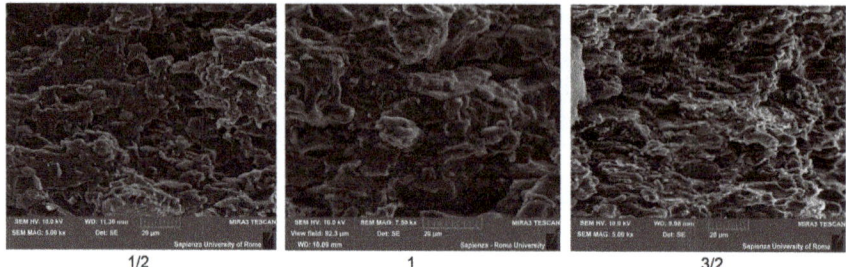

Figure 10. Occurrence of differential deformation in the various composites.

In general terms, the material proved suitable for the application envisaged, despite the fact that further mechanical fragmentation of leather waste e.g., by milling, had been purposely avoided. As expected, the dimensional variation of the particles resulted in some further limitation of the plasticization effect provided by the corn starch-glycerol-citric acid interaction, though reduced already by isinglass, as common in TPS [22]. It needs to be considered in any case that the mechanical properties obtained and the thermal stability offered can be considered acceptable: further refinements of the production process and of the film drying uniformity, leading to a stricter dimensional tolerance, would follow in future investigations.

Regarding the material that proved to be the most suitable for the envisaged application, hence TPS_Leather 1, further characterization was carried out. In particular, a number of peaks were observed from FTIR analysis, whose attribution is reported in Table 3, mainly from the comprehensive and reference study of FTIR spectra on starch, carried out in [23]. Some of these are also corroborated by other information offered in the respective references, and also the peaks referring to the other non-polysaccharide components in the materials, hence glycerol, citric acid and collagen from leather are attributed according to the other references quoted in Table 3. The investigation did not allow the attribution of FTIR peaks to vegetable tannins, despite some studies on this also being available, such as [24], probably due to their amount of leather wastes being too low.

Moreover, FTIR analysis suggested that peaks correlated to polysaccharides (starch), to interaction between starch and glycerol, in particular the 1104 cm^{-1}, to amide I (collagen) and amino-acid, namely the 998 cm^{-1}, only showed, as from Figure 11, a slight degradation after ageing for eight days to simulated exposure to the sunlight. Post-ageing, a re-heating at 80 °C of the material did lead to a water loss not exceeding 3%, as indicated in Figure 12, which demonstrates that most water was linked to the polymeric structure, which could suggest the material is sufficiently strong and stable during storage and application, possibly with limited shrinkage. In fact, coming back to thermogravimetric tests reported in Figure 6, they also suggested a very limited weight loss up to around 80 °C, indicating that the components are hydrated in a stable form.

Table 3. Attribution of FTIR peaks observed during the scanning.

Peak (cm^{-1})	Attribution
3286–3291	Hydrogen bonds given by hydrolyzing
2925–2928	C–H stretching mode of starch [25,26]
1712–1717	C=O stretching vibration in carboxyl groups due to citric acid [27]
1649–1654	Amide I (collagen) [28]
1335–1336	Polysaccharides C–OH bending [28]
1150–1155	Polysaccharides (starch) C–H bending [29]
1104	C–O stretching vibration peak of glycerol [30]
1074–1076	C–O–H stretching vibration [30]
998	Out-of-plane OH-vibrations in carboxyl (amino-acid) [31]

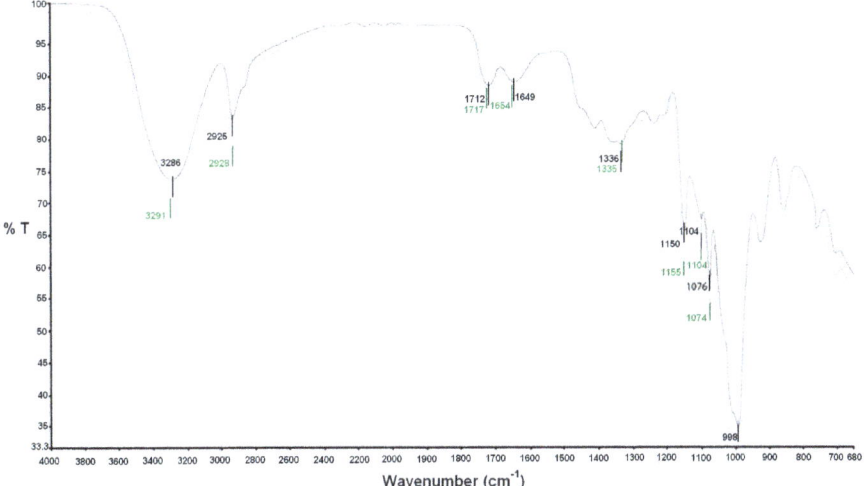

Figure 11. FTIR spectra carried out on the material newly produced (darker curve) and then subjected to UV ageing (lighter curve).

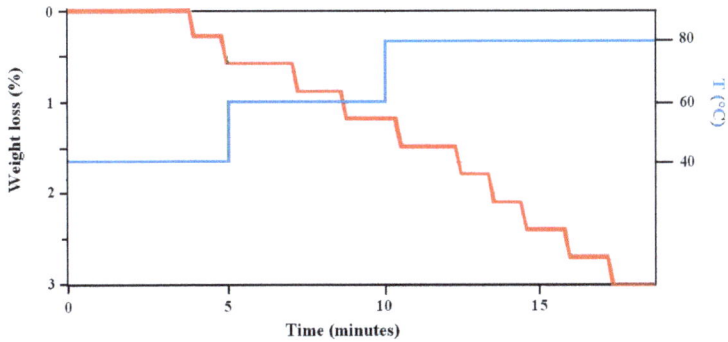

Figure 12. Weight loss of the material during heating up to 80 °C.

4. Conclusions

The introduction of leather fragments in three different quantities in a self-produced thermoplastic starch (TPS) based on starch plasticized with glycerol and cross-linked using citric acid proved to be promising. This can be regarded as a sustainable procedure for introducing in soil a biodegradable composite e.g., for lawn growth support, since leather is chrome-free, being smoketree-tanned. The material with intermediate amount of leather fragments did not suffer any significant degradation below 80 °C, which makes it basically acceptable for the application envisaged.

The positive aspects are also in the formation of a sound matrix-filler interface, though with possible improvements based on a more accurate selection of particle distribution. As expected, collagen-based reinforcement does result in hindering the plasticization of TPS, whereas it also provides more controllable mechanical properties. As far as thermal properties are concerned, the introduction of leather fragments does not change to a significant extent the degradation patterns of TPS, which take place just below 300 °C. The amount of leather waste introduced could be possibly increased by working on particle size with more accurate fragmentation methods, while, in the present work, only basic cutting operations have been performed.

Author Contributions: Conceptualization, S.P., F.S., and C.S.; methodology, S.P., J.T., and F.S.; measurements, S.P., J.T., and F.S.; investigation, S.P. and C.S.; writing—original draft preparation, S.P. and C.S.; writing—review and editing, F.S. All authors have read and agreed to the published version of the manuscript.

Funding: This research received no external funding.

Conflicts of Interest: The authors declare no conflict of interest.

References

1. Covington, A.D. Modern tanning chemistry. *Chem. Soc. Rev.* **1997**, *26*, 111–126.
2. Kumar, J.K.; Sinha, A.K. Resurgence of natural colourants: A holistic view. *Nat. Prod. Lett.* **2004**, *18*, 59–84. [CrossRef] [PubMed]
3. Kanth, S.V.; Venba, R.; Madhan, B.; Chandrababu, N.K.; Sadulla, S. Cleaner tanning practices for tannery pollution abatement: Role of enzymes in eco-friendly vegetable tanning. *J. Clean. Product.* **2009**, *17*, 507–515. [CrossRef]
4. Gil, R.R.; Ruiz, B.; Lozano, M.S.; Martín, M.J.; Fuente, E. VOCs removal by adsorption onto activated carbons from biocollagenic wastes of vegetable tanning. *Chem. Eng. J.* **2014**, *245*, 80–88. [CrossRef]
5. Park, H.M.; Li, X.; Jin, C.Z.; Park, C.J.; Cho, W.J.; Chang, S.H. Preparation and properties of biodegradable thermoplastic starch/clay hybrids. *Macromol. Mater. Eng.* **2002**, *287*, 553–558.
6. Scognamiglio, F.; Mirabile Gattia, D.; Roselli, G.; Persia, F.; De Angelis, U.; Santulli, C. Thermoplastic starch films added with dry nopal (Opuntia Ficus Indica) fibres. *Fibers* **2019**, *7*, 99. [CrossRef]
7. Troiano, M.; Santulli, C.; Roselli, G.; Di Girolami, G.; Cinaglia, P.; Gkrilla, A. DIY bioplastics from peanut hulls waste in a starch-milk based matrix. *FME Trans.* **2018**, *46*, 503–512. [CrossRef]
8. Galentsios, C.; Santulli, C.; Palpacelli, M. DIY bioplastic material developed from banana skin waste and aromatised for the production of bijoutry objects. *J. Basic Appl. Res. Int.* **2017**, *23*, 138–150.
9. Ortega-Toro, R.; Collazo-Bigliardi, S.; Talens, P.; Chiralt, A. Influence of citric acid on the properties and stability of starch-polycaprolactone based films. *J. Appl. Polym. Sci.* **2015**. [CrossRef]
10. Zeleznak, K.J.; Hoseney, R.C. The glass transition in starch. *Cereal Chem.* **1987**, *64*, 121–124.
11. Wang, K.; Wang, W.; Ye, R.; Liu, A.; Xiao, J.; Liu, Y.; Zhao, Y. Mechanical properties and solubility in water of corn starch-collagen composite films: Effect of starch type and concentrations. *Food Chem.* **2017**, *216*, 209–216. [CrossRef] [PubMed]
12. Ghanbarzadeh, B.; Almasi, H.; Entezami, A.A. Improving the barrier and mechanical properties of corn starch-based edible films: Effect of citric acid and carboxymethyl cellulose. *Ind. Crops Prod.* **2011**, *33*, 229–235. [CrossRef]
13. Scognamiglio, F.; Mirabile Gattia, D.; Roselli, G.; Persia, F.; De Angelis, U.; Santulli, C. Thermoplastic Starch (TPS) films added with mucilage from Opuntia Ficus Indica: Mechanical, microstructural and thermal characterization. *Materials* **2020**, *13*, 1000. [CrossRef] [PubMed]
14. Martins, I.M.G.; Magina, S.P.; Oliveira, L.; Freire, C.S.R.; Silvestre, A.J.D.; Pascoal Neto, C.; Gandini, A. New biocomposites based on thermoplastic starch and bacterial cellulose. *Compos. Sci. Technol.* **2009**, *69*, 2163–2168. [CrossRef]
15. Cyras, V.P.; Manfredi, L.B.; Ton-That, M.T.; Vázquez, A. Physical and mechanical properties of thermoplastic starch/montmorillonite nanocomposite films. *Carbohydr. Polym.* **2008**, *73*, 55–63. [CrossRef]

16. Ardakani, K.M.; Navarchian, A.H.; Sadeghi, F. Optimization of mechanical properties of thermoplastic starch/clay nanocomposites. *Carbohydr. Polym.* **2010**, *79*, 547–554. [CrossRef]
17. Liu, Z.Q.; Yi, X.-S.; Feng, Y. Effects of glycerin and glycerol monostearate on performance of thermoplastic starch. *J. Mater. Sci.* **2001**, *36*, 1809–1815. [CrossRef]
18. Popescu, C.; Budrugeac, P.; Wortmann, F.-J.; Miu, L.; Demco, D.E.; Baias, M. Assessment of collagen-based materials which are supports of cultural and historical objects. *Polym. Degrad. Stab.* **2008**, *93*, 976–982. [CrossRef]
19. Ramaraj, B. Mechanical and thermal properties of ABS and leather waste composites. *J. Appl. Polym. Sci.* **2006**, *101*, 3062–3066.
20. Kaushik, A.; Singh, M.; Verma, G. Green nanocomposites based on thermoplastic starch and steam exploded cellulose nanofibrils from wheat straw. *Carbohydr. Polym.* **2010**, *82*, 337–345. [CrossRef]
21. Su, F.Y.; Bushong, E.A.; Deerinck, T.J.; Seo, K.; Herrera, S.; Graeve, O.A.; Kisailus, D.; Lubarda, V.A.; Mc Kittrick, J. Spines of the porcupine fish: Structure, composition, and mechanical properties. *J. Mech. Behav. Biomed. Mater.* **2017**, *73*, 38–49. [CrossRef] [PubMed]
22. Santulli, C.; Vita, A.; Scardecchia, S.; Forcellese, A. A material proposed for re-use of hemp shives as a waste from fiber production. *Mater. Today Proceed.* **2019**. [CrossRef]
23. Galat, A. Study of the Raman scattering and infrared absorption spectra of branched polysaccharides. *Acta Biochim. Pol.* **1980**, *27*, 135–142. [PubMed]
24. Falcão, L.; Araújo, M.E.M. Tannins characterization in historic leathers by complementary analytical techniques ATR-FTIR, UV-Vis and chemical tests. *J. Cult. Herit.* **2013**, *14*, 499–508. [CrossRef]
25. Shi, R.; Zhang, Z.; Liu, Q.; Han, Y.; Zhang, L.; Chen, D.; Tian, W. Characterization of citric acid/glycerol co-plasticized thermoplastic starch prepared by melt blending. *Carbohydr. Polym.* **2007**, *69*, 648–655. [CrossRef]
26. Herrera Brandelero, R.P.; Grossmann, M.V.; Yamashita, F. Films of starch and poly (butylene adipate co-terephthalate) added of soybean oil (SO) and Tween 80. *Carbohydr. Polym.* **2012**, *90*, 1452–1460. [CrossRef]
27. De Campos Vidal, B.; Mello, M.L.S. Collagen type I amide I band infrared spectroscopy. *Micron* **2011**, *42*, 283–289.
28. Shukur, M.F.; Ithnin, R.; Kadir, M.F.Z. Ionic conductivity and dielectric properties of potato starch-magnesium acetate biopolymer electrolytes: The effect of glycerol and 1-butyl-3-methylimidazolium chloride. *Ionics* **2016**, *22*, 1113–1123. [CrossRef]
29. Campos, A.; Marconcini, J.M.; Imam, S.H.; Klamczynski, A.; Ortis, W.J.; Wood, D.H.; Williams, T.G.; Martins-Franchetti, S.M.; Mattoso, L.H.C. Morphological, mechanical properties and biodegradability of biocomposite thermoplastic starch and polycaprolactone reinforced with sisal fibers. *J. Reinf. Plast. Compos.* **2012**, *31*, 573–581. [CrossRef]
30. Dai, L.; Zhang, J.; Cheng, F. Effects of starches from different botanical sources and modification methods on physicochemical properties of starch-based edible films. *Int. J. Biol. Macromol.* **2019**, *132*, 897–905. [CrossRef] [PubMed]
31. Pandiarajan, S.; Umadevi, M.; Rajaram, R.K.; Ramakrishnan, V. Infrared and Raman spectroscopic studies of l-valine l-valinium perchlorate monohydrate. *Spectrochim. Acta A* **2005**, *62*, 630–636. [CrossRef] [PubMed]

 © 2020 by the authors. Licensee MDPI, Basel, Switzerland. This article is an open access article distributed under the terms and conditions of the Creative Commons Attribution (CC BY) license (http://creativecommons.org/licenses/by/4.0/).

Article

Effect of Fique Fibers in the Behavior of a New Biobased Composite from Renewable Mopa-Mopa Resin

José Herminsul Mina Hernandez [1,*], Edward Fernando Toro Perea [2], Katherine Caicedo Mejía [1] and Claudia Alejandra Meneses Jacobo [1]

[1] Grupo Materiales Compuestos, Universidad del Valle, Calle 13 No. 100-00, 76001 Cali, Colombia; kathe.caicedo@hotmail.com (K.C.M.); alejandrameneses91@gmail.com (C.A.M.J.)
[2] Grupo Sistemas de Gestión Científica y Tecnológica, Universidad Nacional Abierta y a Distancia, Av. Roosevelt No. 36-60, 110311 Cali, Colombia; edward.toro@unad.edu.co
* Correspondence: jose.mina@correounivalle.edu.co; Tel.: +57-2-3212170; Fax: +57-2-3392450

Received: 17 May 2020; Accepted: 13 July 2020; Published: 16 July 2020

Abstract: A fully biobased composite was developed using a natural resin from the *Elaeagia Pastoensis Mora* plant, known as *Mopa-Mopa* reinforced with fique fibers. Resin extraction was through solvent processing reaching an efficient extraction process of 92% and obtaining a material that acted as a matrix without using any supplementary chemical modifications as it occurs with most of the biobased resins. This material was processed by the conventional transform method (hot compression molding) to form the plates from which the test specimens were extracted. From physicochemical and mechanical characterization, it was found that the resin had obtained a tensile strength of 15 MPa that increased to values of 30 MPa with the addition of 20% of the fibers with alkalization treatment. This behavior indicated a favorable condition of the fiber-matrix interface in the material. Similarly, the evaluation of the moisture adsorption in the components of the composite demonstrated that such adsorption was mainly promoted by the presence of the fibers and had a negative effect on a plasticization phenomenon from humidity that reduced the mechanical properties for all the controlled humidities (47%, 77% and 97%). Finally, due to its physicochemical and mechanical behavior, this new biobased composite is capable of being used in applications such as wood–plastic (WPCs) to replace plastic and/or natural wood products that are widely used today.

Keywords: Mopa-Mopa resin; biobased composite; fique fibers; wood–plastic

1. Introduction

Due to the environmental impact caused by the conventional synthetic polymers when they are not properly disposed of at the end of their life cycle, research studies are currently being carried out in the field of polymeric materials that focused on the development of polymers characterized by being biobased, generated from renewable sources such as starches, proteins, hydroxy alkanoates, among others, and by presenting complete biodegradability under composting conditions [1–8]. Unlike traditional synthetic polymers, these materials are not oil-dependent; therefore, they have an added value as a potential alternative to produce eco-friendly materials. Among this family of polymer materials, the natural resins extracted from plants stand out due to the potential use of a wide variety of plants in the ecosystem that constitute a renewable source for polymer obtention [6,7]. These resins are currently being employed for the development of biobased composites, which, in most cases, are used as a partial substitute of reagents on the synthesis of polymers, such as canola oil for the obtention of polyols in order to react with isocyanates for the production of polyurethane adhesives and foams [8], tannin-furfuryl alcohol for thermoset resins [9], soya as polyol for polyurethanes [10], starches, wood, and other natural materials as sources for synthesizing reagents to produce epoxy bio-resins [11], such as modified vegetable oils, sugars, polyphenols, terpenes, colophony, natural rubber, and lignin for chemical synthesis of resins and curing agents for epoxy polymers [12], among others. The Mopa-Mopa resin forms the base of the varnish that is extracted from the *Elaeagia Pastoensis Mora* wild shrub, which belongs to the *Rubiaceae* family and grows in the Department of Putumayo, Amazon region in the Colombian jungle. Twice a year, the plant produces a gelatinous paste, which, through an artisanal process, is transformed into a thin sheet that can be molded to make decorative drawings on pre-painted wood [13]. The resin has been extracted and used by generations of farmers to commercialize it as a raw material mainly for manufacturing and/or restoring handicrafts [14,15]. Although the knowledge of the Mopa-Mopa has been limited to the development of new techniques and products on an artisanal level, there has been a growing interest within the scientific community to analyze and expand the understanding of this polymeric material through research about its physical, mechanical, thermal and chemical properties. Among those studies, Insuasty et al. [16] started the chemical characterization of the Mopa-Mopa resin from solubility tests and several spectroscopic techniques and identified that ethanol and methanol were the best solvents for the resin without losing its physical properties, especially its elasticity. Other studies [17,18] have shown both some physicochemical and thermal properties of the Mopa-Mopa resin and the effect of add polycaprolactone on the properties of binary mixtures using Differential Scanning Calorimetry (DSC), X-ray Diffraction (XRD), Thermogravimetry Analysis (TGA), and Fourier Transform Infrared Spectroscopy (FTIR) tests as well as in determining the semi-crystalline nature of the resin.

Considering the aforementioned aspects, this project developed an efficient extraction process of Mopa-Mopa resin to be implemented as a fully biobased matrix for the production of a biobased composite reinforced with short fibers of fique (25 mm); fique fibers are available in Southwestern Colombia, they are mainly used in packaging and cordage [19]. It is also noted that, due to their good mechanical properties and the technological development associated with its extraction process, fique fibers have also been considered as reinforcement in plastic matrix composite materials. [20–23]. In the same manner, this work contemplated the obtained biobased composite of Mopa-Mopa resin with 10 and 20% (m/m) of fique fibers, the evaluation effect of the fiber superficial modification process (by alkalization) and the exposure of the material to three different relative humidities (47, 77 and 97%), in its physicochemical and mechanical properties. Furthermore, the new material developed was characterized because it can be processed through conventional transformation methods and because its behavior means that it can also can be used in other applications, such as wood–plastic.

2. Materials

The Mopa-Mopa resin used in this research was extracted from the buds of the *Elaeagia Pastoensis Mora*, a plant native to the Department of Putumayo (Mocoa, Colombia) (typical taxonomic classification is provided in Table 1). Furthermore, the obtained resin showed some physical characteristics, such as ρ = 1.108 g/cm^3, T_m = 117 °C, and T_g = 34 °C. The fique fiber used belongs to the *Furcraea* genus of the Uña de Águila (white variety plant), which was provided by the Empaques del Cauca Company located in the city of Popayán, Colombia. The raw material was used to make a short reinforcement with an average length of 25 mm randomly located, and with and without an alkaline treatment inside a Mopa-Mopa resin matrix. Finally, sodium hydroxide used in the alkaline treatment of the fibers and ethanol employed in obtaining Mopa-Mopa resin [24] were reactive grade acquired by the company Técnica Química S.A. (Cali, Colombia).

Table 1. Taxonomic classification of the Mopa-Mopa resin.

	Description
Division	*Trachelophyta*
Subdivision	*Angiospermae*
Class	*Dicotyledonae*
Order	*Rubiales*
Family	*Rubiaceae*
Genus	*Elaeagia*
Species	*Elaeagia pastoensis Mora*
Shapes	*Elaeagia pastoensis Mora fma pastoensis* / *Elaeagia pastoensis Mora fina acuminata Mora*
Synonym	*Elaeagia pastoganomophora*

3. Experimental Procedure

3.1. Obtaining the Mopa-Mopa Resin

Through the use of a manual disk coffee milk, a physical treatment of comminution was carried out to reduce the buds of the Mopa-Mopa tree that were initially agglomerated (Figure 1a), thus achieving an efficient size reduction (Figure 1b). Subsequently, for the obtention processing, 300 mL of a solution of the Mopa-Mopa resin in ethanol was prepared at a concentration of 20% (m/v) by using a flat-bottom three-necked flask with a capacity of 500 mL and set up in a closed distillation system, keeping the solution under heating at a temperature of 75 °C for 25 min. The remaining fluid was cooled to approximately 40 °C and the solid residues were separated from the mixture, which corresponds to the remains of leaves, seed husks, and stems, among others, by using a vacuum filtration system composed of a porcelain funnel, an Erlenmeyer that received the filtered solution and a vacuum pump Welch model BS-8000 Fisher Scientific (Gardner, MA, USA). This procedure was carried out in two stages, the first stage with absorbent towels, eliminating the larger residues, and the second stage, with qualitative filter paper, to ensure cleaning of the solution. The already filtered mixture was heated again at a temperature between 76 and 78 °C for 45 min, to evaporate as much as possible, that is to say, to concentrate the Mopa-Mopa, taking special care to not degrade the resin; this procedure was conducted in aid of a fractional distillation system to condense the solvent and to regain around 80% of the initial ethanol. Simultaneously to the previous procedure, a volume of 100 to 150 mL of distilled water was heated to a temperature of 100 °C, in which it was added to the Mopa-Mopa concentrated solution. In the making of this mixture, the resin kept suspended on the water surface due to the immiscibility of these two phases (Figure 1c). Subsequently, constant agitation was maintained using a spatula lab until the resin precipitates and can be separated from the liquid. This step corresponded to the 20 min after mixing with distilled water. Similarly, it is important to specify that in this step the remaining solvent that was left in the previous stage is evaporated. The Mopa-Mopa resin obtained was deposited in a watch glass, where it was left to cool for 5 min until reaching a temperature of 25 °C to register the mass. (Figure 1d). A two-stage size reduction process was necessary to carry out the elaboration of test specimens. In a first stage, the material was crushed in a low-speed granulator of the SG-16/20 series until obtaining a particle size of 3 mm, in a second stage, using a manual coffee grinder, until reaching a particle size between 0.3 and 0.5 mm. It is important to note that this resin Mopa-Mopa is characterized by containing a variety of metabolites, including alkaloids, flavonoids, and cardiotonic aglycones [16]. The test specimens were formed in a Carver press model MH 4389-4021 (Wabash, IN, USA) equipped with heating plates and a forced circulation water system) using a pressure of 25,000 lb, and a temperature of 165 °C for 30 min; after that, the obtained plates were die-cut following the shape of the Type IV probes in accordance with ASTM D 638 [25].

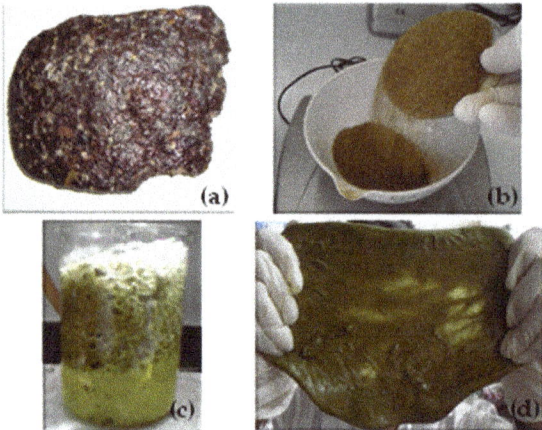

Figure 1. Mopa-Mopa resin: (**a**) in the raw state; (**b**) after the grinding process; (**c**) during the separation process; (**d**) after the extraction process.

3.2. Surface Modification by Alkalization of Fique Fibers

Following a similar methodology to that proposed by Valadez et al. [26] for henequen fibers, the alkalization of the fique fibers consisted of pre-drying the material at 100 °C for 24 h, to then carry out a surface treatment by immersing the fibers in an aqueous solution of NaOH (2% *m/v*) for 1 h at 25 °C; later, the fibers were washed with distilled water acidified with acetic acid, until reaching a neutral pH, that is, until the washing water had no residual alkaline solution. Finally, the fibers were dried at 60 °C for 24 h.

3.3. Preparation of the Biobased Composite

For the elaboration of the biobased composite, the components, the Mopa-Mopa resin and the native and alkaline fique fibers, were pressed separately and 25,000 lb of pressure was applied, forming nonwoven fique mats (Figures 2a and 3b). Finally, the components were molded using a sandwich-type layout (Mopa-Mopa sheet–fiber mat–Mopa-Mopa sheet) at a temperature of 170 °C, under the following pressure scheme: 10,000 lb for 3 min, 20,000 lb for 3 min and 25,000 lb for 30 min (Figure 3c,d); reinforcing proportions of 10 and 20% concerning the total mass were maintained.

Figure 2. Biobased composite preparation: (**a**) Mopa-Mopa compacted sheet; (**b**) nonwoven fique mats; (**c**) Mopa-Mopa/nonwoven fique mats/Mopa-Mopa laminate; (**d**) molding the biobased composite.

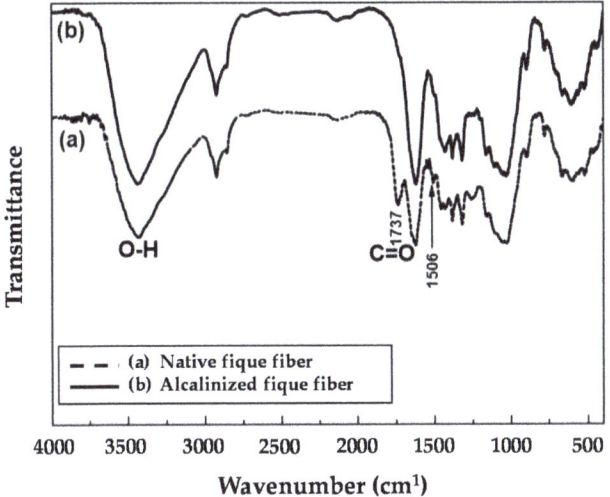

Figure 3. Fourier Transform Infrared Spectroscopy (FTIR) for the fique fibers: (a) native; (b) alkalized.

3.4. Fourier Transform Infrared Spectroscopy (FTIR)

For the analysis of the materials, a Fourier Transform Infrared Spectrum 100 model was used. In the case of fique fibers with and without alkaline treatment, the Attenuated Total Reflectance (ATR) technique was followed, working at 100 sweeps and a resolution of 2 cm^{-1}. For the Mopa-Mopa resin, the infrared analysis was performed on conditioned specimens at three relative humidity ranges (47, 77 and 97%), as well as on an additional sample that was dried in an oven at 60 °C for 2 h. The Diffuse Reflectance (DRIFT) technique was used in these materials, working at 200 scans and a resolution of 2 cm^{-1}.

3.5. Moisture Adsorption

Salts of potassium carbonate, sodium chloride, and distilled water were added to desiccators for maintaining constant the relative humidity at 47%, 77%, 97%, respectively, in accordance with ASTM E 104 [27]. Humidity meters were placed in the desiccators to monitor such relative humidity. For the evaluation of the fique fibers, three bundles of the fibers were cut, with masses between 3 and 6 g, both for the native and those that had the alkaline treatment and for each relative humidity. Subsequently, the samples were dried in an oven at 100 °C for 24 h and placed in the respective desiccators, registering their mass. In the case of the Mopa-Mopa resin and the biobased composite, three type IV test specimens were prepared according to ASTM D 638 [25] for each relative humidity and each condition of the composite (proportions of 10 and 20%, with and without treatment alkaline), to then analyze the effect of relativity humidity and exposure time on mechanical properties by applying a tensile test. The data of the mass gain as a function of time (Mt) were taken in each case and the percentage of moisture adsorption (H) was determined based on the model presented in Equation (1), considering the mass after drying in the oven (Ms).

$$H = \left(\frac{Mt - Ms}{Ms}\right) \times 100 \tag{1}$$

3.6. Density Determination

This test was carried out on a Mettler Toledo AG245 equipment with a maximum capacity of 210 g and a deviation of 0.001 g. The density value for the Mopa-Mopa resin and the biobased composite in all the study conditions (proportions of 10 and 20% with native and alkalinized fiber) was determined by method A of the ASTM D 792 [28]. In the case of fique fibers with and without alkaline treatment, the density was estimated following method B of the same standard. The theoretical density was estimated with the aid of the rule of mixture model shown in Equation (2)

$$\rho_b = \rho_f v_f + \rho_m(1 - v_f) \qquad (2)$$

where ρ is the density, and the subindices b, f, and m refer to the biobased composite, the fiber, and the matrix, respectively; v_f is the volume fraction of fiber incorporated into the biobased composite.

3.7. Tension Test

The determination of the tensile mechanical properties of the fique fibers with and without alkaline treatment, the Mopa-Mopa resin, and the biobased composite was carried out by a Tinius Olsen model H50KS universal testing machine. An experimental setup was used in the fique fibers, which consisted of the fixing of filaments in cardboard frames, allowing for a better grip of these to the clamp, in accordance to the ASTM D 3822 [29] standard, with a clamp displacement rate of 3 mm/min and taking into account a previous analysis of the distribution of average diameters in each fiber tested. A sample of 140 fibers (70 native and 70 alkalized) was selected at random and estimating about 8 diameter values set for each of the filaments, which generated a standard deviation of 0.03 for the native fibers and 0.02 for the alkalized fibers. For Mopa-Mopa resin and biobased composite, ASTM D638 [25] was followed with type IV specimens and a speed displacement rate of 5 mm/min.

3.8. Scanning Electron Microscopy (SEM)

The morphological characterization of the fique fibers, the Mopa-Mopa resin, and the biobased composite were carried out by a Electron Microscope model JMS 6490 LV (Jeol, Mexico D.F., Mexico), in the secondary electrons mode (SEM) and under an accelerating voltage of 20 kV. The chemical microanalysis was performed on several inspection areas using an Energy Dispersive X-ray Spectroscopy (EDS) model IncaPentaFETx3 (Oxford Instrument, Belfast, UK). This technique allowed us to qualitatively know the surface effect of the treatment of the fique fibers, the ductile nature of the Mopa-Mopa resin, and the interface of the biobased composite. The samples were previously bonded on a carbon tape and then metalized with a gold layer between 10 and 50 nm (thin film deposition equipment Model Desk IV at a pressure of 50 m Torr and a time of 60 s, (Denton Vacuum, Moorestown, NJ, USA) in order to generate a conductive surface, analyzing the cross-section of the resin, the biobased composite and the contour of the fique fibers.

4. Results and Discussion

4.1. Fourier Transform Infrared Spectroscopy (FTIR)

The infrared spectroscopy test of the Mopa-Mopa resin was carried out carefully on four conditioned samples: one dried sample in the oven at 60 °C for 2 h (Reference sample), the remaining samples, at three relative humidities of 47, 77 and 97%. The purpose of this procedure was to learn the effect of water adsorption on the molecular interactions of the resin. Consequently, the following bands for the dry sample could be observed: an intense signal in the region below 3000 cm^{-1}, with two bands at 2979.6 and 2951.0 cm^{-1} assignable to the vibration in tension (asymmetric and symmetrical) of the CH bond; at 1751.0 cm^{-1}, an intense signal corresponding to the vibration to a tension of the carbonyl group C=O; at 1656.6 and 1474.9 cm^{-1}, associated with the vibration in tension C=C, between 1400 and 1000 cm^{-1}, other slightly widened bands were observed, most likely related to CO bonds and the deformation vibration of CH bonds. The results found were similar to those reported for the Mopa-Mopa resin in other studies [17,18]. On the other hand, the samples conditioned at 97%, 77%, and 47% relative humidity presented a new band associated with interactions of the hydroxyl group with water molecules at 3284.3 cm^{-1}, but with a difference in the peak's intensity seen in the axis of ordinates that varied in values of 0.017, 0.009 and 0.007, respectively, being greater for the higher humidities. This fact can be correlated with a possible plasticization of the Mopa-Mopa by the influence of the humidity. This phenomenon is similar to the one that happens in the thermoplastic starch [30], whereby incorporating a plasticizer to the cassava starch means that the interactions of the hydroxyl groups are modified within the material and new second-order intermolecular associations (hydrogen bonds) are established with fewer steric hindrances. Table 2 shows the spectral bands for each sample used with the characteristic link type.

Table 2. Characteristic bands and type of bond for the Mopa-Mopa resin.

	Wavenumber (cm^{-1})			Type of Link
Dry Sample	HR 47%	HR 77%	HR 97%	
2979.59	2976.70	2985.38	2977.66	Tension C-H
2951.04	2951.07	2952.45	2974.32	Tension C-H
1751.01	1751.00	1750.81	1751.11	Tension C=O
1656.55	1658.62	1656.55	1654.62	Tension C=C
1474.86	1473.96	1474.21	1475.38	Tension C=C
1394.64	1394.69	1394.80	1394.27	C-O; C-H
1293.28	1293.85	1294.20	1292.83	C-O; C-H
1203.09	1201.79	1203.06	1200.74	C-O; C-H
1123.33	1123.15	1123.70	1123.29	C-O; C-H
1055.42	1056.63	1054.87	1055.59	C-O; C-H

Figure 3 shows the infrared spectra obtained in the native and alkalized fibers; in these spectra, the representative bonds of the fundamental constituents of the fiber, such as cellulose, lignin, and hemicellulose, stood out. With the treatment of the fibers in the alkaline solution, it was observed that the peak corresponding to the tension stretch of the carbonyl C=O, which occurred at a wavenumber of 1736 cm^{-1}, disappeared from the spectrum, this peak is associated with Ester-type bonds that usually appear in the hemicellulose structure; therefore, the loss of this band indicated that, at least at the surface level, the removal of this component in the fiber was generated. Similarly, it showed a decrease in peaks at 1505 and 2862 cm^{-1} related to the aromatic skeleton and stretching of the -OCH$_3$ bond of lignin, respectively, indicating that the lignin surface concentration decreased with alkaline treatment. Similar results were reported by Mina [31] who additionally included the chemical composition for the fique and other natural fibers, finding that the percentage related to lignin and cellulose in the fique was under 14 and 61.2%, respectively.

4.2. Moisture Adsorption

This study was carried out considering the conditioning of the material at relative humidities of 47, 77 and 97%, finding in specimens a reduced increase in mass due to low water adsorption in all the different relative humidities controlled. The humidity adsorption values of the equilibrium corresponded to 1.9, 0.64, and 0.09% for the humidity of 97, 77, and 47%, respectively. These data indicated that the Mopa-Mopa resin presented low humidity adsorption compared to other natural polymers, such as thermoplastic starch, that, according to reported results [30], presents about 7% adsorption when reaching equilibrium; for a relative humidity of 43% and 25 °C, these conditions were lower than those used in the present study and with very high adsorption results that were not reached by the Mopa-Mopa resin even at the highest relative humidity studied (97%). The general behavior of the humidity adsorption of the Mopa-Mopa resin in the different atmospheres is shown in Figure 4a, where it becomes evident both the relationship of proportionality between the relative humidity to which the resin was exposed and its adsorption percentage for the humidity of 97 and 77%. Moreover, it must be considered that, during the initial times, the adsorption of the resin occurred at a faster rate (first stage of the curve), followed by an almost constant behavior defined as balance. In a particular case, the curve corresponding to the humidity of 47% showed a decrease in the humidity adsorption because the specimens were not completely dry before being introduced into the desiccator since the resin is very vulnerable to oxidation in prolonged healing periods; therefore, when these presented a higher humidity than the exposure medium, the salts generate a drying effect to balance their humidity, represented in the decreasing behavior of the curve. Fique fibers' ability to absorb moisture could be influenced by the superficial modification generated by the alkalization treatment, since, with it, some hydrophobic groups are removed from the fiber, such as waxes and pectins, generating a concentration of cellulose on its surface that has hydroxyl groups and a great affinity with water, forming hydrogen bonds and thus increasing its adsorption moisture. This behavior was reflected in the increase in the mass of the fibers when they were exposed to different relative humidities (97, 77, and 47%), with higher adsorption at higher humidities compared to native fibers. Figure 4b shows the curves corresponding to the adsorption isotherms of native fique fibers and of the alkalized fibers, conditioned to the three relative humidities of 47, 77 and 97%. Due to the type of bonds it has, Mopa-Mopa is a material with the proper functionality to form hydrogen bonds with water molecules present in the humidity of the environment, behavior that is similar to that of the fique fibers used as reinforcement and that turn to the increase in the level of moisture adsorption of the Mopa-Mopa/Fique biobased composite concerning the individual behavior of the components, as can be seen in Figure 5a,b. The increase in mass due to the adsorbed water is directly affected by the content of the fibers used, being greater in those with 20% of them. Besides, the alkaline treatment generated a greater absorption of moisture on the part of the biobased composite; this was attributed, as mentioned above, to the fact that the treatment caused an increase in cellulose on the surface of the fiber, which has hydroxyl groups and a great affinity with water, forming hydrogen bonds and increasing their ability to absorb moisture. According to the above, the increase in the mass by water adsorption in the compounds with alkalized fibers could be produced through a diffusion mechanism, due to the presence of micro-spaces between the fiber and the matrix that allowed the water filtration in the material [32]. This phenomenon becomes more important when the amount of fiber increases in the matrix. It is important to mention that, as in the mechanical characterization of the Mopa-Mopa resin, the adsorption of humidity in the same three exposure times was studied in the biobased material, showing a proportional increase in the adsorption of water throughout the exposure time and generating a negative effect on the mechanical performance of the material.

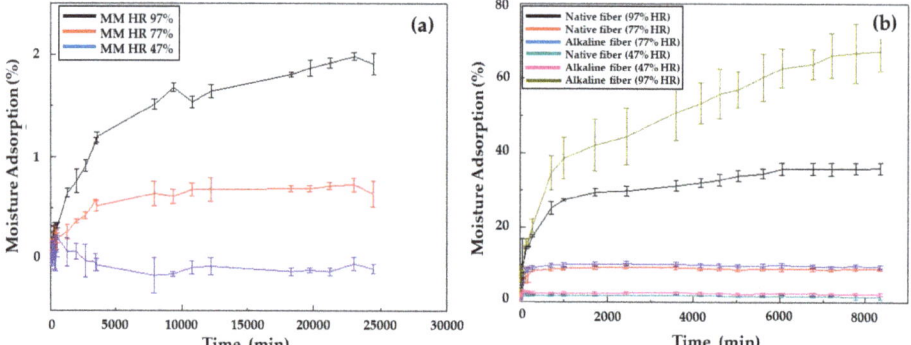

Figure 4. Adsorption isotherms at a relative humidity of 97, 77, and 47% for (**a**) Mopa-Mopa resin; (**b**) native and alkalized fique fibers.

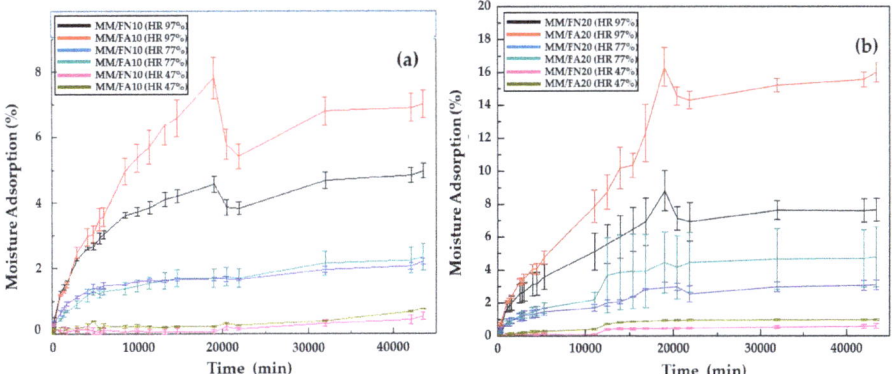

Figure 5. Adsorption isotherms at relative humidities of 97, 77, and 47 for (**a**) biobased composite with 10% native and alkalized fique fibers; (**b**) biobased composite with 20% native and alkalized fique fibers.

4.3. Density Estimation

This test was performed on all the manufactured Mopa-Mopa/Fique biobased composite and the Mopa-Mopa resin matrix to establish the effect of fiber incorporation and the surface treatment of the fique on the density of the material, the used values for the theoretical determination of the different compounds were $\rho_m = 1.108 \pm 0.03$ g/cm^3, $\rho_f = 1.393 \pm 0.04$ and 1.308 ± 0.06 g/cm^3 for the native and alkalize fibers, respectively. Figure 6 shows the results obtained and that was compared with the estimated values employing the rule of mixture in Equation (2). By incorporating 10% of fique fibers with and without alkaline treatment, it was found that the density of the composite decreased concerning the density of the matrix. However, the density of the composite with the alkalized fibers was higher than the one with the native fibers. This positive resulting effect from the treatment of the fique fibers in the density of the composite is due to the reduction of voids and/or cavities between the fiber and the matrix that also affects the improvement of the interfacial zone as will be discussed later with the help of SEM images. On the other hand, in the composite made with 20% of native fique fibers and previously treated with NaOH, it was observed that the density increase regarding that of the Mopa-Mopa resin and the compounds with 10% of the fibers is greater than that of the material with 20% alkalinized fibers, presenting a similar trend to the one estimated from the theoretical values. The materials that presented higher values of density also showed greater resistance in tension because they had a better mechanical anchorage; the relationship between the density, the interfacial zone, and the properties of the compounds has been documented in the literature [33]. It is also important to highlight that the dispersion of data decreased with the increase in the volume of fiber, being closer to the theoretical values than the density of the biobased composite reinforced with alkalized fiber.

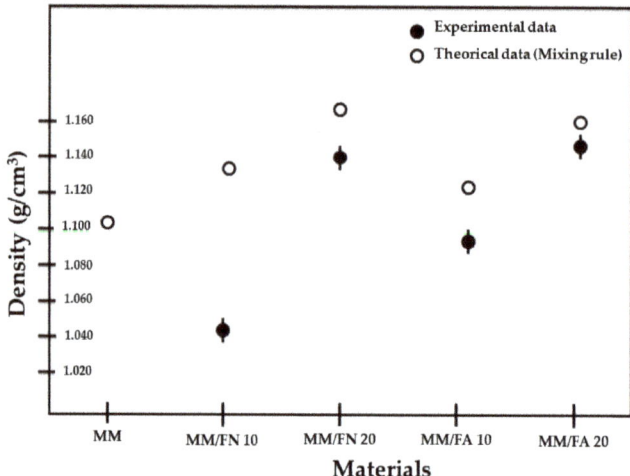

Figure 6. Density data for Mopa-Mopa resin and biobased composite with 10 and 20% native and alkalized fique fibers.

4.4. Tensile Strength

The specimens used to determine the mechanical properties were tested at three time points (t_1: before exposure to the desiccator; t_2: 3 days of exposure; t_3: 15 days) of conditioning, and a relative humidity of 47, 77 and 97%. Table 3 shows the results of the tensile properties for the natural Mopa-Mopa resin at the different relative humidities studied. Here, it can be seen that the material is characterized by having a tensile strength of 10.42 MPa for the conditioning time 1. This value turns out to be low when compared to some conventional synthetic polymers that show strengths around 22 and 30 MPa, as in the case of hight density polyethylene (HDPE) and polypropylene (PP), respectively [34]. However, it has an interesting behavior when compared to materials such as low density polyethylene (LDPE), for which there had been reported values from 5 MPa [35] and some bio-based polymers such as thermoplastic starch that reaches values between 0.23 and 5.5 MPa [31]. On the other hand, it is important to consider that the mechanical properties can be increased with the use of reinforcing materials such as natural fibers, in this case, those of fique. The mechanical properties of the Mopa-Mopa resin varied with the relative humidity to which it was exposed, mainly due to the phenomena of plasticization by water that the material undergoes and which allowed decreases of 69.78, 62.57 and 39.05% for the relative humidities of 97, 77 and 47%, respectively. Such a phenomenon influenced the secondary interactions that were previously discussed through an infrared analysis that is also a condition that occurs in biobased polymers such as thermoplastic starch, for example [30–36]. When incorporating the fibers in the matrix based on the Mopa-Mopa resin, an increase in the tensile strength of 43.35% for the composite with native fibers and 53.95% for the composite with alkalized fibers was evident, while in the case of the Young's modulus of the biobased composite, the increase was 34.87 and 37.88% for the fibers without and with alkaline treatment, respectively. These values were a direct function of the content of incorporated fibers, corroborating that the use of these had a positive impact on the mechanical properties of the material. In return, the alkaline surface treatment on the fibers improved the mechanical behavior of the biobased composite by 24.75 and 38.32% compared to that which was reinforced with 10 and 2% fibers without any treatment (native fibers), respectively, due to the increase in roughness and generation of mechanical anchorage. It is important to highlight that the mechanical properties were negatively affected by factors such as humidity and conditioning time, causing a considerable decrease in tensile strength, as can be seen in the data reported in Table 3. This behavior was attributed to the plasticizing effect due to the humidity that the biobased composite adsorbs due to the hydrophilic character of both the matrix and the reinforcement, as previously mentioned. The best mechanical performance in biobased composite associated with the material that was reinforced with 20% alkalized fibers, for which a tensile strength and Young's modulus of 32.73 and 2128.36 MPa was obtained, respectively; the above for the first evaluation time that corresponded to the specimens before conditioning.

Table 3. Tension mechanical properties of Mopa-Mopa resin and biobased composites with native and alkalized fique fibers at 10 and 20% incorporation.

	Time 1 (Before Conditioning)			Time 2 (3 Days of Conditioning)			Time 3 (15 Days of Conditioning)		
	Tensile Strength (MPa)	Strain (mm/mm)	Young's Modulus (MPa)	Tensile Strength (MPa)	Strain (mm/mm)	Young's Modulus (MPa)	Tensile Strength (MPa)	Strain (mm/mm)	Young's Modulus (MPa)
MM HR 97%				3.15 ± 0.65	1.36 ± 0.35	3.27	2.36 ± 0.82	0.89 ± 0.30	3.18
MM HR 77%	10.42 ± 1.83	1.44 ± 0.63	35.78 ± 5.22	5.51 ± 1.00	1.37 ± 0.37	11.44	3.9 ± 1.43	1.11 ± 0.36	7.10
MM HR 47%				7.21 ± 1.59	1.43 ± 0.39	23.03	11.6 ± 2.53	0.89 ± 0.34	20.09
MM/FN 10 (HR 97%)				7.21 ± 2.15	0.06 ± 0.03	627.21	6.36 ± 1.47	0.09 ± 0.03	267.04
MM/FN 10 (HR 77%)	11.34 ± 4.44	0.09 ± 0.02	739.8	10.51 ± 1.92	0.13 ± 0.03	512.84	8.53 ± 1.34	0.14 ± 0.03	325.83
MM/FN 10 (HR 47%)				15.13 ± 1.77	0.07 ± 0.02	920.29	12.16 ± 1.08	0.19 ± 0.04	857.65
MM/FA 10 (HR 97%)				6.73 ± 1.04	0.13 ± 0.04	545.12	4.25 ± 0.68	0.17 ± 0.04	143.37
MM/FA 10 (HR 77%)	15.07 ± 2.68	0.01 ± 0.02	1321.98	11.7 ± 1.34	0.10 ± 0.03	790.57	8.05 ± 0.87	0.14 ± 0.06	325.14
MM/FA 10 (HR 47%)				19.57 ± 2.25	0.1 ± 0.04	817.75	10.2 ± 1.24	0.14 ± 0.03	590.32
MM/FN 20 (HR 97%)				7.52 ± 1.55	0.11 ± 0.03	288.54	6.09 ± 0.94	0.25 ± 0.06	201.62
MM/FN 20 (HR 77%)	20.02 ± 7.99	0.05 ± 0.01	1135.86	10.96 ± 2.25	0.09 ± 0.02	688.83	10.87 ± 2.56	0.04 ± 0.02	621.69
MM/FN 20 (HR 47%)				16.17 ± 4.55	0.04 ± 0.01	1022.74	15.22 ± 2.54	0.02 ± 0.01	872.45
MM/FA 20 (HR 97%)				12.85 ± 2.48	0.08 ± 0.03	483.14	5.98 ± 0.74	0.12 ± 0.04	388.62
MM/FA 20 (HR 77 %)	32.73 ± 10.13	0.02 ± 0.01	2128.36	16.65 ± 4.14	0.06 ± 0.02	1667.61	7.43 ± 2.76	0.09 ± 0.02	400.55
MM/FA 20 (HR 47 %)				28.85 ± 6.63	0.05 ± 0.01	1667.61	18.49 ± 4.23	0.03 ± 0.01	1417.07

These results were superior to those reported by Delgado et al. [37], who developed wood–plastic composites from a low density polyethylene (LDPE)/hight impact polystyrene (HIPS) matrix reinforced with natural fibers, reaching values below 6 and 70 MPa for the parameters of interest. Additionally, the results reported by Fajardo et al. [38], who developed a PP matrix reinforced with bamboo fibers, that contained values of 12 and 1400 MPa in a tensile strength and Young's modulus, respectively. In the case of the biobased composites studied in the present study, the mechanical properties achieved were greater, independent of the humidity and conditioning times employed.

4.5. Scanning Electron Microscopy (SEM)

From SEM micrographs, it was possible to evaluate the quality of the fiber–matrix interface for the biobased composite that was reinforced with the fibers with and without surface treatment. In Figure 7 it can be seen some gaps in the areas located between the matrix and the fiber for the two case studies and that could give an indication of the formation of a weak interface between the phases of the compound. However, thanks to the intrinsic adhesion exhibited by the matrix, the mechanical properties showed increases in the resistance and modulus with the incorporation of the fibers and performing greater values with the fiber alkalized by the mechanical anchorage promoted by the roughness generated by the treatment, as mentioned in other studies [32]. The interface and the mechanical anchorage of the biobased composite are themselves affected by the relative humidity in which the material was exposed (47%, 77% and 97%). It can be seen in Figure 7 that at a higher relative humidity the mechanical anchorage decreases, due to the plasticization of the material caused by the adsorption of water. The lack of adherence between the fibers and the matrix can be seen in the reduction of the amount of matrix on the fibers and the increase in the empty spaces between them (see red circle), being observed a greater intensity at the higher relative humidity. This fact was also visible in the decrease in the mechanical properties of the biobased composites.

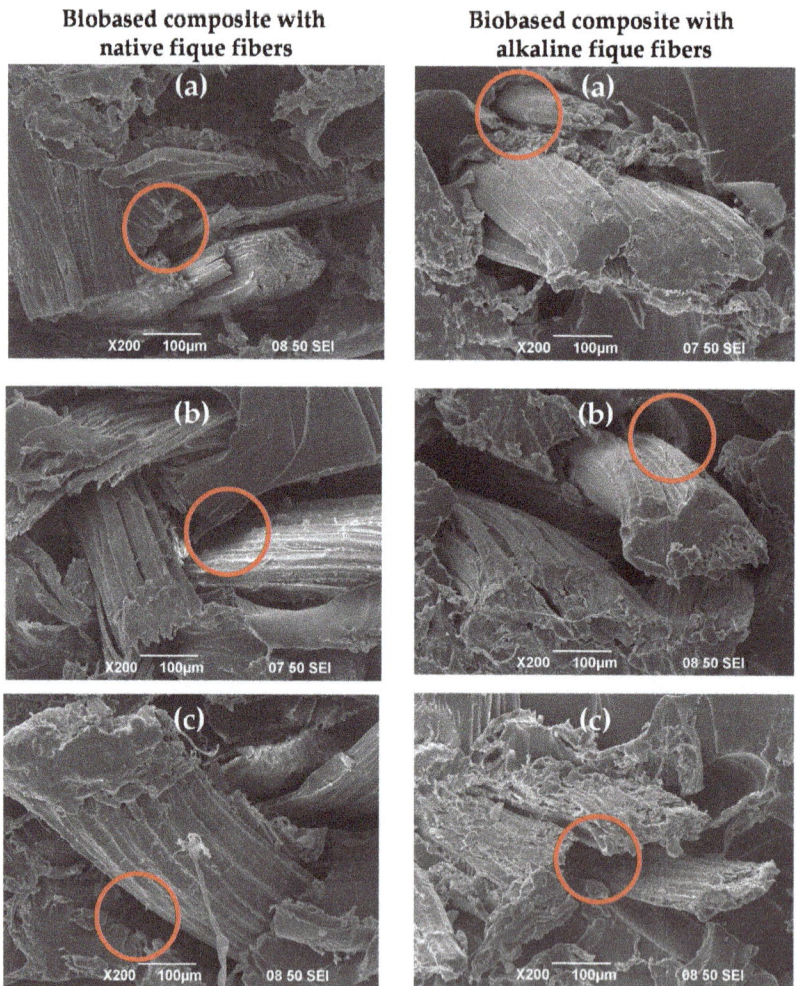

Figure 7. Scanning Electron Images of fracture surfaces of the biobased composite with native and alkaline fibers conditioned at (**a**) a relative humidity of 47%; (**b**) a relative humidity of 77%; (**c**) a relative humidity of 97%. The red circles show voids between the fiber and the matrix.

5. Conclusions

It was possible to develop a new composite material in which the matrix corresponded to the Mopa-Mopa resin, extracted from the bud of the *Elaeagia Pastoensis Mora* plant, characterized by being a fully biobased composite and, unlike most materials reported in the literature, only requiring a physical process of extraction through a closed distillation system that provided a significant yield (92%). In addition, this biobased material could be processed by hot compression molding, which is a conventional transformation process, through which it is possible to manufacture plates of fique fiber-reinforcement and from which test specimens for physical-chemical and mechanical characterizations were produced.

The biobased composite presented a positive synergy between the Mopa-Mopa resin and the fique fibers that were evidenced in the quality of the interfacial zone and from a macro-mechanical perspective, in an increase of the tensile strength and the Young's modulus as a function of the content of fibers in the material; much more significant increase in those properties was shown when a 20% of fibers superficially modified by an alkalization treatment were incorporated.

On the other hand, it was found that the relative humidity and the conditioning time of the material played an important role in a plasticization phenomenon that generated important reductions in the mechanical properties, especially to 97%, albeit the amount of moisture absorption in the equilibrium was low compared to what was reported for other biocomposites based on natural polymers such as thermoplastic starch. Because of the ease of processing and the physicochemical and mechanical characteristics evaluated that are evident when comparing the results achieved in this new material with those reported LDPE, HIPS and PP biocomposites reinforced with natural fibers, it is possible that the Mopa-Mopa resin/fique fibers biobased composite developed can be used as wood–plastic for the substitution of plastic and/or natural wood, increasing this viability if in further research the effect of integrating additives (lubricants, fillers, pigments, among others) in the composition of the material is studied.

Author Contributions: Conceptualization, J.H.M.H., E.F.T.P., K.C.M., and C.A.M.J.; Formal analysis, J.H.M.H., and E.F.T.P; Investigation, J.H.M.H., E.F.T.P., K.C.M., and C.A.M.J.; Methodology, J.H.M.H., E.F.T.P., K.C.M., and C.A.M.J.; Writing—original draft, J.H.M.H., and E.F.T.P.; Writing—review & editing, J.H.M.H., and E.F.T.P. All authors have read and agreed to the published version of the manuscript.

Funding: This research received no external funding.

Acknowledgments: The authors acknowledge the Escuela de Ingeniería de Materiales of the Universidad del Valle, for funding this research.

Conflicts of Interest: The authors declare no conflicts of interest.

References

1. Hatti-Kaul, R.; Nilsson, L.J.; Zhang, B.; Rehnberg, N.; Lundmark, S. Designing Biobased Recyclable Polymers for Plastics. *Trends Biotechnol.* **2019**, *38*, 50–67. [CrossRef]
2. Ilyas, R.A.; Sapuan, S.M.; Ishak, M.R.; Zainudin, E.S. Development and characterization of sugar palm nanocrystalline cellulose reinforced sugar palm starch bio-nano composites. *Carbohydr. Polym.* **2018**, *200*, 186–202. [CrossRef] [PubMed]
3. Yatigala, N.S.; Bajwa, D.S.; Bajwa, S.G. Compatibilization improves physico-mechanical properties of biodegradable biobased polymer composites. *Compos. Part A Appl. Sci. Manuf.* **2018**, *107*, 315–325. [CrossRef]
4. Maqsood, M.; Seide, G. Development of biobased socks from sustainable polymer and statistical modeling of their thermo-physiological properties. *J. Clean. Prod.* **2018**, *197*, 170–177. [CrossRef]
5. Huang, M.; Yu, J.; Ma, X. Ethanolamine as a novel plasticizer for thermoplastic starch. *Polym. Degrad. Stab.* **2005**, *90*, 501–507. [CrossRef]
6. Guan, J.; Eskridge, K.M.; Hanna, M. Acetylated starch-polylactic acid loose-fill packaging materials. *Ind. Crops Prod.* **2005**, *22*, 109–123. [CrossRef]
7. Lai, S.M.; Don, T.M.; Huang, Y.C. Preparation and properties of biodegradable thermoplastic starch/poly(hydroxybutyrate) blends. *J. Appl. Polym. Sci.* **2006**, *100*, 2371–2379. [CrossRef]
8. Kong, X.; Omonov, T.S.; Curtis, J.M. The development of canola oil based bio-resins. *Lipid Technol.* **2012**, *24*, 7–10. [CrossRef]

9. Lagel, M.; Hai, L.; Pizzi, A.; Basso, M.; Delmotte, L.; Abdalla, S.; Zahed, A.; Al-Marzouki, F. Automotive brake pads made with a bioresin matrix. *Ind. Crops Prod.* **2016**, *85*, 372–381. [CrossRef]
10. Jiang, L.; Walczyk, D.; McIntyre, G.; Bucinell, R.; Li, B. Bioresin infused then cured mycelium-based sandwich-structure biocomposites: Resin transfer molding (RTM) process, flexural properties, and simulation. *J. Clean. Prod.* **2019**, *207*, 123–135. [CrossRef]
11. Rad, E.R.; Vahabi, H.; De Anda, A.R.; Saeb, M.R.; Thomas, S. Bio-epoxy resins with inherent flame retardancy. *Prog. Org. Coat.* **2019**, *135*, 608–612. [CrossRef]
12. A Baroncini, E.; Yadav, S.K.; Palmese, G.R.; Stanzione, I.J.F. Recent advances in bio-based epoxy resins and bio-based epoxy curing agents. *J. Appl. Polym. Sci.* **2016**, *133*, 44103. [CrossRef]
13. Ávila, M.P.; Ochoa, I.C.; Rodríguez, C.E. *Memoria de oficio: Mopa Mopa Pasto*; Atesanías de Colombia: Bogotá, Colombia, 2016.
14. Cortés, C. La lógica de la representación en lo artesanal; un asunto de identidad. *Revista Estudios Latinoamericanos.* **2000**, *1*, 42–62.
15. Newman, R.; Kaplan, E.; Derrick, M. Mopa mopa: Scientific analysis and history of an unusual South American resin used by the inka and Artisans in pasto, Colombia. *J. Am. Inst. Conserv.* **2015**, *54*, 123–148. [CrossRef]
16. Insuasty, B.; Argoti, J.; Altarejos, J.; Cuenca, G.; Chamorro, E. Caracterización Fisicoquímica preliminar de la resina del Mopa-Mopa (Elaeagia pastoensis Mora), Barniz de Pasto. *Scientia et Technica* **2007**, *1*, 365–368.
17. Mina, J.; Anderson, S.; Bolaños, C.; Toro, E. Preparación y caracterización físico-química y térmica de mezclas binarias de resina Mopa-Mopa (elaegia pastoensis Mora) y policaprolactona (PCL). *Revista Latinoamericana de Metalurgía y Materiales* **2012**, *32*, 176–184.
18. Toro, E.; Mina, J.; Bolaños, C. Estudio físico-mecánico y térmico de una resina natural Mopa-Mopa acondicionada a diferentes humedades relativas. *Biotecnología en el Sector Agropecuario y Agroindustrial* **2013**, *11*, 30–36.
19. Gañán, P.F.; Mondragon, I. Surface modification of fique fibers. Effect on their physico-mechanical properties. *Polym. Compos.* **2002**, *23*, 383–394. [CrossRef]
20. Mina, J.; Valadez, A.; Muñoz, M. Micro- and Macro mechanical Properties of a Composite with a Ternary PLA–PCL–TPS Matrix Reinforced with Short Fique Fibers. *Polymers* **2020**, *12*, 58. [CrossRef]
21. Hoyos, C.G.; Vazquez, A. Flexural properties loss of unidirectional epoxy/fique composites immersed in water and alkaline medium for construction application. *Compos. Part B Eng.* **2012**, *43*, 3120–3130. [CrossRef]
22. Pereira, A.C.; De Assis, F.S.; Filho, F.D.C.G.; Oliveira, M.S.; Demosthenes, L.C.D.C.; Lopera, H.A.C.; Monteiro, S.N. Ballistic performance of multilayered armor with intermediate polyester composite reinforced with fique natural fabric and fibers. *J. Mater. Res. Technol.* **2019**, *8*, 4221–4226. [CrossRef]
23. Salazar, M.A.H.; Correa, J.P.; Hidalgo-Salazar, A.M.; Juan, P. Mechanical and thermal properties of biocomposites from nonwoven industrial Fique fiber mats with Epoxy Resin and Linear Low Density Polyethylene. *Results Phys.* **2018**, *8*, 461–467. [CrossRef]
24. Botina, R.; Mejía, M. Contribución al Conocimiento del Barniz o "Mopa-Mopa", Elaeagia pastoensis Mora. *Acta Agron.* **1987**, *37*, 56–65.
25. ASTM D638-14. *Standard Test Method for Tensile Properties of Plastics*; ASTM: West Conshohocken, PA, USA, 2014.
26. Valadez-Gonzalez, A.; Cervantes-Uc, J.; Olayo, R.; Franco, P.J.H. Chemical modification of henequén fibers with an organosilane coupling agent. *Compos. Part B Eng.* **1999**, *30*, 321–331. [CrossRef]
27. ASTM E104-02. *Standard Practice for Maintaining Constant Relative Humidity by Means of Aqueous Solutions*; ASTM: West Conshohocken, PA, USA, 2012.
28. ASTM D792-13. *Standard Test Methods for Density and Specific Gravity (Relative Density) of Plastics by Displacement*; ASTM: West Conshohocken, PA, USA, 2013.
29. ASTM D3822-07. *Standard Test Method for Tensile Properties of Single Textile Fibers*; ASTM: West Conshohocken, PA, USA, 2007.
30. Mina, J.H.; Valadez, A.; Franco, P.J.H.; Toledano, T. Influencia del tiempo de almacenamiento en las propiedades estructurales de un almidón termoplástico de yuca (TPS). *Ingeniería y competitividad* **2011**, *11*, 53–61. [CrossRef]
31. Mina, J. Preparación y caracterización fisicoquímica y mecánica de materiales compuestos de PLA/PCL/TPS reforzados con fibras de fique. Ph.D. Thesis, Universidad del Valle, Cali, Colombia, 2010.

32. Muñoz, M. Desarrollo y caracterización fisicoquímica y mecánica de un material compuesto de matriz PEBD/Al reforzado con fibras cortas de fique. Master's Thesis, Universidad del Valle, Cali, Colombia, 2012.
33. Muñoz-Vélez, M.F.; Salazar, M.A.H.; Hernandez, J.H.M. Effect of Content and Surface Modification of Fique Fibers on the Properties of a Low-Density Polyethylene (LDPE)-Al/Fique Composite. *Polymers* **2018**, *10*, 1050. [CrossRef]
34. Dikobe, D.; Luyt, A. Thermal and mechanical properties of PP/HDPE/wood powder and MAPP/HDPE/wood powder polymer blend composites. *Thermochim. Acta* **2017**, *654*, 40–50. [CrossRef]
35. Dehghani, S.; Peighambardoust, S.H.; Hosseini, S.V.; Regenstein, J.M. Improved mechanical and antibacterial properties of active LDPE films prepared with combination of Ag, ZnO and CuO nanoparticles. *Food Packag. Shelf Life* **2019**, *22*, 1–8. [CrossRef]
36. Shogren, R. Effect of moisture content on the melting and subsequent physical aging of cornstarch. *Carbohydr. Polym.* **1992**, *19*, 83–90. [CrossRef]
37. Delgado, A.; Aperador, W.; Gómez, W. Mejoramiento de las Propiedades de Tensión en WPC de LDPE: HIPS/Fibra Natural Mediante Entrecruzamiento con DCP. *Polímeros Ciência e Tecnología.* **2014**, *24*, 291–299.
38. Lima, L.D.P.F.C.D.; Santana, R.M.C.; Chamorro, C.D. Influence of Coupling Agent in Mechanical, Physical and Thermal Properties of Polypropylene/Bamboo Fiber Composites: Under Natural Outdoor Aging. *Polymers* **2020**, *12*, 929. [CrossRef] [PubMed]

 © 2020 by the authors. Licensee MDPI, Basel, Switzerland. This article is an open access article distributed under the terms and conditions of the Creative Commons Attribution (CC BY) license (http://creativecommons.org/licenses/by/4.0/).

Article

Characterization Study of Empty Fruit Bunch (EFB) Fibers Reinforcement in Poly(Butylene) Succinate (PBS)/Starch/Glycerol Composite Sheet

Rafiqah S. Ayu [1], Abdan Khalina [2,*], Ahmad Saffian Harmaen [1], Khairul Zaman [3], Tawakkal Isma [2], Qiuyun Liu [4], R. A. Ilyas [5] and Ching Hao Lee [1,*]

1. Laboratory of Biocomposite Technology, INTROP, Universiti Putra Malaysia, Serdang 43400, Selangor, Malaysia; ayu.rafiqah@yahoo.com (R.S.A.); harmaen@upm.edu.my (A.S.H.)
2. Engineering Faculty, UPM, Serdang 43400, Selangor, Malaysia; intanamin@upm.edu.my
3. Polycomposite Sdn Bhd, Jalan Maharajalela, Hilir Perak 36000, Perak, Malaysia; dr.khairulz@gmail.com
4. The BioComposites Centre, Bangor University, Bangor LL57 2UW, UK; q.liu@bangor.ac.uk
5. Advanced Engineering Materials and Composites Research Centre (AEMC), Department of Mechanical and Manufacturing Engineering, Universiti Putra Malaysia, Serdang 43400, Selangor, Malaysia; ahmadilyasrushdan@yahoo.com
* Correspondence: khalina@upm.edu.my (A.K.); leechinghao@upm.edu.my (C.H.L.)

Received: 24 May 2020; Accepted: 22 June 2020; Published: 15 July 2020

Abstract: In this study, a mixture of thermoplastic polybutylene succinate (PBS), tapioca starch, glycerol and empty fruit bunch fiber was prepared by a melt compounding method using an industrial extruder. Generally, insertion of starch/glycerol has provided better strength performance, but worse thermal and water uptake to all specimens. The effect of fiber loading on mechanical, morphological, thermal and physical properties was studied in focus. Low interfacial bonding between fiber and matrix revealed a poor mechanical performance. However, higher fiber loadings have improved the strength values. This is because fibers regulate good load transfer mechanisms, as confirmed from SEM micrographs. Tensile and flexural strengths have increased 6.0% and 12.2%, respectively, for 20 wt% empty fruit bunch (EFB) fiber reinforcements. There was a slightly higher mass loss for early stage thermal decomposition, whereas regardless of EFB contents, insignificant changes on decomposition temperature were recorded. A higher lignin constituent in the composite (for high natural fiber volume) resulted in a higher mass residue, which would turn into char at high temperature. This observation indirectly proves the dimensional integrity of the composite. However, as expected, with higher EFB fiber contents in the composite, higher values in both the moisture uptake and moisture loss analyses were found. The hydroxyl groups in the EFB absorbed water moisture through formation of hydrogen bonding.

Keywords: empty fruit bunch fiber (EFB); polybutylene succinate (PBS); starch; glycerol; characterizations; biocomposite; polymer Blends

1. Introduction

The development of biodegradable materials has attracted much research interest by scientists on worldwide. Aliphatic polyesters are among the most promising materials for the production of high-performance biodegradable plastics. One of the polyesters, polybutylene succinate (PBS) which is commercially available in the market, has very high fame as a high-performed bioplastic [1]. Many recent studies have selected PBS as the composite matrix for various applications and purposes [2–4].

PBS is synthesized from succinic acid and 1,4-butanediol (BDO) via a polycondensation process, and exhibits balanced performance in thermal and mechanical properties as well as processability [5]. It is more thermally stable than PLA polymer [6]. PBS is able to undergo biodegradation and even disposal in compost, moist soil, fresh water (by activated sludge), or sea water. It also can be composted by microorganism activities to convert it into CO_2, H_2O, and inorganic products under aerobic conditions, or CH_4, CO_2, and inorganic products under anaerobic conditions. The biodegradability of PBS depends mainly on its chemical structure and especially on its hydrolysable ester bond in the main chain, which is susceptible to microbial attack [7,8]. One study prepared a reactive-PBS polymer (RPBS) with insertion of toluene-2,4,diisocyanate (TDI) chemical in different ratios and blends with starch. The properties of the blended specimens were found to be significantly improved, even with only 10 wt% of RPBS. The TDI chemical insertion smoothened the PBS/starch polymer blend's surface, showing better miscibility of the two phases [9]. However, PBS has some negative properties such as slow crystallization rate, low melt viscosity, and softness. These have restricted its processing condition and potential applications. Polymer mixing with other materials is commonly used, to develop new blend materials that are suitable for specific working environments or specific purposes. However, most of the polymers are not miscible with each other and tend to phase-separate in a melt state [10]. Besides, although a fast crystallization reaction can happen when mixing with other materials, this may cause deterioration of PBS composite's strength [11]. Therefore, plasticizers such as glycerol were added to overcome and improve the flexibility of PBS polymer [12]. The council of the IUPAC (International Union of Pure and Applied Chemistry) has defined a plasticizer as "a substance or material incorporated in a material (usually a plastic or elastomer) to increase its flexibility, and workability by lowering glass transition temperature (T_g)" [13]. Glycerol is a pure anhydrous structure and has a specific gravity of 1.261 g·mL^{-1}, melting point of 18.2 °C and boiling point of 290 °C under normal atmospheric pressure [14]. On the other hand, grafting is another method to improve a compatibilizer between two materials. Suchao-in et al., 2013, have grafted PBS on tapioca starch blends. Results revealed a strong interfacial adhesion of the blend and enhanced modulus properties, as evidenced from SEM micrographs [15].

Starch is one of the materials that is readily available, low cost and one of the important bioresources used in the food industry, e.g., as a thickener and gelling agent. It also possesses good physical, mechanical and oxygen barrier properties, that give it potential to become active film [6,16]. It is much more reliable and chemically stable than other spacers [17]. Starch is a natural polymeric product and is found in almost every plant. Usually the main sources of starch come from tapioca, potato, maize, rice and wheat [18]. Starch contains two different molecular structures, linear (1,4)-linked α-D-glucan amylose and highly (1,6)-branched α-D-glucan amylopectin. The starch molecules are tied by van der Waals bonds and strong intermolecular hydrogen bonds. Common native starch granules have a semi-crystalline, radially oriented spherulitic structure. They contain water on different structural levels [19]. Amylopectin consist of a branching chain that forms double helices and produce crystalline structure of the granules, whereas amylose is amorphous and interspersed among amylopectin molecules [20]. Some starch polymers form helical structures due to the existence of α linkages, which contribute to its extraordinary properties and enzyme digestibility [21]. The relative amounts of amylose and amylopectin depend upon the plant source. Corn starch granules typically contain approximately 70% amylopectin and 30% amylose [22]. However, native starch itself cannot be satisfactorily used due to its hydrophilicity and brittleness which lead to the poor mechanical properties, so it requires some chemical modification to overcome this drawback [23]. Blending thermoplastic starch with PBS is one of the frequently selected options by researchers. Higher water resistance, good processability, fully biodegradable, and superior mechanical properties were being claimed for PBS/corn starch blend with glycerol plasticizers [24].

On the other hand, extensive investigation has been carried out to study the effects of natural fiber reinforcement on polymer composites [25–27]. The majority of outcomes have agreed that reinforced natural fiber has a better performing load transfer mechanism, and results in higher mechanical properties [28,29]. Empty fruit bunch (EFB) fibers have shown comparable quality to high strength kenaf bast fibers [30]. However, the hydrophilic nature of the EFB fiber is found to be incompatible with the hydrophobic polymer matrix. This caused poor interfacial adhesion between the fiber and matrix, leading to lower performances. Chemically treated EFB fibers had greater thermal and morphologies properties [31]. Moreover, it consists of wood-like constituents (cellulose, hemicellulose and lignin), showing lower thermal stability towards high heat environments, yet producing high residue at high temperature [32]. Furthermore, the hydrophilic behavior is expected to have higher moisture absorption, leading to swelling of the EFB fiber. Nevertheless, the extremely low cost of EFB fiber as a byproduct and its 100% biodegradable properties have created a high interest in it [33].

This study is a continuation of previous study, which investigated the characterization of high volume contents of EFB fiber reinforced in PBS/tapioca starch composite [34]. The high volume of fiber reinforcement found deterioration of mechanical properties due to poor interfacial bonding, evidenced from SEM micrograph and this is not accepted by the market, and similar findings were reported that show a lower tensile strength when alkaline treated-sugarcane fibers were inserted without any plasticizers [35]. Hence, in the present study, a lower volume of EFB fiber was added into the PBS/starch composite sheet with glycerol plasticizers to improve compatibility. This study has filled the knowledge-of-gap on low EFB fiber reinforcement in PBS/starch composite sheet with plasticizer fillers. The outcomes of this investigation (mechanical, morphological and thermal characterization) could serve as valuable knowledge for future developments on EFB fiber reinforcement in polymer composite.

2. Experimental

2.1. Materials

PBS in the form of pallets were bought from PTT Public Company Limited in Thailand. Density of PBS is 1.26 g/cm^3. Tapioca starch in form of powder was obtained from PT Starch solution in Indonesia. Empty fruit bunch fiber (EFB) was used and obtained from Polycomposite Sdn Bhd in Negeri Sembilan. The EFB were chopped using a grinder machine and sieved to get an average 300–600 μ in size. Meanwhile, glycerol was purchased from Duro Kimia Sdn Bhd in Selangor. The properties of materials as tabulated in Table 1.

Table 1. Properties of polybutylene succinate (PBS), starch and empty fruit bunch (EFB) fiber.

Properties	PBS	Starch	Properties	EFB Fiber
Density (g/cm^3)	1.26 g/cm^3	0.63	Density	0.98 g/cm^3
MFR	5 g/10 min	None	Cellulose (%)	45
Color	White	White	Lignin (%)	23
Odor	No Odor	No Odor	Hemicellulose (%)	21
Melting Point	115 °C	None	Size Mesh (μ)	300–600
Molecular Weight	65,000 g/mol	692.7 g/mol	Moisture (%)	9.41

2.2. PBS Composite Preparation

The PBS pallets and EFB fiber was first dried in an oven at 80 °C to prevent excessive hydrolysis which can compromise physical properties of the polymer. Starch, glycerol and EFB were dry mixed in an industrial mixer machine and sieved to remove excessive lumps during the mixing process. Then, PBS and the mixed compound of starch/EFB/glycerol were added into an industrial counter rotating extruder feeder for a total of 300 kg per processing. After that, the compound was melted in an industrial extruder machine comprising 10 heat zones, which were set temperatures in between 115–145 °C with rotation speed of 80 RPM. As a result of the shear stress imposed on fibers during compounding, homogenization of PBS/starch/fiber/glycerol was carried out by cycling the mixture in the extruder for 15 min and then extruded through a 2 mm gauge strand die at a rate of 10 mm/s. The melted compound was then passed through a calendaring machine before producing a sheet. Then, the sheets were cut into shapes according to specific characterization testing. The image of the extruded compound is shown in Figure 1.

Figure 1. Sheet extrusion process.

2.2.1. Mechanical Properties (Tensile Properties)

The tensile testing of the composite was conducted using a 5 kN Bluehill INSTRON Universal Testing Machine. The test was carried out according to ASTM standard D-638. The specimens were cut into dog bone shape by a plastic molder machine with the specifications of 120 × 120 × 2 mm^3 of length, width and thickness respectively. The composites were gripped at a 30 mm gauge length and the crosshead speed was set at 2.0 mm/min. All specimens were kept in a conditioning room and the test was run at 22 °C and relative humidity (RH) at 55%. Seven specimens were tested per test condition.

2.2.2. Mechanical Properties (Flexural Properties)

Flexural test of the composite was performed using 5 kN Bluehill INSTRON Universal Testing Machine. Test samples were cut to the dimension of 70 × 15 × 2 mm^3 and three-point bending tests were performed according to ASTM D790 standard. The crosshead speed was set at 2 mm/min with a support span-to-depth ratio of 16:1. All specimens were kept in a conditioning room and the test was run at 22 °C with the relative humidity (RH) at 55%. Seven specimens were tested per test condition.

2.2.3. Morphological Analysis

Morphology of the samples was observed using Hitachi S-3400N scanning electron microscope (SEM) equipped with energy dispersive X-ray (EDX) under an accelerating voltage of 15 kV and at an emission current of 58 µA. The tensile-tested-samples were gold sputtered before observation to avoid the charging effect during sample examination. SEM helps to analyze the microscopic structure and characterization of the compound on the basis morphology and structural changes.

2.2.4. Thermal Analysis

The thermal stability of the samples was characterized using a TA Instruments Q500 thermogravimetric analyzer, TGA. About 6 mg of the sample was scanned from 30 to 700 °C at a heating rate of 20 °C min^{-1} under a nitrogen gas atmosphere.

2.2.5. Moisture Absorption and Moisture Loss Analysis

Sample sheets of rectangular shape with dimensions of 15 × 15 × 0.5 mm^3 were dried in a vacuum oven at 60 °C for 24 h and weighed prior to testing. The vacuum dried rectangular sheets were immersed in distilled water at 20 °C to determine the water absorption and soluble ratio. The sample was taken out to measure the water absorption and soluble ratio in a certain time, and then the same sample was vacuum dried to measure the weight loss of the sample. The weights of the original sample and the sample after water absorption were designated as W_0 and W_1, and the dry weight of the water extracted sample was designated as W_2. The value of moisture absorption was obtained by Equation (1):

$$\text{Moisture uptake} = \frac{W_1 - W_2}{W_2} \times 100\% \qquad (1)$$

with the value of the soluble ratio derived from Equation (2):

$$\text{Soluble ratio} = \frac{W_0 - W_2}{W_0} \times 100\% \qquad (2)$$

Three measurements were performed for each sample, and the result was reported as the average value. This procedure followed the short-term immersion standard method ASTM D570-98.

On the other hand, seven samples were prepared for the moisture content evaluation. The samples were placed in normal climatic conditions at room temperature (27 ± 2 °C) with 65% relative humidity of air for 24 h before being weighed. Percentages of moisture content were determined by using Equation (3). The samples were heated in the oven for 24 h at 105 °C. Before heating the samples were measured as M_0. After 24 h in the oven, the fiber was weighed again as M_1. Therefore:

$$\text{Moisture content (\%)} \frac{M_1 - M_0}{M_0} \times 100\% \qquad (3)$$

3. Result and Discussion

3.1. Mechanical Testing

Filler reinforcement is an important factor in determining mechanical properties of the composite. The most crucial factor that affects the mechanical properties of the fiber reinforced materials is its fiber/matrix interfacial adhesion. The strength of the interfacial bonding was determined by several factors, such as the nature of the fiber and polymer components, fiber aspect ratio and processing procedure [36,37]. The mechanical properties of the PBS composite are presented and illustrated in Table 2 and Figures 2 and 3, respectively. It was clearly shown that the tensile and flexural strength of specimens were decreased for fiber reinforcement up to 8 wt%. This is due to poor dispersion and incompatibility between fillers and the PBS matrix according to previous studies [38,39]. Fibers are unable to disperse evenly in the PBS matrix, creating high stress concentration spots, resulting in a dramatic reduction in tensile strength [40]. However, increments in tensile and flexural strength were observed, indicating that the reinforcing ability of the natural fibers has overcome the shortage from the interfacial adhesion factor. A previous study reported the same trend, that higher fiber contents led to an improvement in the tensile strength of the matrix due to the interaction related to the fiber contents [41].

On the contrary, there were relatively higher mechanical properties for a 0% EFB specimen (which contained 30 wt% of starch/glycerol with a 2:1 ratio) in a current study, when compared to a previous study, which only gave 16.12 and 21.78 MPa for tensile and flexural strength, respectively, for pure PBS polymer [34]. The insertion of starch supposedly reduces the composite's strength performance due to low compatibility [6]. However, the addition of glycerol has the adverse effect of strength deterioration by localization of a compatibilizer at the interface for a stable morphology from a SEM micrographic [42].

Accoding to Thirmizir et al., the flexural strength of PBS composites was higher than neat PBS polymer. [8]. Higher fiber loadings have improved the flexural strength due to mechanical interlocks found between fiber and matrix. The fiber/matrix mechanical interlocking was expected to act as a mechanism to withstand the bending force in flexural testing. On the other hand, flexural strength was reduced by 6% for 8 wt% EFB fiber reinforcement composites. This may be attributed to interruption of the continuous long polymer chain by the presence of hydrophilic lignocellulose. Similarly, higher flexural strength values were recorded for higher EFB fiber reinforcement specimens. EFB fibers work as a carrier of loads in the matrix, synchronized with the tensile performance.

Table 2. Mechanical properties of PBS/starch/glycerol and EFB blends.

Formulation [a]	Specimen	Tensile Strength (MPa)	Tensile Modulus (MPa)	Flexural Strength (MPa)	Flexural Modulus (MPa)
PBS 70% Starch/Glycerol 30%	0% EFB	19.04 ± 1.27	392.76 ± 28.16	29.08 ± 0.64	1049.13 ± 67.15
PBS 70%, EFB 8%, Starch/Glycerol 22%	8% EFB	15.96 ± 0.63	360.40 ± 26.17	27.17 ± 1.21	872.10 ± 42.36
PBS 70%, EFB 12%, Starch/Glycerol 18%	12% EFB	17.38 ± 1.29	376.33 ± 31.06	27.19 ± 0.56	845.17 ± 41.17
PBS 70%, EFB 16%, Starch/Glycerol 14%	16% EFB	19.19 ± 1.08	466.84 ± 29.14	29.20 ± 1.05	954.35 ± 51.25
PBS 70%, EFB 20%, Starch/Glycerol 10%	20% EFB	20.18 ± 0.72	497.95 ± 30.17	32.63 ± 1.14	1029.15 ± 54.15

[a] Starch/Glycerol in 2:1 ratio.

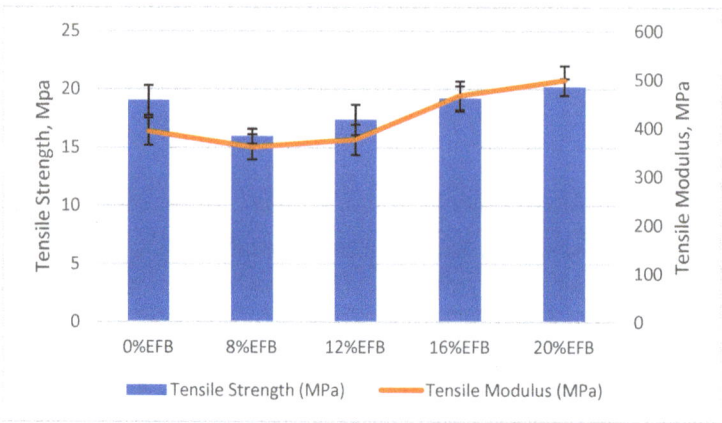

Figure 2. Tensile strength and tensile modulus of PBS composites.

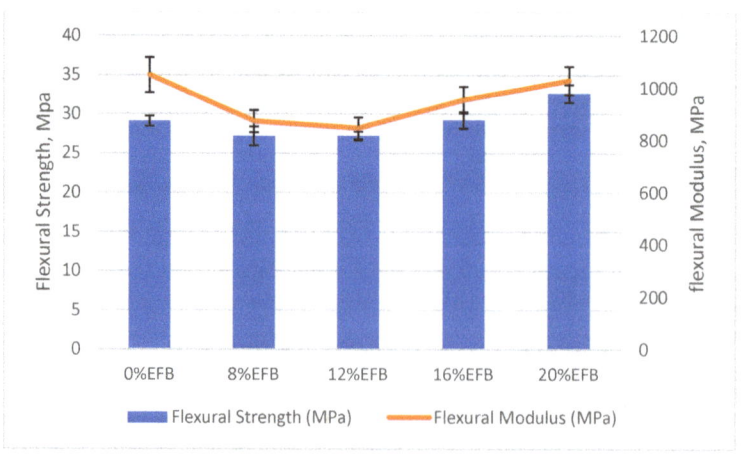

Figure 3. Flexural strength and flexural modulus of PBS composites.

3.2. Morphological Analysis

Figure 4 shows SEM images for specimens' surface morphology, under 500× magnification. The strength performances of the composites are directly affected by morphology status. Figure 4a shows a smooth and regular PBS surface, while Figure 4b shows the image of modified tapioca starch granules on the surface. Figure 4c,d shows the presence of EFB fiber, which consists of long fibers surrounded on the PBS/starch matrix. The poor adhesion of fibers on the matrix shows correlation with the reduction of mechanical properties for "8 wt% EFB" specimens. The poor impregnation makes it easier for the fiber to be pulled out, and causes a lower strength performance for the composite. This trend was also reported by a previous researcher [36]. For Figure 4e,f, it can be observed that the fibers are adhered to the matrix. The longitudinal fibrous shapes of the fibers were evenly mixed and evenly distributed on the matrix surface. The fibers mix homogenously with the matrix and are not clearly seen on the surface morphology analysis. This indicates the good fiber/matrix adhesion. On the other hand, the void between the EFB fiber and matrix is less, which gives a better fiber/matrix bonding and increased mechanical strength to the composite.

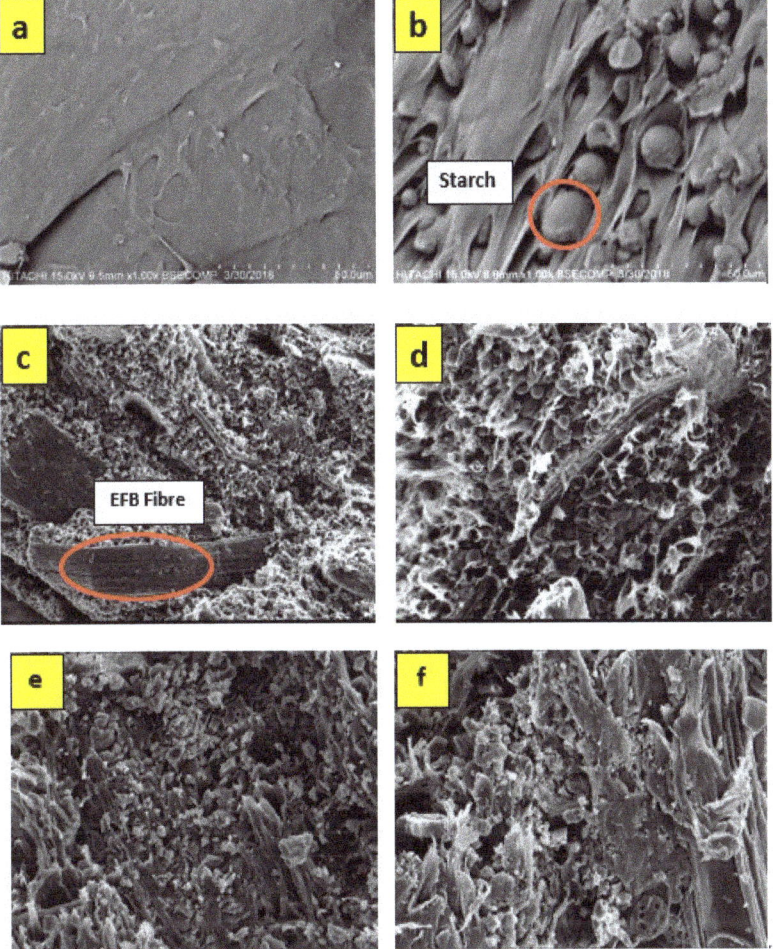

Figure 4. SEM micrographs for: (**a**) raw PBS, (**b**) 0% EFB, (**c**) 8% EFB, (**d**) 12% EFB, (**e**) 16% EFB and (**f**) 20% EFB.

3.3. Thermal Analysis

Thermogravimetric analysis (TGA) is a useful method for quantitative determination of the degradation behavior, thermal stability and mass change in a composite. The appearance of starch and glycerol in the PBS polymer composite has reduced the thermal stability of the specimen in generally. Excess amounts of glycerol have taken part in the reaction with hydroxyl groups of the PBS polymer, which promoted a lower thermal stability [43]. However, with more starch/glycerol contents replaced by EFB fibers, the effects of glycerol are lesser and gradually dominated by EFB fibers.

Figure 5 shows the TGA profiles of the EFB composites, while Table 3 lists the mass loss in every stage with peak temperature until sample reach 600 °C. There is a small but noticeable step between 75–95 °C, which was due to the presence of free water in the composite. Other researchers also have reported that this is due to water removal, as starch has a higher tendency to absorb moisture [6]. It also was reported that at the initial stage weight loss may be ascribed to the evaporation of water in the fiber [32,44,45]. Sharp transitions at peak 2 and 3 between 200–265 °C is due to decomposition of polysaccharide components in the starches. At higher temperatures, hemicellulose degradation occurs, followed by cellulose degradation [46]. Both degradation processes involve complex reactions (dehydration, decarboxylation, among others) as well as breakage of C-H, CO and C-C bonds [47]. Apart from this, lignin starts to degrade at a temperature range between 250–450 °C. Lignin degradation generates water, methanol, carbon monoxide and carbon dioxide [48,49]. PBS matrix is a thermally stable biopolymer and it begins to degrade near 300 °C with high degradation rates, as similarly found by Lee et al. [50]. In this analysis, there was slightly higher mass loss for early stage thermal decomposition whereas insignificant changes on decomposition temperature, regardless of EFB contents. However, with the higher lignin constituent in the composite there was a higher mass residue, which would turn into char at high temperature. This observation indirectly proves the dimensional integrity of the composite. Besides, the better mechanical performance for the high natural fiber reinforcement could offer wider the applications for this composite material.

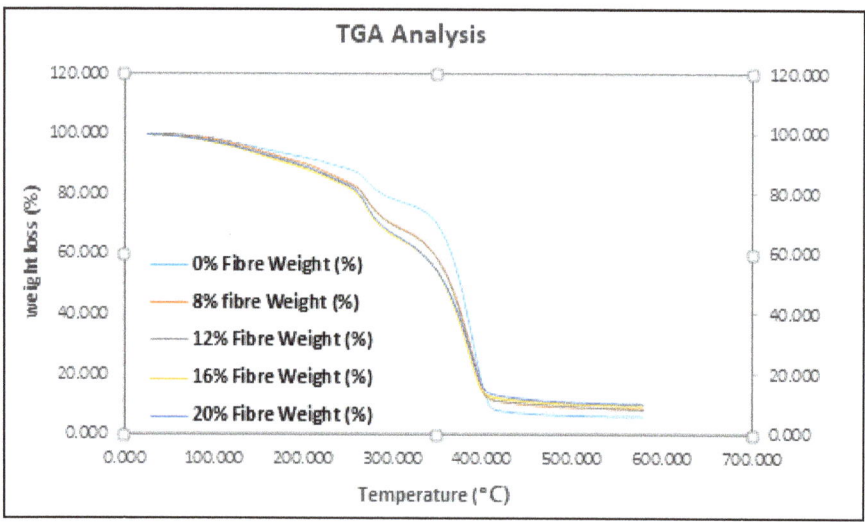

Figure 5. TGA profiles of EFB composites.

Table 3. A summary of peak temperatures for EFB composites.

Specimens	Peak 1, °C	Mass loss, %	Peak 2, °C	Mass loss, %	Peak 3, °C	Mass loss, %	Peak 4, °C	Mass loss, %	Mass Residue, %
0% EFB	78.13	7.161	214.45	3.404	261.43	12.19	362.05	71.22	5.995
8% EFB	95.13	8.086	209.19	6.774	262.82	16.63	358.92	59.95	8.492
12% EFB	87.89	9.534	209.61	6.129	262.12	16.42	358.85	59.63	8.176
16% EFB	88.50	9.885	-	-	254.79	25.07	357.29	55.38	9.568
20% EFB	84.87	5.940	201.68	6.634	260.63	21.82	357.93	55.32	10.16

3.4. Moisture Uptake and Average Loss of Moisture Contents

The amount of water absorbed in the composite was calculated by weight difference between before and after samples exposed to water. Figure 6 shows moisture uptake over the time and average loss of moisture contents for EFB composites. The moisture uptake test was conducted to identify the amount of water absorbed by the composites while the average loss of moisture content is to measure the mass loss after being subjected to heat. Generally, the moisture uptake was depending on several factors such as volume fraction of fiber, voids, viscosity of matrix, humidity and temperature [51].

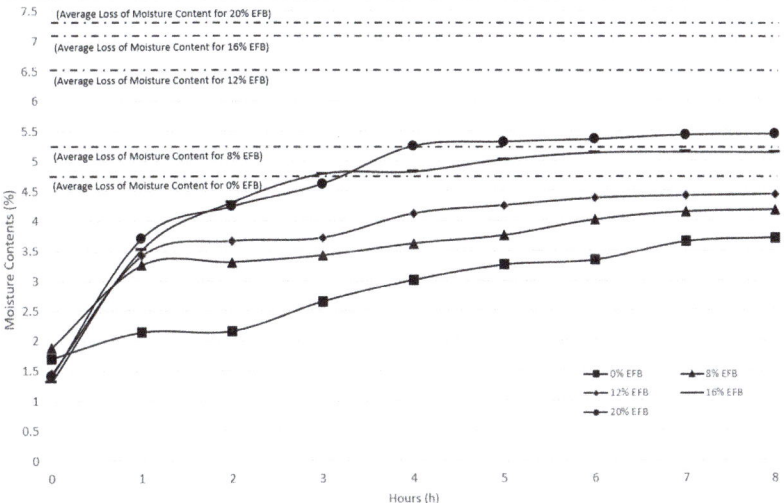

Figure 6. Moisture uptake analysis and average loss of moisture contents for EFB composites.

Water absorption is one of the disadvantages of applying lignocellulosic materials. Insertion of starch components into PBS polymer (0% EFB), comes with expected higher water absorption [52,53], as the starch component may take up to 300% of water absorption, as reported previously [9]. However, when a portion of the starch/glycerol is replaced by EFB fiber, higher values are found in both moisture uptake and loss analyses. This is because of the hydrophilic properties of the natural fibers in the poor interfacial bonding, leading to higher increments of moisture uptake, due to the presence of hydroxyl groups. Hence, it was observed that 20% EFB composite has the highest moisture uptake. The hydroxyl groups absorbed water moisture through formation of hydrogen bonding. The higher moisture content of the natural fiber may result in a weak interfacial bonding between the fiber and matrix [54]. The water molecules were absorbed in the inter-fibrillar space of the cellulosic structure that exists in the fiber and causes cracks and micro voids in the composite surface [55]. During immersion of the samples in water, capillarity action conducts water molecules to fill the voids, causing cracks and dimensional change. Swelling of fiber also leads to interfacial debonding and thereby reduction of mechanical strength [56,57]. In this study, at 6 to 8 h immersion, samples reached stable moisture contents, showing a saturation point, where no more water was absorbed. Similarly, when subjected to heat, the high EFB loadings composite loses a higher amount of water content. This shows that fiber reinforcement improves strength profiles, yet may cause higher susceptibility to moisture attack, thereby reducing overall composite properties.

4. Conclusions

In this study, the effect of fiber content on the mechanical and thermal properties of polybutylene succinate (PBS) composites were mainly evaluated. The control specimen (0% EFB) was compared with PBS polymer to discuss the changes affected by the appearance of starch/glycerol components. Generally, insertion of starch/glycerol provided better strength performance, but worse thermal and water uptake to all specimens.

On the other hand, it was found that there was poor interfacial adhesion between the EFB and PBS matrix, leading to lower mechanical properties. Fortunately, this was overcome and improved by higher fiber reinforcement, that regulated a better load transfer mechanism. Higher fiber loadings have improved the flexural strength due to mechanical interlocks found between the fiber and matrix. As a result, the tensile and flexural strength had increases of 6.0% and 12.2%, respectively, for 20 wt% EFB reinforcements.

In the SEM micrographic, it shows a smooth surface for PBS, while appearances of the EFB fiber show poor adhesion on the matrix, and was found to correlate with the mechanical properties analysis. On the other hand, the void between the EFB fiber and matrix was less and gave better fiber/matrix for a high fiber volume content composite.

A total of four thermal degradation peaks were recorded in the TGA analysis. The first peak was observed at 75–95 °C, due to the presence of free water in the composite. Sharp transitions at peak 2 and 3 between 200–265 °C were due to decomposition of the polysaccharide components in the starches and natural fibers. The last thermal decomposition peak was recorded at around 350 °C, which was responsible for the degradation of the PBS matrix. In this analysis, there was a slightly higher mass loss for early stage thermal decomposition, whereas insignificant changes on decomposition temperature were recorded, regardless of EFB contents. However, the higher lignin constituent in the composite had a higher mass residue, which would turn into char at high temperature. This observation indirectly proves the dimensional integrity of the composite. Moreover, the better mechanical performance of the high natural fiber reinforcement could offer wider applications for this composite material.

The moisture uptake over time and average loss of moisture contents for EFB composites were analyzed in this study. The higher the EFB fiber content in the composite, the higher values in both moisture uptake and loss data were found. This is expected due to the hydrophilic properties of the natural fibers that lead to higher increments of moisture uptake, due to the presence of hydroxyl groups. Hence, it was observed that 20% EFB composite has the highest moisture uptake. In this study, at 6 to 8 h immersion, samples reached a stable moisture content, showing a saturation point, where no more water was absorbed. Similarly, when subjected to heat, the high EFB loadings composite loses a higher amount of water content. This shows that fiber reinforcement improves the strength profile yet may cause higher susceptibility to moisture attack, thereby reducing overall composite properties.

As concluding remarks, the present results suggest that the use of 20% EFB fiber contents in the composite may be a potential candidate for effectively improving the properties and performances of the composite for future application. Nevertheless, the content of starch/glycerol may need to strategically planned to obtain a balance between performance and costing.

Author Contributions: Conceptualization, A.K. and R.A.I.; methodology, R.S.A.; validation, C.H.L.; formal analysis, Q.L.; investigation, R.A.I.; resources, A.S.H.; data curation, T.I.; writing—original draft preparation, R.S.A.; writing—review and editing, C.H.L.; visualization, K.Z.; supervision, A.K.; project administration, A.K.; funding acquisition, A.K. All authors have read and agreed to the published version of the manuscript.

Funding: This research was funded by Malaysian Industry Government Group of High Technology (MIGHT) under Newton Engku Omar Fund Grant no 6300873 (Safebiopack food packaging).

Conflicts of Interest: The authors declare no conflict of interest.

References

1. Sinha, R.S.; Okamoto, K.; Okamoto, M. Structure—Property relationship in biodegradable poly (butylene succinate)/layered silicate nanocomposites. *Macromolecules* **2003**, *36*, 2355–2367. [CrossRef]
2. Jiang, S.; Wei, Y.; Hu, Z.; Ge, S.; Yang, H.; Peng, W. Potential application of bamboo powder in PBS bamboo plastic composites. *J. King Saud Univ. Sci.* **2020**, *32*, 1130–1134. [CrossRef]
3. Li, M.; Wang, X.; Ruan, H.; Zhang, Q.; Wu, Z.; Liu, Y.; Lu, Z.; Hai, J. A facile cation-exchange approach to 2D PbS/amorphous MoSx heterojunction composites with enhanced photocatalytic activity. *J. Alloys Compd.* **2018**, *768*, 399–406. [CrossRef]
4. Nanni, A.; Messori, M. Thermo-mechanical properties and creep modelling of wine lees filled Polyamide 11 (PA11) and Polybutylene succinate (PBS) bio-composites. *Compos. Sci. Technol.* **2020**, *188*, 107974. [CrossRef]
5. Xu, J.; Guo, B.H. Poly (butylene succinate) and its copolymers: Research, development and industrialization. *Biotechnol. J.* **2010**, *5*, 1149–1163. [CrossRef]
6. Ayu, R.; Khalina, A.; Harmaen, A.; Zaman, K.; Jawaid, M.; Lee, C. Effect of modified tapioca starch on mechanical, thermal, and morphological properties of PBS blends for food packaging. *Polymers* **2018**, *10*, 1187. [CrossRef]
7. Tserki, V.; Matzinos, P.; Pavlidou, E.; Vachliotis, D.; Panayiotou, C. Biodegradable aliphatic polyesters. Part I. Properties and biodegradation of poly (butylene succinate-co-butylene adipate). *Polym. Degrad. Stab.* **2006**, *91*, 367–376. [CrossRef]
8. Thirmizir, M.A.; Ishak, Z.M.; Taib, R.; Rahim, S.; Jani, S.M. Kenaf-bast-fiber-filled biodegradable poly (butylene succinate) composites: Effects of fiber loading, fiber length, and maleated poly (butylene succinate) on the flexural and impact properties. *J. Appl. Polym. Sci.* **2011**, *122*, 3055–3063. [CrossRef]
9. Zeng, J.-B.; Jiao, L.; Li, Y.-D.; Srinivasan, M.; Li, T.; Wang, Y.-Z. Bio-based blends of starch and poly(butylene succinate) with improved miscibility, mechanical properties, and reduced water absorption. *Carbohydr. Polym.* **2011**, *83*, 762–768. [CrossRef]
10. Manson, J.A. *Polymer Blends and Composites*; Springer Science & Business Media: Berlin, Germany, 2012.
11. Ojijo, V.; Sinha Ray, S.; Sadiku, R. Role of specific interfacial area in controlling properties of immiscible blends of biodegradable polylactide and poly [(butylene succinate)-co-adipate]. *ACS Appl. Mater. Interfaces* **2012**, *4*, 6690–6701. [CrossRef]
12. Lavorgna, M.; Piscitelli, F.; Mangiacapra, P.; Buonocore, G.G. Study of the combined effect of both clay and glycerol plasticizer on the properties of chitosan films. *Carbohydr. Polym.* **2010**, *82*, 291–298. [CrossRef]
13. Vieira, M.G.A.; da Silva, M.A.; dos Santos, L.O.; Beppu, M.M. Natural-based plasticizers and biopolymer films: A review. *Eur. Polym. J.* **2011**, *47*, 254–263. [CrossRef]
14. Pagliaro, M.; Rossi, M. Glycerol: Properties and production. *Future Glycerol.* **2010**, *1*, 20–21.
15. Suchao-in, K.; Koombhongse, P.; Chirachanchai, S. Starch grafted poly(butylene succinate) via conjugating reaction and its role on enhancing the compatibility. *Carbohydr. Polym.* **2014**, *102*, 95–102. [CrossRef]
16. Fabunmi, O.O.; Tabil, L.G.; Chang, P.R.; Panigrahi, S. Developing biodegradable plastics from starch. In Proceedings of the ASABE/CSBE North Central Intersectional Meeting, Las Vegas, NV, USA, 27–30 July 2003; p. 1.
17. Mansourighasri, A.; Muhamad, N.; Sulong, A.B. Processing titanium foams using tapioca starch as a space holder. *J. Mater. Process. Technol.* **2012**, *212*, 83–89. [CrossRef]
18. Naz, M.; Sulaiman, S.; Ariwahjoedi, B.; Shaari, K.Z.K. Characterization of modified tapioca starch solutions and their sprays for high temperature coating applications. *Sci. World J.* **2014**, *2014*, 1–10. [CrossRef]
19. Xiaofei, M.; Jiugao, Y.; Jin, F. Urea and formamide as a mixed plasticizer for thermoplastic starch. *Polym. Int.* **2004**, *53*, 1780–1785. [CrossRef]
20. Ai, Y.; Jane, J.l. Gelatinization and rheological properties of starch. *Starch Stärke* **2015**, *67*, 213–224. [CrossRef]
21. Sanyang, M.L.; Ilyas, R.; Sapuan, S.; Jumaidin, R. Sugar palm starch-based composites for packaging applications. In *Bionanocomposites for Packaging Applications*; Springer: Berlin, Germany, 2018; pp. 125–147.
22. Avella, M.; De Vlieger, J.J.; Errico, M.E.; Fischer, S.; Vacca, P.; Volpe, M.G. Biodegradable starch/clay nanocomposite films for food packaging applications. *Food Chem.* **2005**, *93*, 467–474. [CrossRef]
23. Ariyanti, S.; Man, Z.; Bustam, M.A. Improvement of hydrophobicity of urea modified tapioca starch film with lignin for slow release fertilizer. *Adv. Mater. Res.* **2012**, *626*, 350–354. [CrossRef]
24. Li, J.; Luo, X.; Lin, X.; Zhou, Y. Comparative study on the blends of PBS/thermoplastic starch prepared from waxy and normal corn starches. *Starch Starke* **2013**, *65*, 831–839. [CrossRef]

25. Lee, C.H.; Sapuan, S.M.; Lee, J.H.; Hassan, M.R. Melt volume flow rate and melt flow rate of kenaf fibre reinforced Floreon/magnesium hydroxide biocomposites. *SpringerPlus* **2016**, *5*, 1680. [CrossRef]
26. Dashtizadeh, Z.; Khalina, A.; Cardona, F.; Lee, C.H. Mechanical characteristics of green composites of short kenaf bast fiber reinforced in cardanol. *Adv. Mater. Sci. Eng.* **2019**, *2019*, 6. [CrossRef]
27. Samadi, M.; Zainal Abidin, Z.; Yoshida, H.; Yunus, R.; Awang Biak, D.R.; Lee, C.H.; Lok, E.H. Subcritical water extraction of essential oil from Aquilaria malaccensis leaves. *Sep. Sci. Technol.* **2019**, 1–20. [CrossRef]
28. Sumesh, K.R.; Kanthavel, K.; Kavimani, V. Peanut oil cake-derived cellulose fiber: Extraction, application of mechanical and thermal properties in pineapple/flax natural fiber composites. *Int. J. Biol. Macromol.* **2020**, *150*, 775–785. [CrossRef]
29. Jeyapragash, R.; Srinivasan, V.; Sathiyamurthy, S. Mechanical properties of natural fiber/particulate reinforced epoxy composites—A review of the literature. *Mater. Today Proc.* **2020**, *22*, 1223–1227. [CrossRef]
30. Anuar, N.I.S.; Zakaria, S.; Gan, S.; Chia, C.H.; Wang, C.; Harun, J. Comparison of the morphological and mechanical properties of oil Palm EFB fibres and kenaf fibres in nonwoven reinforced composites. *Ind. Crop. Prod.* **2019**, *127*, 55–65. [CrossRef]
31. Khalili, P.; Tshai, K.Y.; Kong, I. Comparative thermal and physical investigation of chemically treated and untreated oil palm EFB fiber. *Mater. Today Proc.* **2018**, *5*, 3185–3192. [CrossRef]
32. Lee, C.H.; Sapuan, S.M.; Hassan, M.R. Thermal analysis of kenaf fiber reinforced floreon biocomposites with magnesium hydroxide flame retardant filler. *Polym. Compos.* **2018**, *39*, 869–875. [CrossRef]
33. Peças, P.; Carvalho, H.; Salman, H.; Leite, M. Natural fiber reinforced composites in the context of biodegradability: A review. In *Encyclopedia of Renewable and Sustainable Materials*; Hashmi, S., Choudhury, I.A., Eds.; Elsevier: Oxford, UK, 2020; pp. 160–178.
34. Ayu, R.S.; Khalina, A.; Harmaen, A.S.; Zaman, K.; Mohd Nurrazi, N.; Isma, T.; Lee, C.H. Effect of empty fruit brunch reinforcement in polybutylene-succinate/modified tapioca starch blend for agricultural mulch films. *Sci. Rep.* **2020**, *10*, 1166. [CrossRef]
35. Huang, Z.; Qian, L.; Yin, Q.; Yu, N.; Liu, T.; Tian, D. Biodegradability studies of poly(butylene succinate) composites filled with sugarcane rind fiber. *Polym. Test.* **2018**, *66*, 319–326. [CrossRef]
36. Khalid, M.; Ratnam, C.T.; Chuah, T.; Ali, S.; Choong, T.S. Comparative study of polypropylene composites reinforced with oil palm empty fruit bunch fiber and oil palm derived cellulose. *Mater. Des.* **2008**, *29*, 173–178. [CrossRef]
37. Pochiraju, K.; Tandon, G.; Pagano, N. Analyses of single fiber pushout considering interfacial friction and adhesion. *J. Mech. Phys. Solids* **2001**, *49*, 2307–2338. [CrossRef]
38. Siyamak, S.; Ibrahim, N.A.; Abdolmohammadi, S.; Yunus, W.M.Z.B.W.; Rahman, M.Z.A. Enhancement of mechanical and thermal properties of oil palm empty fruit bunch fiber poly (butylene adipate-co-terephtalate) biocomposites by matrix esterification using succinic anhydride. *Molecules* **2012**, *17*, 1969–1991. [CrossRef]
39. Wu, C.-S.; Liao, H.-T.; Jhang, J.-J. Palm fibre-reinforced hybrid composites of poly (butylene succinate): Characterisation and assessment of mechanical and thermal properties. *Polym. Bull.* **2013**, *70*, 3443–3462. [CrossRef]
40. Then, Y.Y.; Ibrahim, N.A.; Zainuddin, N.; Ariffin, H.; Wan Yunus, W.M.Z. Oil palm mesocarp fiber as new lignocellulosic material for fabrication of polymer/fiber biocomposites. *Int. J. Polym. Sci.* **2013**, *2013*, 1–7. [CrossRef]
41. Ma, X.; Yu, J.; Kennedy, J.F. Studies on the properties of natural fibers-reinforced thermoplastic starch composites. *Carbohydr. Polym.* **2005**, *62*, 19–24. [CrossRef]
42. Wang, W.; Zhang, G.; Zhang, W.; Guo, W.; Wang, J. Processing and thermal behaviors of poly (butylene succinate) blends with highly-filled starch and glycerol. *J. Polym. Environ.* **2013**, *21*, 46–53. [CrossRef]
43. Yun, I.S.; Hwang, S.W.; Shim, J.K.; Seo, K.H. A study on the thermal and mechanical properties of poly (butylene succinate)/thermoplastic starch binary blends. *Int. J. Precis. Eng. Manuf. Green Technol.* **2016**, *3*, 289–296. [CrossRef]
44. Ok Han, S.; Muk Lee, S.; Ho Park, W.; Cho, D. Mechanical and thermal properties of waste silk fiber-reinforced poly (butylene succinate) biocomposites. *J. Appl. Polym. Sci.* **2006**, *100*, 4972–4980. [CrossRef]
45. Lee, C.H.; Sapuan, S.M.; Hassan, M.R. Mechanical and thermal properties of kenaf fiber reinforced polypropylene/magnesium hydroxide composites. *J. Eng. Fibers Fabr.* **2017**, *12*. [CrossRef]
46. Lee, C.H.; Salit, M.S.; Hassan, M.R. A review of the flammability factors of kenaf and allied fibre reinforced polymer composites. *Adv. Mater. Sci. Eng.* **2014**, *2014*, 8. [CrossRef]

47. Liminana, P.; Garcia-Sanoguera, D.; Quiles-Carrillo, L.; Balart, R.; Montanes, N. Development and characterization of environmentally friendly composites from poly (butylene succinate)(PBS) and almond shell flour with different compatibilizers. *Compos. Part B Eng.* **2018**, *144*, 153–162. [CrossRef]
48. Frollini, E.; Bartolucci, N.; Sisti, L.; Celli, A. Poly (butylene succinate) reinforced with different lignocellulosic fibers. *Ind. Crop. Prod.* **2013**, *45*, 160–169. [CrossRef]
49. Huang, M.; Luo, J.; Fang, Z. Solvent-thermal degradation of waste lignin over Ni-based catalysts in the presence of homogeneous acids. *Chem. Ind. For. Prod.* **2015**, *35*, 126–132.
50. Lee, S.M.; Cho, D.; Park, W.H.; Lee, S.G.; Han, S.O.; Drzal, L.T. Novel silk/poly (butylene succinate) biocomposites: The effect of short fibre content on their mechanical and thermal properties. *Compos. Sci. Technol.* **2005**, *65*, 647–657. [CrossRef]
51. Jawaid, M.; Khalil, H.A.; Khanam, P.N.; Bakar, A.A. Hybrid composites made from oil palm empty fruit bunches/jute fibres: Water absorption, thickness swelling and density behaviours. *J. Polym. Environ.* **2011**, *19*, 106–109. [CrossRef]
52. Phiriyawirut, M.; Mekaroonluck, J.; Hauyam, T.; Kittilaksanon, A. Biomass-based foam from crosslinked tapioca starch/polybutylene succinate blend. *J. Renew. Mater.* **2016**, *4*, 185–189. [CrossRef]
53. Dash, B. Water absorption, Xrd and ftir analysis of pbs-starch blended halloysite composites. *J. Drug Deliv. Ther.* **2017**, *7*. [CrossRef]
54. Yusoff, M.Z.M.; Salit, M.S.; Ismail, N. Tensile properties of single oil palm empty fruit bunch (OPEFB) fibre. *Sains Malays.* **2009**, *38*, 525–529.
55. Khalil, H.A.; Jawaid, M.; Bakar, A.A. Woven hybrid composites: Water absorption and thickness swelling behaviours. *BioResources* **2011**, *6*, 1043–1052.
56. Ahmed, K.S.; Vijayarangan, S. Experimental characterization of woven jute-fabric-reinforced isothalic polyester composites. *J. Appl. Polym. Sci.* **2007**, *104*, 2650–2662. [CrossRef]
57. Abdul Khalil, H.P.S.; Ismail, H.; Ahmad, M.N.; Ariffin, A.; Hassan, K. The effect of various anhydride modifications on mechanical properties and water absorption of oil palm empty fruit bunches reinforced polyester composites. *Polym. Int.* **2001**, *50*, 395–402. [CrossRef]

 © 2020 by the authors. Licensee MDPI, Basel, Switzerland. This article is an open access article distributed under the terms and conditions of the Creative Commons Attribution (CC BY) license (http://creativecommons.org/licenses/by/4.0/).

Article

Mechanical and Rheological Behaviour of Composites Reinforced with Natural Fibres

Mariana D. Stanciu [1,*], Horatiu Teodorescu Draghicescu [1], Florin Tamas [2] and Ovidiu Mihai Terciu [1]

1. Department of Mechanical Engineering, Transilvania University of Brasov, Eroilor 29, 500036 Brasov, Romania; draghicescu.teodorescu@unitbv.ro (H.T.D.); terciu.ovidiu.mihai@gmail.com (O.M.T.)
2. Department of Civil Engineering, Transilvania University of Brasov, Eroilor 29, 500036 Brasov, Romania; florin.tamas@unitbv.ro
* Correspondence: mariana.stanciu@unitbv.ro

Received: 23 May 2020; Accepted: 17 June 2020; Published: 22 June 2020

Abstract: The paper deals with the mechanical behaviour of natural fibre composites subjected to tensile test and dynamic mechanical analysis (DMA). Three types of natural fibre composites were prepared and tested: wood particle reinforced composites with six different sizes of grains (WPC); hemp mat reinforced composites (HMP) and flax reinforced composite with mixed wood particles (FWPC). The tensile test performed on universal testing machine LS100 Lloyd's Instrument highlights the elastic properties of the samples, as longitudinal elasticity modulus; tensile rupture; strain at break; and stiffness. The large dispersion of stress–strain curves was noticed in the case of HMP and FWPC by comparison to WPC samples which present high homogeneity of elastic–plastic behaviour. The DMA test emphasized the rheological behaviour of natural fibre composites in terms of energy dissipation of a material under cyclic load. Cole–Cole plots revealed the connection between stored and loss heat energy for studied samples. The mixture of wood particles with a polyester matrix leads to relative homogeneity of composite in comparison with FWPC and HMP samples which is visible from the shape of Cole–Cole curves. The random fibres from the hemp mat structure lead to a heterogeneous nature of composite structure. The elastic and viscous responses of samples depend on the interface between fibres and matrix.

Keywords: natural fibre composites; mechanical properties; elastic behaviour; viscous response

---

1. Introduction

The natural fibres, vegetal, animal or mineral, consist of sustainable resources for composite materials used both in industrial applications and building structures. The source of vegetal fibres is different parts of plants such as bust for jute, hemp, flax, ramie, kenaf, leaf for sisal, banana and manila hemp (abaca), seeds in the case of cotton, coir and oil palm, wood and grass stem [1–4]. All vegetal fibres contain cellulose, hemicellulose and lignin in various proportions. For composite materials, reinforcement can be done with continuous fibres or with short fibres, in yarn or mat form. The ratio of mechanical properties and low weight, the possibilities to design different volumetric composition with effects on the mechanical, thermal, optical and electrical properties, and the environment protection and biodegradable properties of some natural composites are demonstrated by numerous pieces of research [5–7].

The main criteria for assessing new natural fibre composites were usually [8–10]: the degree of capitalization of the vegetal raw material and of other materials; the efficiency in use of raw material sources; volume or surface density; the limit values of the resistances to different mechanical stresses (traction parallel and perpendicular to/on the surface of the plate, bending, shearing in different planes and directions, detachment, fatigue, etc.); the rigidity and elasticity of the products, expressed by the values of the longitudinal and transversal modulus of elasticity; the physical properties of lignocellulose composite materials (profile density, swelling coefficient and water absorption in different environments and periods, degree of penetration of different substances, humidity, content in crystalline substances, etc.); the ecological properties (volatile emissions, etc.); and biodegradation and recyclability. The natural fibre reinforced polymers are characterized by two stress–strain states: an elastic behaviour as a response to a fast applied force and a viscous behaviour as response to a slowly applied force. The fibre reinforcement plays an important role in elastic behaviour of polymeric composite, as the matrix contributes to the viscous behaviour. For instance, the most efficient volume fractions of fiber content is maximum 50–65 m% according to [11]. The tests to determine the flow and relaxation phenomena provide important information about the dimensional stability of the polymer [12,13]. Some studies have shown that the natural fibres used for reinforcement are thermally unstable above 200 °C, for which reason the used matrices are based on polyethylene (PE), PP, polyolefin, polyvinylchloride and polystyrene and thermosets (which can be cured below this temperature) [11–16].

Most studies on wood reinforced composites (plastic wood composites) have analysed either the influence of the particle volume fraction/resin, or the influence of the size of wood particles, or the influence of the fibre type used as reinforcing elements on the mechanical properties of composites [14,15]. Mechanical properties of wood plastic composites (WPC), determined by tensile and bending tests, show an increase of the elasticity modulus E and yield modulus at the same time with the increasing of the particle size between 0.1 mm and 0.2 mm. These mechanical parameters are lower for composites with larger particles (1.00–4.00 mm) [8,12–16]. The mechanical characteristics of plastic composites with wood sawdust as filler material were studied in previous research by [14–18]. They carried out a comparative analysis of the effects of water/seawater absorption on the degradation of mechanical properties obtained at bending, in the case of hybrid composites made of polyester resin reinforced with fiberglass E and wood flour as filler. Thus, the maximum flexural stress decreases by almost 33% after immersion in sea water of wood sawdust mixture and epoxy resin during 6572 h [15,16]. Other studies are addressed to degradation of lignocellulosic composites both by weathering, ageing or humidity, because, from a mechanical point of view, the exposure of lignocellulosic composites to different aggressive factors can produce modifications of the elastic characteristics, and can induce stress concentrations at the surface of the polymeric material (which can lead to its premature failure) [19–24]. The modifications of the elastic-dynamic properties of a composite made of small waste oak particles of various sizes and a polyester resin subjected to photo degradation by UV radiation and thermal degradations were studied by [16,25–31]. From the point of view of the change of the energy loss capacity (internal friction) after the exposure to UV radiations, the most stable are the specimens reinforced with 0.4 mm particles, whereas the most unstable are those reinforced with 1 mm particles [16]. Ref. [32] studied new lignocellulose composites with carbon nanotube, having improved the mechanical properties, stability and fire retardment. The mechanical behaviour of fibre-reinforced composites depends on the size of the fibres (diameter and length), their distribution in the composite structure, the strength and elasticity of the fibres, the chemical stability and thermal resistance of the matrix and the nature of the fibre-matrix interface. Depending on the main directions of stress in the final product, the layers can be oriented differently so as to ensure a multidirectional distribution of stress during applications [33,34]. Numerous studies have shown that the method of reinforcement, length and nature of the fibres influence the mechanical properties of the composite. There is a linear relationship between the increase of the fibre content and the increase of the elastic modulus of the composite [35–37]. The complexity of the composite behaviour consists of the differences between the rigidity of the components that lead to the development of shear stress at the interface between the matrix and the fibres [38–40]. The determination of the mechanical and dynamic properties of composites reinforced with natural fibres plays a key role in the different structures and applications for both the proper design of the structures and the prediction of their lifetime cycle. The finished products from natural fibre reinforced composites obtained from wood waste or hemp or flax and polyester resin are exposed to environmental and technological risks such as high relative humidity, temperature, UV radiation, vibrations, etc., reducing their resistance to different loads. The mechanical and dynamical characterisation of lignocelluloses composites based on wood particles, flax woven and polyester resin have not been investigated yet.

The aim of this paper is to examine the mechanical and dynamical properties of three different types of three categories of composites (wood particle reinforced composites, hemp mat reinforced composites and flax reinforced composite with mixed wood particles). The mechanical characterization of natural fibre composites is very important from the design and analysis as well as from the life prediction point of view, and it was obtained by a uni-axial tension test according to SR EN ISO 527-4, where the longitudinal Young's modulus, tensile of rupture and strain at break are obtained. The polyester resin as a matrix in the studied composites is used for outdoor applications due to its high resistance to environment factors. The natural fibre composites proposed and studied in this piece of research have a good potential to be used in different parts of automotive structures as door panels, ornaments or for the indoor parts of boats. The effects of thermal degradation in terms of glass transition and heat deflection temperature were detecting by thermal dynamical mechanical analysis. Dynamic mechanical analysis (DMA) was carried out in order to provide quantitative information about the performance of material to cyclic stress and variation of temperature. Based on dynamic properties determined by DMA (stiffness, energy dissipation), the viscous-elastic properties of manufactured natural composites can be improved in order to increase the quality of composites. For the producers of composite materials and for the users of the products made from them, the present research offers information about the possibilities of use versus limitations depending on the determined mechanical properties.

2. Materials

Because the aim of this study is to compare the mechanical characteristics of composites reinforced with natural fibres, the sample preparation is presented for each type of specimens: wood particle reinforced composites abbreviated as WPC, hemp mat and polyurethane resin abbreviated as HMP (in longitudinal and transversal direction of mat fibres), flax and wood particle reinforced composites (abbreviated as FWPC). The specimens have the specific shape and dimensions of tensile test composite materials, according to ASRO SR EN ISO 527 and were made by a hand lay-up process.

2.1. Natural Fibre Composites

2.1.1. Wood Particle Reinforced Composite (WPC)

The WPC specimens were obtained by mixing wood particles with polyester resin. The oak wood particles resulted from the mechanical processing of the wood logs from a Romanian factory. Oak particles were conditioned at 5–6% moisture content and their specific gravity was established by a pycnometer [41]. For the production of the lignocellulosic composites, in the first stage, the wood particles were sorted according to their size using granulometric sieves. Five classes of oak wood particles were obtained: less than 0.04 mm (coded WPC 0.04); between 0.04 ÷ 0.1 mm (coded WPC 0.10); between 0.1 ÷ 0.2 mm (WPC 0.20); between 0.2 ÷ 0.4 mm ((WPC 0.40); between 0.4 ÷ 1 mm ((WPC 1.00) and from 1 mm to 2 mm (WPC 2.00) (Table 1). From each type of grain sizes used as reinforcement, five specimens for the tensile test were prepared using the same volume fraction of 25% in a mixture with 440-M888 POLYLITE type polyester resin (Table 2), obtaining a total of 25 WPC type samples) (Figure 1a). For the DMA test, 2 specimens for the test with constant temperature and 2 specimens for the test with temperature variation (Figure 1b) were prepared from each type of wood particle reinforced composites The physical features of WPC samples for the tensile test are shown in Table 1 and for the DMA test in Table 3.

Table 1. The physical characteristics of samples for the tensile test.

Samples	No. of Samples	No. of Layers	Thickness [mm]	Width [mm]	Area [mm^2]	Gauge Length [mm]
WPC 0.04	5	1	4.10 ± 0.10	10 ± 0.5	42 ± 1.5	50
WPC 0.10	5	1	4.10 ± 0.10	10 ± 0.5	42 ± 1.5	50
WPC 0.20	5	1	4.25 ± 0.25	10 ± 0.5	43 ± 1.5	50
WPC 0.40	5	1	4.20 ± 0.25	10 ± 0.5	42 ± 1.5	50
WPC 1.00	5	1	4.20 ± 0.20	10 ± 0.5	42 ± 1.5	50
WPC 2.00	5	1	4.30 ± 0.20	10 ± 0.5	43 ± 1.5	50
HMPL	5	1	1.80 ± 0.20	10 ± 0.5	18 ± 1.5	50
HMPT	5	1	1.80 ± 0.20	10 ± 0.5	18 ± 1.5	50
FWPC_L	5	6	6.80 ± 0.50	10 ± 0.3	69 ± 4.8	50
FWPC_T	5	6	6.60 ± 0.60	10 ± 0.4	68 ± 5.5	50

Table 2. The characteristics of the polyester resin type 440-M888 Polylite, at 23 °C.

Properties	Units	Value	Tested Method
Brookfield Viscosity LVF	mPa·s(cP)	1100–1300	ASTM D 2196-86
Density	g/cm^3	1, 10	ISO 2811-2001
PH (max.)	mgKOH/g	24	ISO 2114-1996
Styrene content	% of weight	43 ± 2	B070
Curing time: 1% NORPOL PEROXIDE 1	Minutes	35–45	G020
Tensile Strength	MPa	50	ISO 527-1993
Longitudinal Elasticity Modulus	MPa	4600	ISO 5271993
Elongation	%	1.6	ISO 527-1993
Bending strength	MPa	90	ISO 178-2001
Elasticity modulus at bending	MPa	4000	ISO 178-2001
The shock resistance P4J	mJ/mm^2	5.0–6.0	ISO 179-2001
Volume contraction	%	5.5–6.5	ISO 3521-1976
Glass transition temperature	°C	62	ISO 75-1993

Table 3. The physical characteristics of the samples for the DMA test (Legend: DMA operated under constant temperature T = const.; DMA operated under temperature variation T).

Samples	No. of Samples T = const./T var.	Thickness [mm]	Width [mm]	Gauge Length [mm]
WPC 0.04	2/2	4.10 ± 0.10	10 ± 0.2	40
WPC 0.10	2/2	4.10 ± 0.10	10 ± 0.5	40
WPC 0.20	2/2	4.25 ± 0.25	10 ± 0.5	40
WPC 0.40	2/2	4.20 ± 0.25	10 ± 0.5	40
WPC 1.00	2/2	4.20 ± 0.20	10 ± 0.5	40
WPC 2.00	2/2	4.30 ± 0.20	10 ± 0.5	40
HMPL	2/2	1.80 ± 0.20	10 ± 0.5	40
HMPT	2/2	1.80 ± 0.20	10 ± 0.5	40
FWPC_L	2/2	6.80 ± 0.50	10 ± 0.3	40
FWPC_T	2/2	6.60 ± 0.60	10 ± 0.4	40

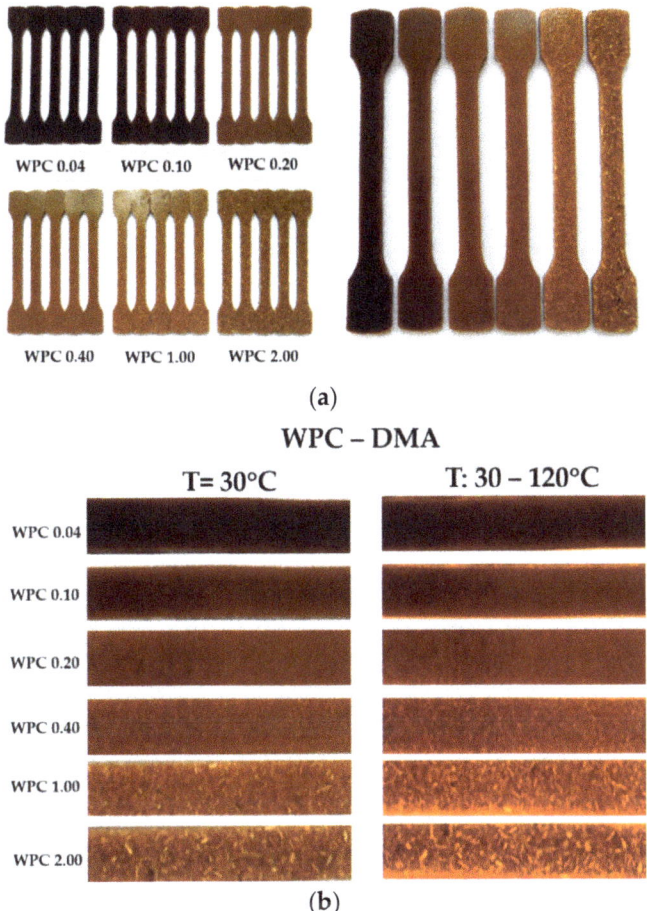

Figure 1. The specimens made from wood particles and polyester resin: (**a**) specimens for the tensile test; (**b**) samples for DMA.

2.1.2. Hemp Mat Reinforced Composites

The analysed composite material contains hemp mat and polyurethane resin (RAIGITHANE 8274/RAIGIDUR CREM), with 50% percent of reinforcing natural fibres. This type of natural fibre composite is used in the automotive industry for the interior panel of the car door. To evaluate the mechanical behaviour of composite reinforced with hemp fibres subjected to the tensile test, 5 samples on the longitudinal direction of the mat, coded HMPL were cut from the plate and 5 samples on the transversal direction of the mat, coded HMPT (Figure 2a). For the DMA test, 2 specimens for the test with constant temperature and 2 specimens for the test with temperature variation (Figure 2b) were prepared from each type of hemp mat reinforced composites. The physical features of hemp mat samples for tensile testing in Table 1 and for the DMA test in Table 3.

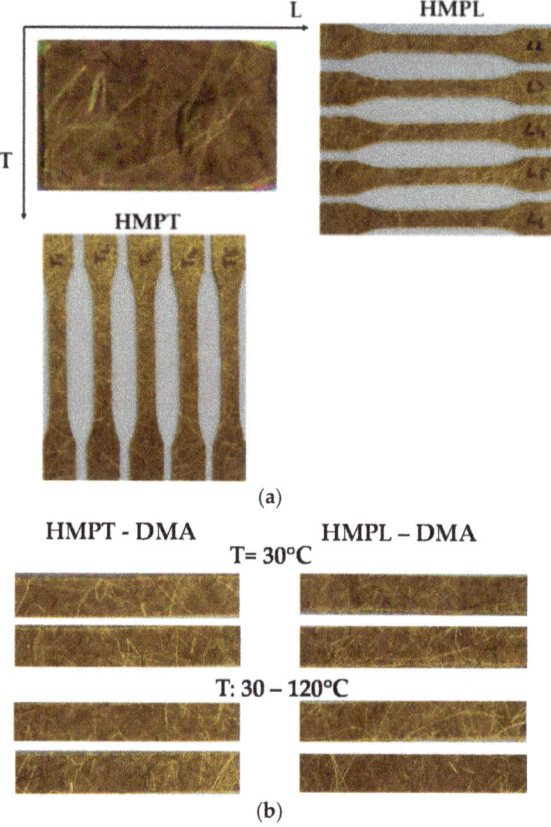

Figure 2. The specimens made from hemp mat reinforced and polyurethane resin: (**a**) geometry for tensile test: HMPL—sample cut in longitudinal direction; HMPT—sample cut in transversal direction; (**b**) geometry for the DMA test: HMPL—sample cut in longitudinal direction; HMPT—sample cut in transversal direction.

2.1.3. Flax Reinforced Composites

For these tests, the specimens were made of polyester resin reinforced with 6 layers of flax fabric and oak wood particles with dimensions between 0.1 ÷ 0.2 mm, arranged between layers as it can be seen in Figure 3. The 6 layers of fabric have the same orientation of the warp and weft threads, respectively. The total volume percentage of reinforcement with natural fibres, in this case is approx. 30%. Samples for the tensile test were cut from the composite plate on the two main directions of the fabrics, respectively the warp direction (named FWPC_L) and the weft direction (named FWPC_T) (Figure 3a). For the DMA test, from each type of FWPC 2 specimens for the test with constant temperature and 2 specimens for the test with temperature variation were prepared (Figure 3b). The geometrical characteristics of FWPC for the tensile test are presented in Table 1 and for the DMA test in Table 3.

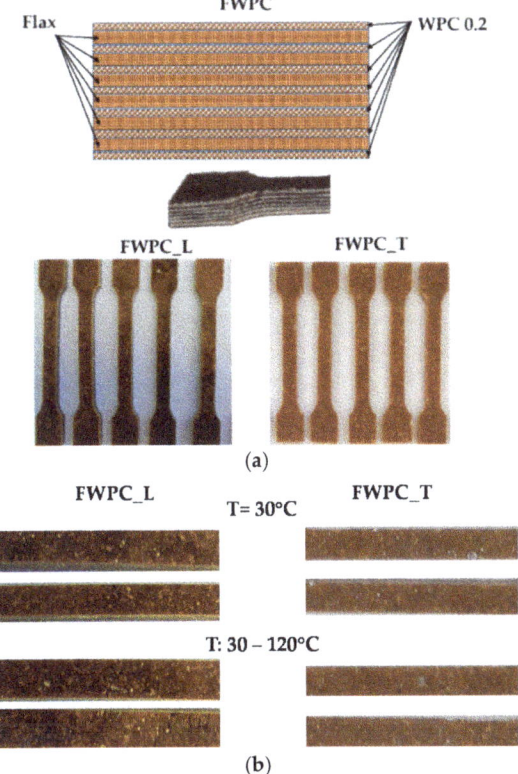

Figure 3. The specimens made from flax and wood particles reinforced and polyester resin: (**a**) samples for tensile test: FWPC_L—samples cut in longitudinal direction; FWPC_T—samples cut in transversal direction; (**b**) samples for DMA test.

2.2. Experimental Set-Up

2.2.1. Tensile Test

To determine the elastic characteristics of a material, the samples were subjected to a static tensile test. In this study, for the analysis of the mechanical behaviour of the composites, the specimens were tested on the universal testing machine LS100 Lloyd's Instrument belonging to the Mechanical Engineering Department of Transilvania University of Brașov. The specimens were loading with a constant speed of 1 mm/min until breaking. The elongation was measured simultaneously with loading using extension device (Figure 4a).

For data acquisition, the Nexygen Plus software was used. After the tensile tests (according to SR EN ISO 527-4), the characteristic curve, the specific deformation, the longitudinal elastic modulus, the rupture tension of each reinforced composite were determined, and on the basis of the load curves, the average deformation energy for each type of sample was calculated. The fracture of samples was analysed with optical devices.

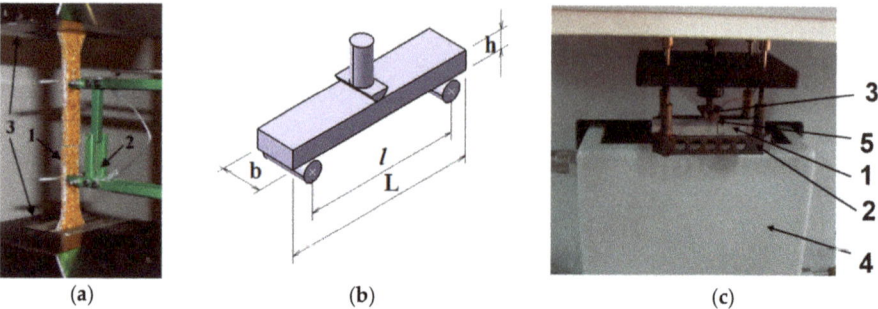

Figure 4. The experimental devices: (**a**) sample WPC 2.00 during the tensile test (Legend: 1—sample; 2—extension device for the elongation measurement; 3—tensile test machine jaws); (**b**) experimental set-up for the flexural test (3 points bending); (**c**) the DMA equipment (Legend: 1—sample; 2—sample supports; 3—loading device; 4—conditioner chamber; 5—temperature sensor).

2.2.2. Dynamic Mechanical Analysis

The rheological characteristics of the natural fibre composites were measured with the dynamic mechanical analysis by using the Dynamic Mechanical Analyzer DMA 242C Netzsch Germany at the Institute of Research and Development for Technical Physics in Iași. The method is based on ASTM D7028-07 which covers the procedure for the determination the glass transition temperature of polymer composites under the flexural oscillation mode. Thus, the complex modulus E*, with its two components (the conservation modulus E' and the loss modulus E") and the damping factor tan δ were determined in two cases: under isothermal conditions (T = 30 °C) for 30 min and with temperature variation between 30 to 120 °C, for 45 min. The specimens having the shape and geometry as it can be seen in Figure 4b were subjected to a flexural test. The input data were set up to 6 N for the applied force with frequency of 1 Hz. In Table 3, the specimen features for the DMA test are presented. For the DMA test, the samples have the same width and thickness as in the case of the tensile test and the length between supports is standard being set-up at 45 mm (Figure 4c).

3. Results and Discussion

The results in terms of qualitative and quantitative values of mechanical properties of natural fibre composites will be presented successively and then as a comparison. In the case of wood particle reinforced composites, it was noticed that, the smaller the reinforcing particles, the better the mechanical properties to traction. Nevertheless, the highest values of the elasticity modulus and of the tensile strength were obtained in the case of the composites reinforced with 0.2 mm particles. This is highlighted in the literature as well in the case of the polypropylene matrix composites (PP) [42]. The mechanical properties of the composites made of polypropylene and wood particles (WPC) in the case of tensile and bending tests show an increase in the values of the elasticity and resistance modulus simultaneous with the increase of the particle size between 0.25 and 2 mm, which is then followed by a slight decline of these values for larger particles (2–4 mm) [42]. The more elongated the particles, the more the mechanical properties increase because the contact between the reinforcing elements and the matrix occur over a larger surface. The first debonding of the matrix and the dispersed fibres occur near the breaking point. The characteristic curves for the same category of composites reinforced with wood particles did not display a large dispersion of values, the mixture between matrix and fibres being homogeneous (Figure 5).

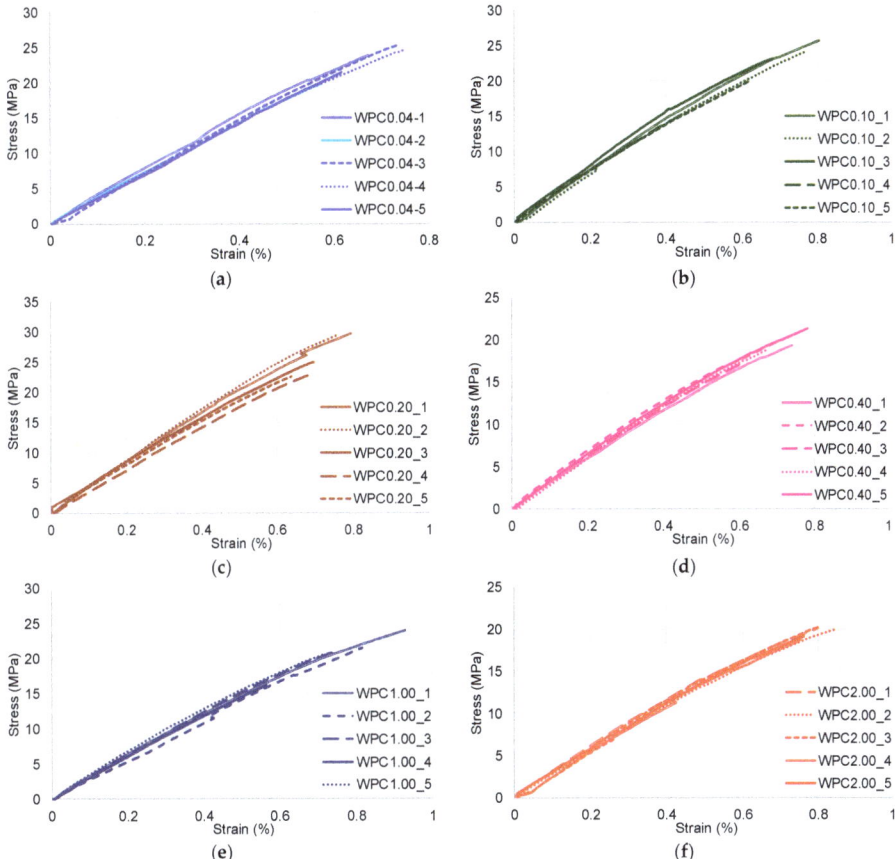

Figure 5. Characteristic curve stress–strain of wood particle reinforced composites: (**a**) stress–strain characteristic curves of WPC0.04 specimens; (**b**) stress–strain characteristic curves of WPC0.10 specimens; (**c**) stress–strain characteristic curves of WPC0.20 specimens; (**d**) stress–strain characteristic curves of WPC0.40 specimens; (**e**) stress–strain characteristic curves of WPC1.00 specimens; (**f**) stress–strain characteristic curves of WPC2.00 specimens.

Tensile tests performed by other researchers have shown that natural fibre fabric reinforced composite materials have major differences in the mechanical properties of traction in the direction of the warp and weft [42–45] as it can be seen in the case of HMP specimens (Figure 6a,b) and FWPC specimens (Figure 6c,d). It is known that the fibre mat is used as reinforcement to assure a quasi-isotropy of composite plates due to the random orientation of the fibres. Despite this assumption, the tests performed on hemp mat composites showed that there are significant differences between the tensile properties in the two directions of the mat (longitudinal and transversal direction). In the case of flax wood particle reinforced composites FWPC, the trend is similar to HMP regarding the mechanical properties in the weft and warp direction. Equally, it can be observed that HMP show a great spread of stress–strain curves in the longitudinal direction (Figure 6a) by comparison to FWPC which indicated a great dispersion of the curves in the transversal direction (Figure 6d).

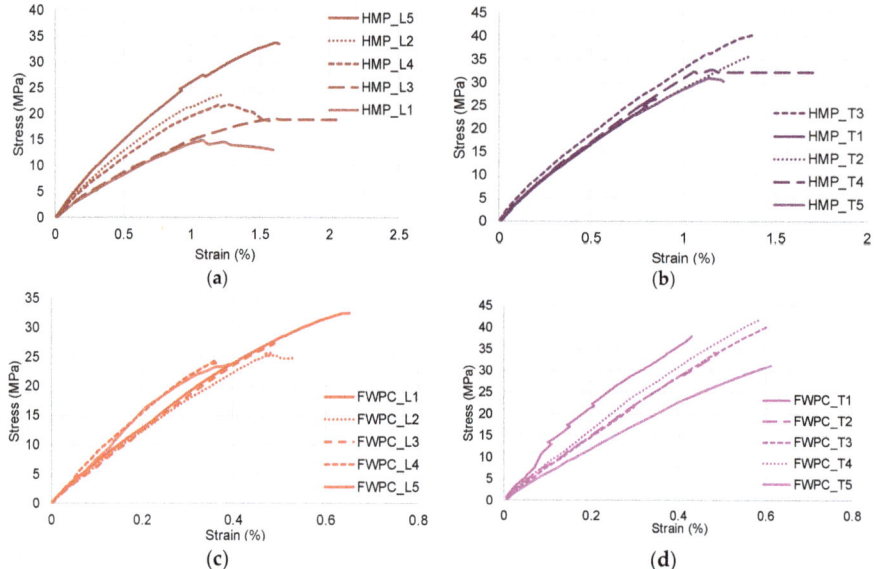

Figure 6. Characteristic curves stress–strain of natural fibres reinforced composites: (**a**) stress–strain curves for HMP_L samples; (**b**) stress–strain curves for HMP_T samples; (**c**) stress–strain curves for FWPC_L samples); (**d**) stress–strain curves for FWPC_T samples.

In Figure 7, it can be noticed that the FWPC samples present a rigid behaviour compared to the HMP_L samples which behave viscously. The WPC0.20 samples are also rigid by comparison to the other wood particle reinforced composite.

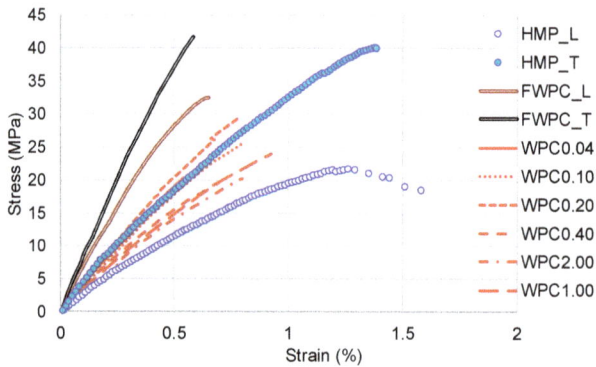

Figure 7. Comparison of stress–strain curves for tested samples.

The mechanical properties of tested samples in terms of the average values of the longitudinal elasticity modulus, stress at break, the strain and stiffness in percent are summarized in Table 4. In the case of WPC specimens, it can be observed that the modulus of elasticity varies from 2877 MPa, for the specimens reinforced with particles about $1 \div 2$ mm, up to the maximum value of 4012 MPa, for the specimens reinforced with particles about $0.1 \div 0, 2$ mm. The minimum values of the tensile strength were noted for specimens reinforced with particles about $1 \div 2$ mm (WPC 2.0), being 19.5 MPa, and the maximum being 26 MPa, for specimens reinforced with particles from 0.1 to 0.2 mm (WPC 0.2).

For the HMP specimens cut in the longitudinal direction, an average tensile strength value of 26 MPa was obtained, and in the weft direction 32 MPa, approximately 23% higher than in the longitudinal direction. In the case of the longitudinal modulus of elasticity, it is observed that, in the transverse direction, its value is approximately 63% higher than in the longitudinal direction (Table 3). For the FWPC specimens cut in the warp direction, tensile strength values between 23.9 and 27.3 MPa were obtained, and in the weft direction between 31.4 and 42 MPa. The longitudinal elasticity modulus is higher in the transversal direction (weft) by almost 95% than in the warp direction (longitudinal). Although the mechanical properties of lignocelluloses samples are relatively low, they can be used in some applications as exterior and interior products, being valuable for the possibility of integrating wood residues from processing operations or from recycled wood in the form of chips and fibres.

Table 4. Average values of elastic characteristics obtained after the tensile test. Legend: E—longitudinal elasticity modulus; STDV—standard deviation; σ_r—tensile of rupture; ε_r—percentage strain at break; k—stiffness.

Samples	E (MPa)	STDV E (MPa)	σ_r (MPa)	STDVσ_r (MPa)	ε_r (%)	k (10^6 N/mm)
WPC 0.04	3626	218	23	2	0.920	0.003052
WPC 0.10	3693	181	22	4	1.261	0.003124
WPC 0.20	4012	328	26	4	0.011	0.003518
WPC 0.40	3109	138	19	2	0.014	0.002762
WPC 1.00	3041	260	21	3	0.011	0.002683
WPC 2.00	2877	85	20	1	0.010	0.002589
HMPL	3086	934	26	9	1.550	1.110
HMPT	5005	569	32	5	1.199	1.802
FWPC_L	8586	1247	27	3	0.377	7.397
FWPC_T	16700	3500	38	3	0.37	8.854

3.1. Dynamic Mechanical Analysis

3.1.1. Isothermal Conditions

The rheological characteristics of the natural fibre composites in terms of complex modulus E *, with its two components (the storage modulus E' and the loss modulus E") and the damping tan δ were determined with the DMA. The storage modulus E' represents the capacity of materials to withstand the applied loading, being the expression of the elastic constant of the composite. The energy dissipation due to the internal friction of the material is called loss modulus E" and it represents the viscous modulus. The ratio between the loss and storage modulus represents damping (tan δ). tan δ is an indicator of how efficiently the material loses energy to molecular rearrangements and internal friction [46]. In Figures 8–10, the elastic and viscous responses of samples to dynamic loading with frequency of 1 Hz can be noticed. The capacity of WPC samples to store the deformation energy decreases slowly while increasing the time of loading. Similar behaviour is noticed in the case of the HMP samples with the mention that there are clear differences between the samples cut in the longitudinal direction compared to those cut in the transverse direction (Figures 8, 9 and 10b).

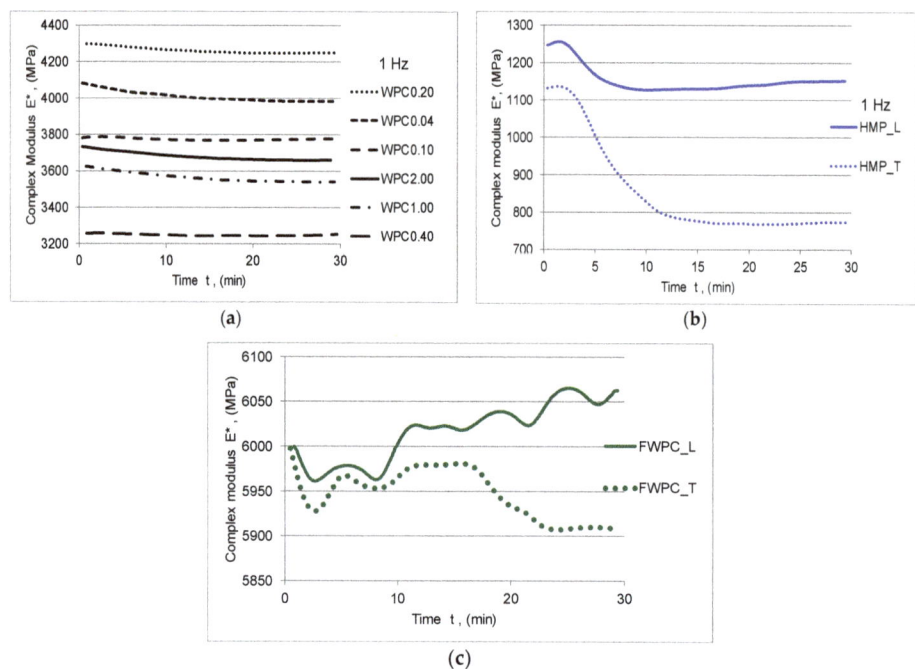

Figure 8. Variation of the complex modulus in time: (**a**) WPC samples; (**b**) HMP samples; (**c**) FWCP samples.

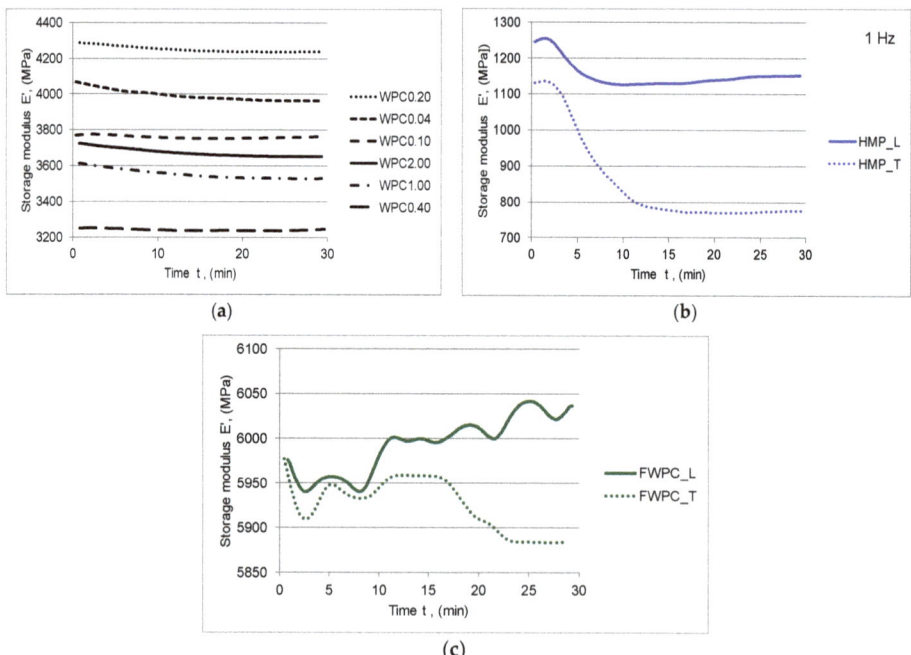

Figure 9. Variation of the storage modulus E' in time: (**a**) WPC samples; (**b**) HMP samples; (**c**) FWCP samples.

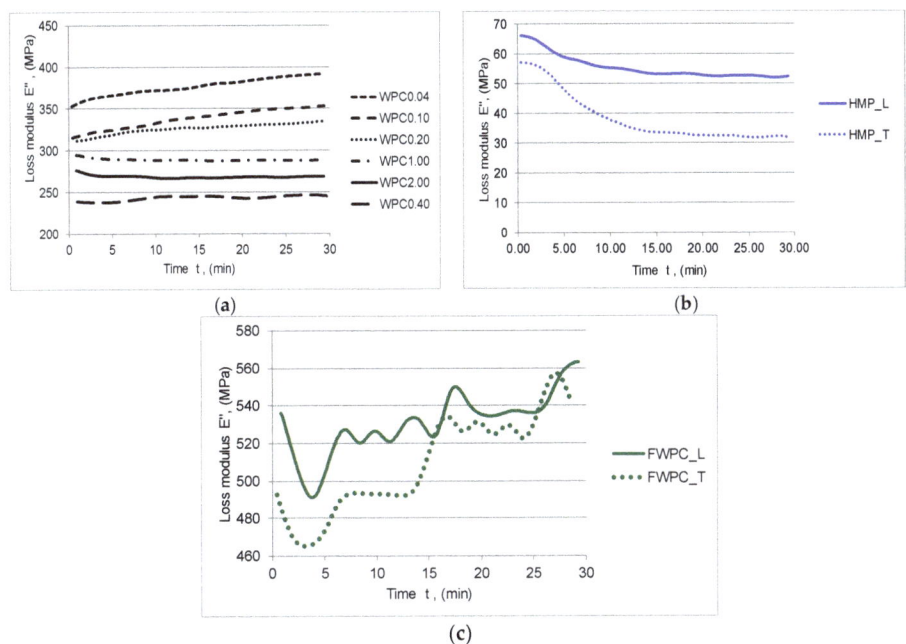

Figure 10. Variation of the loss modulus E" in time: (a) WPC samples; (b) HMP samples; (c) FWCP samples.

Both types of samples have a similar viscous-elastic behaviour at the beginning, so that, after 10 min of stress, the samples cut in the longitudinal direction show a stiffening phenomenon compared to the samples cut in the transverse direction whose elastic behaviour decreases suddenly after 18 min of cyclic loading (Figures 8 and 9c). As far as the values of complex modulus and storage modulus for each type of the tested samples are concerned, it can be noticed that the lower values are noticed in the case of the HMP sample by comparison to WPC and FWPC. Initially, there are 1120–1260 MPa and, after 5 min, they decreased by 12.5% in the case of HMP_L maintaining constant value for remaining time exposure to cyclic loading. The HMP_T indicated a higher decreasing by almost 28% after 10 min and then a stabilization of the values at around 790 MPa. In the case of the WPC samples, the complex modulus ranges between 3250 MPa (WPC0.40) and 4270 MPa (WPC0.20). It can be noticed that the values for the flexural test in a dynamic regime are similar to Young's modulus values obtained in the tensile test. FWPC samples obtained the highest values for the dynamic modulus (5900–6060 MPa) in comparison to the other types of specimens. The loss modulus increases by increasing the time exposure to cyclic loading in the case of WPC and FWPC samples while the HMP samples indicated a decrease of this viscous component (Figure 10). This behaviour is due to the type of matrix: the WPC and FWPC, which contain as matrix polyester resin, indicated a slightly higher network density and it is slightly more cross-linked than the HMP sample based on polyurethane resin. The minimum values of energy dissipation due to internal frictions are indicated in the case of HMP samples. In the longitudinal direction (HMP_L), the viscous modulus decreases by 28% during the cyclic loading and, in the transversal direction (HMP_T), the decrease is 50%. For FWPC, the internal frictions increase by increasing the exposure time to loading; the overall value varied from minimum 465 MPa to maximum 560 MPa (in the case of FWPC_T) and from minimum 490 MPa to maximum 565 MPa (in the case of FWPC_L). For the WPC samples, the loss modulus varied in accordance with the wood particle sizes: the highest value of energy dissipation is recorded for smaller wood particles (WPC0.04). Regarding the variation of loss modulus, two groups can be noticed: WPC0.04, WPC0.10, and WPC0.20 indicated an increase in internal friction and WPC1.00, WPC2.00, WPC0.40 indicated a slight decrease of the viscous modulus.

The damping tan δ as a ratio between the loss and the storage modulus is an expression of the energy dissipation of a material under cyclic load, and it depends on the interface and adhesion between fibres and matrix. Any rigid material is characterized by a high damping value, whereas any ductile material indicates a low damping value [47]. In this sense, the damping tan δ for the HMP samples tends to decrease by increasing the loading time, since, for the WPC and FWPC samples, the damping increases (Figure 11).

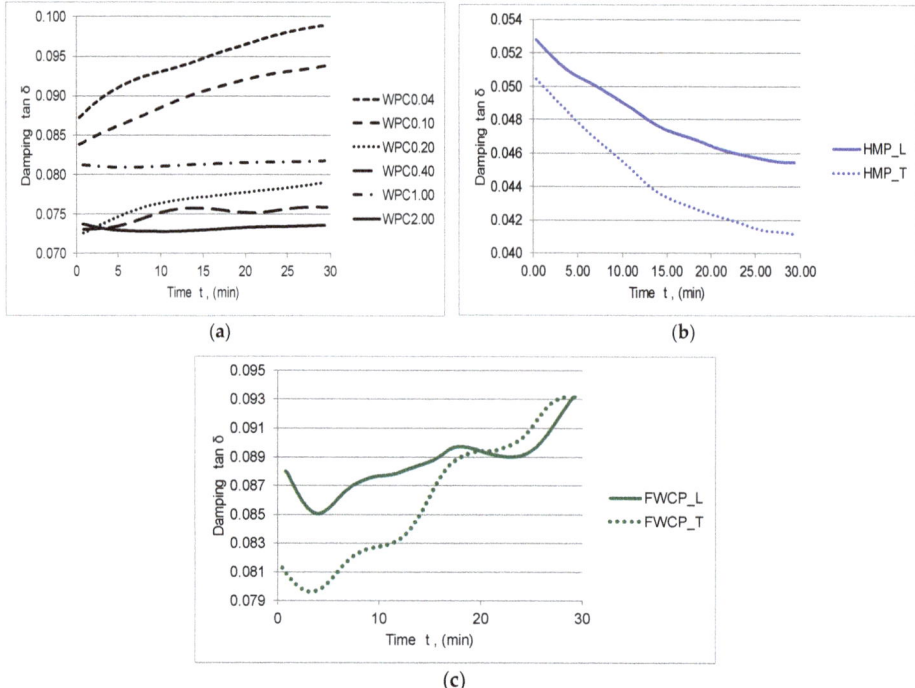

Figure 11. Variation of the damping tan δ in time: (**a**) WPC samples; (**b**) HMP samples; (**c**) FWCP samples.

3.1.2. Temperature Variation

The viscous-elastic behavior of composites is emphasized by the temperature variation during cyclic loading. The stiffness and rigidity stability of composites at a certain temperature can be observed on the storage modulus curves. Thus, similar behavior between the WPC and FWPC samples can be noticed in Figure 12a,c: the glassy region presented between 30 °C and 40 °C characterized the rigidity of composites due to polymeric chains; in the second region (40–80 °C), the storage modulus decreases drastically because by increasing the temperature, the internal friction in polymeric chains is accelerated leading to a rubbery region. The HMP composite behavior differs, their stiffness being affected right from the start of the test and decreasing by temperature increase (Figure 12b). Despite this behavior, the reinforcement with hemp fibers and also the reinforcement with flax fabric lead to higher values of storage modulus by comparison to wood particle reinforced composites.

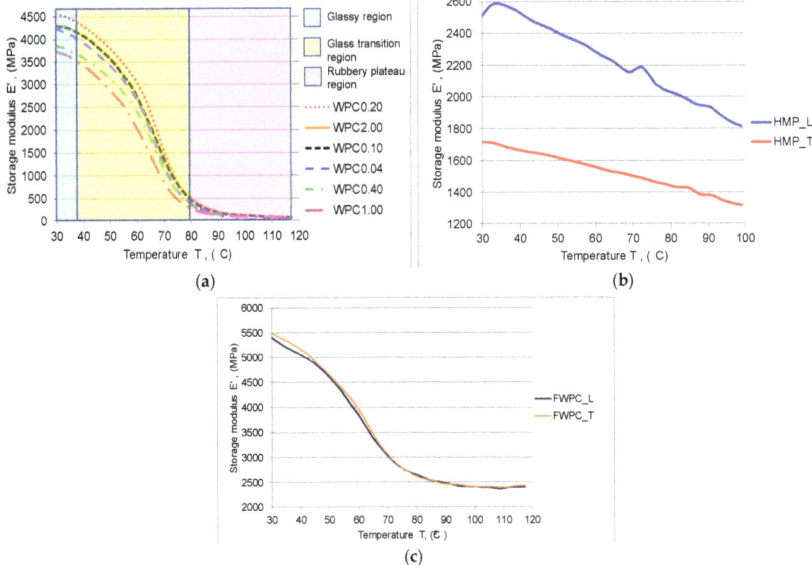

Figure 12. Variation of the storage modulus E' with temperature: (**a**) WPC samples; (**b**) HMP samples; (**c**) FWCP samples.

Figure 13 presents the effect of reinforcement on the damping as a function of temperature at 1 Hz frequency. From this type of chart, the glass transition temperature (T_g) can be extracted from the peak of damping variation curves.

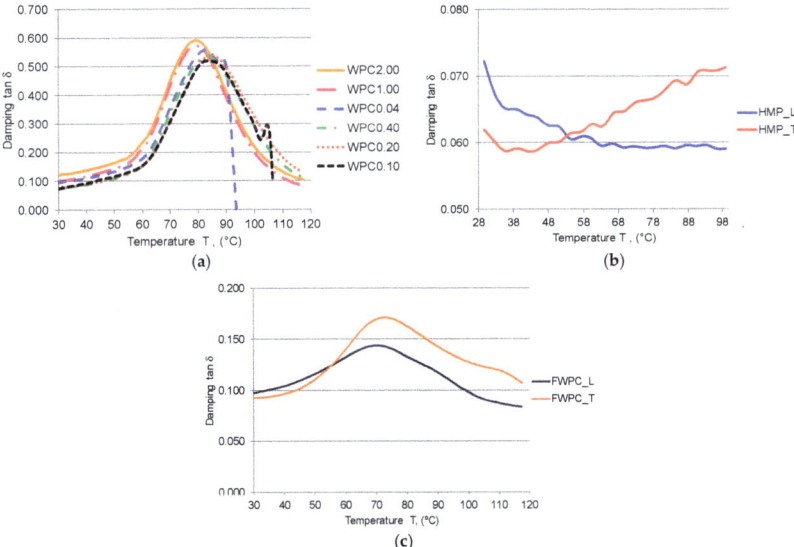

Figure 13. Variation of the damping tan δ with temperature: (**a**) WPC samples; (**b**) HMP samples; (**c**) FWCP samples.

According to [48–50], lower damping values represent the improved interfacial interaction as it can be noticed in the case of FWPC and HMP (Figure 13b,c), while a higher damping value is recorded in the case of poor interfacial adhesion as it can be observed in Figure 13a—the WPC samples. Thus, the T_g for the WPC samples is around 78 °C (for 2.00 mm and 1.00 mm wood particle sizes) and 82 °C for smaller wood particles. The T_g for FWPC is lower than the one for WPC, being 70 °C. The HMP samples indicated a different behaviour. In Figure 14, the Cole–Cole charts are shown in order to analyse the connection between stored and loss heat energy for the studied samples. As [49–51] highlights, this kind of plot is useful to interpret the modification of viscous-elastic material with different reinforcement, as it is illustrated in Figure 14a—the WPC samples. The mixture of wood particles with polyester matrix leads to relative homogeneity of composite by comparison to the FWPC and HMP samples which is visible from the shape of Cole–Cole curves. The random fibres from the hemp mat structure lead to a heterogeneous nature of the composite structure.

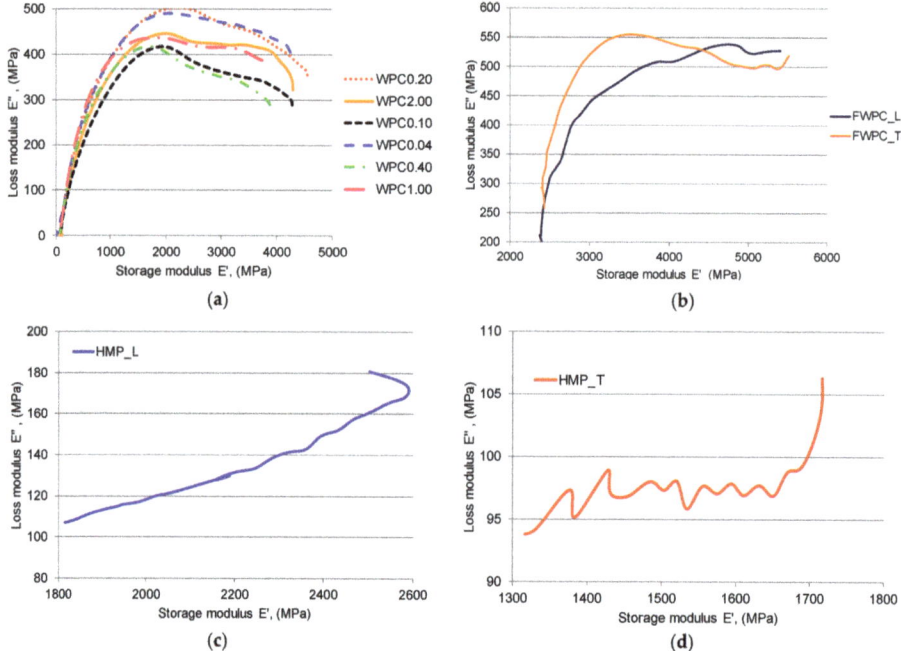

Figure 14. Cole–Cole plot: (**a**) WPC samples; (**b**) FWCP samples; (**c**) HMP_L sample; (**d**) HMP_T sample.

3.2. Fracture Analysing

In Figure 15, the ways of breaking of the tested composites are shown. Thus, it is noticeable that the breakage of the composite reinforced with wood particles (WPC) is produced through the simultaneous destruction of the matrix and of the dispersed fibres (Figure 15). In the case of the HMP samples, the matrix is fractured first; then, the dispersed fibres break. The plane of fracture is obtained on the area with the minimum resistance of the interface between the matrix and the fibres or in the area where the fibres are missing or do not have a good adhesion with the matrix. For the FWPC composites which contain both dispersed fibres (wood particles) and flax fabric, the fracture is produced first in the matrix mixed with wood particles and then the layers of flax fabric fail. It is appreciated that the use of a fabric structure (flax) doubles the tensile strength compared to mat reinforcement. The mechanism of failure differs between short fibres/particles reinforcement and long fibres [52,53]. The tension stress causes interface debonding. [53–56] considered that the final interface between the short fibres and the matrix is easy to debond in the loading process. At the end of the final interface, the stress transfer from the matrix to the fibres depends on the shear stress only on the axial interface (Figure 16a). In the case of long fibre composites, the interface stress (shear stress) is higher due to the adhesion of resin to fibres (Figure 16b).

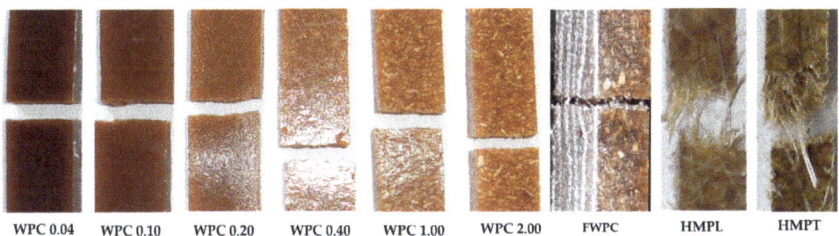

Figure 15. Modes of fractures recorded for all natural tested fibre reinforced composites.

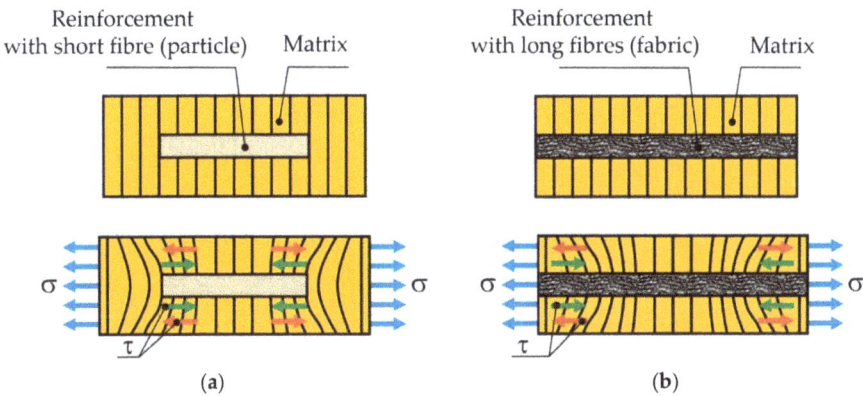

Figure 16. Mechanism of failure modes: (**a**) the case of short fibres reinforced composites; (**b**) the case of long fibres reinforced composites.

4. Conclusions

The mechanical behaviour of composite materials from three types of natural fibres was studied. The results demonstrated that both tensile and rheological behaviour depends on the size of fibres, disposure of fibres (randomly or fabric) and the type of matrix. The mechanical properties of natural fibres composites differ within the same type of composite. For instance, WPC0.20 recorded the higher values of Young's Modulus and tensile strength in comparison with WPC2.00 which have the lower values. It can be concluded that the best wood particles size is 0.20 mm from mechanical point of view. In case of HMP and FWPC, there is strong relation between direction of loading and weft/yarn direction. The longitudinal elasticity modulus is higher in the transversal direction (weft) by almost 80–95% than in the warp direction (longitudinal). The flax woven reinforcement in case of FWPC leads to the best mechanical properties from all types of tested composites.

Regarding DMA results, HMP and also FWPC present higher values of storage modulus by comparison to wood particle reinforced composites although the increasing temperature produces a decreases in viscous-elastic behaviour for all types of samples. The Tg for all tested natural fibres composite range between 65 and 85 °C, the matrix having the main role in modification of polymers stiffness.

The HMP presents a good capacity to absorb the energy of deformation, partially due to polyurethane resin and to the type of dispersed fibres. In the case of a structure made of lignocelluloses composite materials, consisting of layers reinforced with natural fibres as fabrics and layers reinforced with wood particles, by adding an additional layer reinforced with wood particles on the visible surface of the panels. Thus, structures with superior aesthetic properties can be made which no longer require coating with other materials. Although the structure is no longer symmetrical, it has visible surfaces with natural textures whose colour can be changed by using wood particles of different species.

The further work will focus on simulation of mechanical behaviour of complex structures made from natural fibres composites using the elastic properties determined in experimental tests and predicting the stress and strain states of them.

Author Contributions: Conceptualization, M.D.S. and F.T.; methodology, H.T.D. and O.M.T.; software, H.T.D. and O.M.T.; validation, M.D.S. and F.T.; formal analysis, H.T.D. and O.M.T.; investigation, M.D.S. and O.M.T.; resources, F.T.; data curation, H.T.D.; writing—original draft preparation, O.M.T. and M.D.S.; writing—review and editing, H.T.D. and F.T.; visualization, M.D.S.; supervision, O.M.T.; funding acquisition, M.D.S. and F.T. All authors have read and agreed to the published version of the manuscript.

Funding: This research was funded by the Sectorial Operational Program Human Resources Development (SOP HRD), financed from the European Social Fund and by the Romanian Government under the contract number POSDRU/88/1.5/S/59321 and Program partnership in priority domains PNII under the aegis of MECS UEFISCDI, project No. PN-II-PT-PCCA-2013-4-0656.

Acknowledgments: We are grateful to the manager Dorin Rosu, PhD and to the technical staff of S.C. Compozite S.A., Brasov, Romania, manufacturer of composite products for supplying the specimens for the experimental research of the present article. We are grateful to Savin Adriana, head of laboratory "Non-destructive tests" of the National Institute of Research and Development for Technical Physics in Iasi, for the DMA tests.

Conflicts of Interest: The authors declare no conflict of interest.

Abbreviations and Notations

WPC	wood particle reinforced composites;
HMP	hemp mat and polyurethane resin;
FWPC	flax and wood particle reinforced composites;
DMA	dynamic mechanical analysis;
L	length of sample;
b	width of sample;
h	thickness of sample;
l	gauge length between bearings for 3 points bending;
E	longitudinal elasticity modulus;
STDV	standard deviation;
σ_r	tensile rupture;
ε_r	percentage strain at break;
k	stiffness;
E'	storage modulus;
E"	loss modulus;
tan δ	damping ratio;

References

1. Di Bela, G.; Fiore, V.; Valenza, A. Natural Fibre Reinforced Composites (Chapter 2). In *Fiber Reinforced Composites*; Cheng, Q., Ed.; Nova Science Publishers: New York, NY, USA, 2012.
2. Pamuk, G. Natural fibers reinforced green composites. *Tekstilec* **2016**, *59*, 237–243. [CrossRef]
3. Hristov, V.; Vasileva, S. Dynamic mechanical and thermal properties of modified poly (propylene) wood fiber composites. *Macromol. Mater. Eng.* **2003**, *288*, 798–806. [CrossRef]
4. Puglia, D.; Biagiotti, J.; Kenny, J.M. A review on natural fibre-based composites—Part II. *J. Nat. Fibers* **2005**, *1*, 23–65. [CrossRef]
5. Fajrin, J. Mechanical properties of natural fiber composite made of Indonesian grown sisal. *Infoteknik* **2016**, *17*, 69–84.
6. Terciu, O.M.; Curtu, I.; Teodorescu-Draghicescu, H. Effect of wood particle size on tensile strength in case of polymeric composites. In Proceedings of the 8th International Conference of DAAAM Baltic, Industrial Engineering, Tallinn, Estonia, 19–21 April 2012.

7. Stark, N.M.; Berger, M.J. Effect of particle size on properties of wood-flour reinforced polypropylene composites. In Proceedings of the Fourth International Conference on Woodfiber-Plastic Composites, Madison, Wisconsin, 12–14 May 1997.
8. Gozdecki, C.; Wilczynski, A. Effect of wood particle size and test specimen size on mechanical and water resistance properties of injected wood-high density polyethylene composite. *Wood Fiber Sci.* **2015**, *47*, 365–374.
9. Lilholt, H.; Sørensen, B.F. Interfaces between a fibre and its matrix. In *IOP Conference Series: Materials Science and Engineering*; IOP Publishing: Bristol, UK, 2017; Volume 219, p. 012030. [CrossRef]
10. Stanciu, M.D.; Curtu, I.; Groza, M.; Savin, A. The evaluation of rheological properties of composites reinforced with hemp subjected to photo and thermal degradation. In *CONAT 2016 International Congress of Automotive and Transport Engineering*; Chiru, A., Ispas, N., Eds.; Springer: Cham, Switzerland, 2016. [CrossRef]
11. Pickering, K.L.; AruanEfendy, M.G.; Le, T.M. A review of recent developments in natural fibre composites and their mechanical performance. *Compos. Part A* **2016**, *83*, 98–112. [CrossRef]
12. Terciu, O.M.; Curtu, I. New hybrid lignocellulosic composite made of epoxy resin reinforced with flax fibres and wood sawdust. *Mater. Plast.* **2012**, *49*, 114–117.
13. Patel, V.K.; Rawat, N. Physico-mechanical properties of sustainable Sagwan-Teak Wood Flour/Polyester Composites with/without gum rosin. *Sustain. Mater. Technol.* **2017**, *13*, 1–8. [CrossRef]
14. Stanciu, M.D.; Bucur, V.; Valcea, C.S.; Savin, A.; Sturm, R. Oak particles size effects on viscous-elastic properties of wood polyester resin composite submitted to ultraviolet radiation. *Wood Sci. Technol.* **2018**, *52*, 365–382. [CrossRef]
15. Cerbu, C.; Curtu, I.; Ciofoaia, V.; Rosca, I.C.; Hanganu, L.C. Effects of the wood species on the mechanical characteristics in case of some e-glass fibres/wood flour/polyester composite materials. *Mater. Plast.* **2010**, *47*, 109–114.
16. Cerbu, C.; Cosereanu, C. Moisture effects on the mechanical behaviour of Fir Wood Flour/Glass reinforced epoxy composite. *BioResources* **2016**, *11*, 8364–8385. [CrossRef]
17. Araújo, E.M.; Araújo, K.D.; Pereira, O.D.; Ribeiro, P.C.; de Melo, T.J. Fiberglass Wastes/Polyester resin composites: Mechanical properties and water sorption. *Polímeros Ciência e Tecnologia* **2006**, *16*, 332–335.
18. Sinha, A.; Narang, H.K.; Bhattacharya, S. Mechanical properties of natural fibre polymer composites. *J. Polym. Eng.* **2017**, *37*, 879–895. [CrossRef]
19. Tang, H.; Nguyen, T.; Chuang, T.; Chin, J.; Lesko, J.; Wu, F. Fatigue model for fiber-reinforced polymeric composite. *J. Mater. Civil. Eng.* **2000**, *12*, 97–104. [CrossRef]
20. Torabizadeh, M.A. Tensile, compressive and shear properties of unidirectional glass/epoxy composite subjected to mechanical loading and low temperature services. *Indian J. Eng. Mater. Sci.* **2013**, *20*, 299–309.
21. Degrieck, J.; Van Paepegem, W. Fatigue damage modelling of fibre-reinforced composite materials: Review. *Appl. Mech. Rev.* **2001**, *54*, 279–300. [CrossRef]
22. Pepper, T. Polyester Resin. Engineering Materials Handbook. Available online: http://home.engineering.iastate.edu/~{}mkessler/MatE454/Constituent%20Materials%20Chapters%20from%20ASM%20Handbook/%285%29%20Polyester%20Resins.pdf (accessed on 10 October 2016).
23. Aramide, F.O.; Atanda, P.O.; Olorunniwo, O.O. Mechanical properties of a polyester fibre glass composite. *Int. J. Comp. Mater.* **2012**, *2*, 147–151.
24. Cerbu, C. Practical solution for improving the mechanical behaviour of the composite materials reinforced with flax woven fabric. *Adv. Mech. Eng.* **2015**, *74*, 1687814015582084. [CrossRef]
25. Prasad, A.V.R.; Rao, K.M. Mechanical properties of natural fibre reinforcedpolyester composites: Jowar, sisal and bamboo. *Mater. Des.* **2011**, *32*, 4658–4663. [CrossRef]
26. Thakur, V.K.; Thakur, M.K. Processing and characterization of natural cellulose fibers/thermoset polymer composites. *Carbohydr. Polym.* **2014**, *109*, 102–117. [CrossRef]
27. Ndiaye, D.; Diop, B.; Thiandoume, C.; Fall, P.A.; Farota, A.K.; Tidjani, A. Morphology and thermo mechanical properties of wood/polypropylene composites. In *Polypropylene*; INTECH: London, UK, 2012.
28. Ndiaye, D.; Verney, V.; Askanaian, H.; Commereuc, S.; Tidjani, A. Morphology, thermal behaviour and dynamic rheological properties of wood polypropylene composites. *Mater. Sci. Appl.* **2013**, *4*, 730–738.
29. Wambua, P.; Ivens, J.; Verpoest, I. Natural fibres: Can they replace glass in fibre reinforced plastics? *Compos. Sci. Technol.* **2003**, *63*, 1259–1264. [CrossRef]
30. Herrera, F.P.J.; Gonzalez, A. A study of the mechanical properties of shortnatural-fiber reinforced composites. *Compos. Part B* **2005**, *36*, 597–608. [CrossRef]

31. Gassan, J.; Bledzki, A.K. Possibilities to improve the properties of natural fiber reinforced plastics by fiber modification–jute polypropylene composites. *Appl. Comp. Mater.* **2000**, *7*, 373–385. [CrossRef]
32. Espert, A.; Camacho, W.; Karlsson, S. Thermal and thermomechanical properties of biocomposites made from modified cellulose and recycled polypropylene. *J. Appl. Polym. Sci.* **2003**, *89*, 2350–2353. [CrossRef]
33. Seldén, R.; Nyström, B.; Långström, R. UV Aging of Poly(propylene)/Wood-Fiber composites. *Polym. Compos.* **2004**, *25*, 543–553.
34. Gupta, M.K.; Bharti, A. Natural fiber reinforced polymercomposites: A review on dynamic mechanical properties. *Curr. Trends Fash. Technol. Text. Eng.* **2017**, *1*, 1–4. [CrossRef]
35. Ashok, R.B.; Srinivasa, C.V.; Basavaraju, B. Dynamic mechanical properties of natural fiber composites—A review. *Adv. Compos. Hybrid Mater.* **2019**, *2*, 586–607. [CrossRef]
36. Jiang, X.; Gao, Q. Stress-transfer analysis for fibre/matrix interfaces in short-fibre-reinforced composites. *Compos. Sci. Technol.* **2001**, *61*, 1359–1366. [CrossRef]
37. Teacă, C.A.; Roşu, D.; Bodîrlău, R.; Rosu, L. Structural changes in wood under artificial UV light irradiation determined by FTIR spectroscopy and color measurements—A brief review. *BioResources* **2013**, *8*, 1478–1507. [CrossRef]
38. Cosnita, M.; Cazan, C.; Duta, A. The influence of inorganic additive on the water stability and mechanical properties of recycled rubber, polyethylene terephthalate, high density polyethylene and wood composites. *J. Clean. Prod.* **2017**, *165*, 630–636. [CrossRef]
39. Mbarek, T.B.; Robert, L.; Hugot, F.; Orteu, J.J.; Sammouda, H.; Graciaa, A.; Charrier, B. Study of Wood Plastic Composites elastic behaviour using full field measurements. In Proceedings of the EPJ Web of Conferences, Poitiers, France, 4–9 July 2010; Volume 6, p. 28005. [CrossRef]
40. Varganici, C.D.; Rosu, D.; Rosu, L. Life-time prediction of multicomponent polymeric materials. In *Photochemical Behaviour of Multicomponent Polymeric-based Materials*; Rosu, D., Visakh, P.M., Eds.; Springer: Cham, Switzerland, 2016; pp. 227–258. [CrossRef]
41. Facca, A.G.; Kortschot, M.T.; Yan, N. Predicting the tensile strength of natural fibre reinforced thermoplastics. *Comp. Sci. Technol.* **2007**, *67*, 2454–2466. [CrossRef]
42. Romanzini, D.; Lavoratti, A.; Ornaghi, H.L., Jr.; Amico, S.C.; Zattera, A.J. Influence of fiber content on the mechanical and dynamic mechanical properties of glass/ramie polymer composites. *Mater. Des.* **2013**, *47*, 9–15. [CrossRef]
43. Panigrahy, B.S.; Rana, A.; Chang, P.; Panigrahi, S. Overview of flax fibre reinforced thermoplastic composites. *Am. Soc. Agric. Biol. Eng.* **2006**, *6–165*, 1–12. [CrossRef]
44. ISO 527-1:2012 Plastics—Determination of Tensile Properties. Available online: https://www.iso.org/standard/56045.html (accessed on 1 August 2019).
45. ASTM D7028-07. *Standard Test Method for Glass Transition Temperature (DMA Tg) of Polymer Matrix Composites by Dynamic Mechanical Analysis (DMA)*; ASTM International: West Conshohocken, PA, USA, 2015; Available online: www.astm.org (accessed on 10 December 2019).
46. Arputhabalan, J.; Palanikumar, K. Tensile properties of natural fiber reinforced polymers: An overview. In *Applied Mechanics and Materials*; Trans Tech Publications Ltd.: Stafa-Zurich, Switzerland, 2015; Volume 766, pp. 133–139.
47. Menard, K.P. *Dynamic Mechanical Analysis. A practical Introduction*; CRC Press–Taylor &Francis Group: New York, NY, USA, 2008.
48. Cheng, F.; Hu, Y.; Li, L. Interfacial properties of glass fiber/unsaturated polyester resin/poplar wood composites prepared with the prepreg/press process. *Fiber Polym.* **2015**, *16*, 911–917. [CrossRef]
49. Saba, N.; Safwan, A.; Sanyang, M.; Mohammad, F.; Pervaiz, M.; Jawaid, M.; Alothman, O.; Sain, M. Thermal and dynamic mechanical properties of cellulose nanofibers reinforced epoxy composites. *Int. J. Biol. Macromol.* **2017**, *102*, 822–828. [CrossRef] [PubMed]
50. Devi, L.U.; Bhagawan, S.; Thomas, S. Dynamic mechanical analysis of pineapple leaf/glass hybrid fiber reinforced polyester composites. *Polym. Compos.* **2010**, *31*, 956–965. [CrossRef]
51. Jesuarockiam, N.; Jawaid, M.; Zainudin, E.S.; Thariq Hameed Sultan, M.; Yahaya, R. Enhanced thermal and dynamic mechanical properties of synthetic/natural hybrid composites with graphene nanoplateletes. *Polymers* **2019**, *11*, 1085. [CrossRef]
52. Alothman, O.Y.; Alrashed, M.M.; Anis, A.; Naveen, J.; Jawaid, M. Characterization of date palm fiber-reinforced different polypropylene matrices. *Polymers* **2020**, *12*, 597.

53. Awaja, F.; Zhang, S.; Tripathi, M.; Nikiforov, A.; Pugno, N. Cracks, microcracks and fracture in polymer structure: Formation, detection, autonomic repair. *Prog. Mater. Sci.* **2016**, *83*, 536–573. [CrossRef]
54. Dhakal, H.N.; Méner, E.L.; Feldner, M.; Jiang, C.; Zhang, Z. Falling weight impact damage characterisation of flax and flax basalt vinyl ester hybrid composites. *Polymers* **2020**, *12*, 806. [CrossRef] [PubMed]
55. Stanciu, M.D.; Șova, D.; Savin, A.; Ilias, N.; Gorbacheva, G. Physical and mechanical properties of ammonia-treated black locust wood. *Polymers* **2020**, *12*, 377. [CrossRef] [PubMed]
56. Saba, N.; Jawaid, M.; Alothman, O.Y.; Paridah, M.T. A review on dynamic mechanical properties of natural fiber reinforced polymer composites. *Constr. Build. Mater.* **2016**, *106*, 149–159. [CrossRef]

© 2020 by the authors. Licensee MDPI, Basel, Switzerland. This article is an open access article distributed under the terms and conditions of the Creative Commons Attribution (CC BY) license (http://creativecommons.org/licenses/by/4.0/).

Article

Sustainable Micro and Nano Additives for Controlling the Migration of a Biobased Plasticizer from PLA-Based Flexible Films

Laura Aliotta [1,2,*], **Alessandro Vannozzi** [1,2], **Luca Panariello** [1,2], **Vito Gigante** [1,2], **Maria-Beatrice Coltelli** [1,2,*] **and Andrea Lazzeri** [1,2]

[1] Department of Civil and Industrial Engineering, University of Pisa, 56122 Pisa, Italy; alessandrovannozzi91@hotmail.it (A.V.); luca.panariello@ing.unipi.it (L.P.); vito.gigante@dici.unipi.it (V.G.); andrea.lazzeri@unipi.it (A.L.)
[2] Consorzio Interuniversitario Nazionale per la Scienza e Tecnologia dei Materiali (INSTM), 50121 Florence, Italy
* Correspondence: laura.aliotta@dici.unipi.it (L.A.); maria.beatrice.coltelli@unipi.it (M.-B.C.)

Received: 20 May 2020; Accepted: 13 June 2020; Published: 17 June 2020

Abstract: Plasticized poly(lactic acid) (PLA)/poly(butylene succinate) (PBS) blend-based films containing chitin nanofibrils (CN) and calcium carbonate were prepared by extrusion and compression molding. On the basis of previous studies, processability was controlled by the use of a few percent of a commercial acrylic copolymer acting as melt strength enhancer and calcium carbonate. Furthermore, acetyl n-tributyl citrate (ATBC), a renewable and biodegradable plasticizer (notoriously adopted in PLA based products) was added to facilitate not only the processability but also to increase the mechanical flexibility and toughness. However, during the storage of these films, a partial loss of plasticizer was observed. The consequence of this is not only correlated to the change of the mechanical properties making the films more rigid but also to the crystallization and development of surficial oiliness. The effect of the addition of calcium carbonate (nanometric and micrometric) and natural nanofibers (chitin nanofibrils) to reduce/control the plasticizer migration was investigated. The prediction of plasticizer migration from the films' core to the external surface was carried out and the diffusion coefficients, obtained by regression of the experimental migration data plotted as the square root of time, were evaluated for different blends compositions. The results of the diffusion coefficients, obtained thanks to migration tests, showed that the CN can slow the plasticizer migration. However, the best result was achieved with micrometric calcium carbonate while nanometric calcium carbonate results were less effective due to favoring of some bio polyesters' chain scission. The use of both micrometric calcium carbonate and CN was counterproductive due to the agglomeration phenomena that were observed.

Keywords: poly(lactic acid); poly(butylene succinate); plasticizer migration; diffusion

---

1. Introduction

The environmental impact of plastic wastes, also due to the limited disposal methods, is a continuously growing public concern worldwide. This problem has encouraged research and industrial interest on biobased and biodegradable polymers that could overcome sustainability issues and environmental challenges posed by the production and disposal of oil-derived plastics [1,2]. In packaging and agriculture applications, the use of bio-based biodegradable polymers is a strong advantage both for environment and customers [3]. However, it is expected that the demand for these biopolymers will increase and, over the coming few decades, bioplastic materials can complement and gradually substitute the oil-based plastics in different sectors [2,4]. Nowadays different biopolymers can be found not only in food and agriculture applications, but also in the medicine and cosmetic sectors. The development of renewable polymeric modulated materials tailored for specific applications is a subject of active research interest worldwide [5].

Poly(lactic acid) (PLA) is particularly interesting because it exhibits mechanical properties (Young's modulus of about 3 GPa, tensile strength between 50 and 70 MPa with an elongation at break of around 4%, and an impact strength close to 2.5 kJ/m^2) that make it useful for a wide range of applications [6,7]. It can be obtained from renewable resources (e.g., corn, wheat, or rice) and it is not only biodegradable and compostable, but also recyclable. However, in particular, its biocompatibility makes it appealing for biomedical and cosmetic applications. If compared to other commercialized biopolymers (such as poly(hydroxylalkanoates) (PHAs), poly(caprolactone) (PCL), and starch), PLA is easily processable [8]. However, PLA's brittleness and its poor heat resistance limit strongly the application of PLA. The improvement of PLA's toughness can be reached in different ways (that can also be adopted contemporary): plasticization, copolymerization, and melt blending with flexible polymers [9,10]. However, the blending technique is the common method adopted to overcome the PLA brittleness making it useful in those applications (like film production) where high flexibility and toughness are requested. In literature different successful studies can be found where PLA was blended, in different quantities, with other biodegradable polymers such as: polycaprolactone [11], poly(vinyl alcohol) (PVOH) [12,13], polybutylene(succinate) (PBS) [14–16], poly(butylene succinate-co-adipate) [17–19], and poly(butylene adipate-co-terephthalate) (PBAT) [20–22].

Furthermore, the combination of additives such as plasticizers into biopolymers and their blends, is a common practice to further improve the mechanical flexibility, and processability limitations. Generally, plasticizers are a class of low-to-medium molecular weight compounds up to a few thousand, it is expected that their demand will increase reaching approximately the 9.75 million of tons in 2024 [23]. In various applications, like film formation or coating dispersion, the adequate selection of plasticizer greatly improves the processing; however, the choice of these plasticizers, especially in biobased applications, is limited by the required safety, environmental favorability, and chemical and physical properties that dictate their miscibility [2].

It is therefore evident that, unlike classic plastic commodities (e.g., HDPE) which possess good starting physical-chemical properties, biopolyesters need to be improved and in this context biopolymers can release a major amount of additives that are not covalently bound to the polymer, at all stages of the plastic's lifecycle via migration of liquids or solids or via volatilization [24]. The result of this release is the transfer of chemicals (such as plasticizers) that can affect human health and can contaminate soils and water [25,26]. In fact, especially in natural environments, liquid additives can quickly migrate out of plastic and can be absorbed by roots affecting the plant development [27]. On the other hand, a controlled release of beneficial substances contained in the blends are very useful in the cosmetics and biomedical sectors [28,29].

For all of these aspects, the study of migration is very important especially for films containing plasticizers that are not chemically attached to polymeric chains (not reactive plasticizers). Depending on certain conditions, liquid plasticizers can come out from the polymer matrix. During service and storage, this loss is problematic because it leads to unwanted changes of the mechanical properties (loss of flexibility and toughness with an increment of stiffness). Clearly, this decrement of the mechanical properties will be more evident and dangerous in those polymeric matrices that are brittle in the not plasticized state (like PLA). This plasticizer release, as well as creating problems of stiffening, can lead to variations in the crystallization of the samples due to the reorganization over time (aging) of the crystalline structures [30]. Even the plasticizer molecular weight and linearity of the plasticizer influence the migration and so an average plasticizer molecular weight is required to ensure long-term plasticizer retention in the polymeric matrix [31].

According to the type of plasticizer, different strategies can be adopted to prevent the plasticizer migration (such as internal plasticization [32], polymer surface modification [33,34], or addition of nanofillers or ionic liquid [35,36]). The methodologies and the advantages/disadvantages of these strategies are extensively reported in literature [37]. Among the strategies mentioned, the role of micro and nano filler addition on the plasticizer release is particularly interesting.

In fact, the use of fillers has a double effect: they can reduce plasticizer migration thanks to the creation of tortuosity that forces the plasticizer molecules to follow a longer path to leave the polymeric structure [38] and the adsorption mechanism. Fillers, especially nanoparticles or nanofibrils, are extremely small and they possess a large number of groups capable of interactions on their surfaces that make the absorption of other substances easy. On the basis of these characteristics, the plasticizer migration can be reduced both by the absorption on the surface of plasticizer molecules and by the steric resistance that makes the passage of the plasticizer difficult [37]. On the other hand, filler addition can improve the mechanical properties of the materials (such as toughness) according to the rigid filler toughening mechanism [39]. Various nano-scale and micro-scale fillers with different geometries (such as: montmorillonite, silica, calcium carbonate, and aluminum oxide) are reported to improve not only the properties of polymers such as toughness, stiffness, and heat resistance [40,41] but they also can limit the plasticizer migration [35,38]. For example, it has been observed that organic montmorillonite (OMMT), nano-SiO_2, and nano-$CaCO_3$ have a strong adsorption force and diffusion inhibition [35,37]. Therefore, the presence on the fillers surface of functional groups which are active and can easily bond with plasticizer molecules is another important aspect that influences the migration of the plasticizer [42]. It was observed that the migration rates decrease with the increasing of filler content [43]. However, it is essential to reach a homogeneous dispersion of the particles in the polymeric matrix to hinder the plasticizer migration. In fact, the increase of fillers content does not lead always to better barrier properties if there is a poor dispersion of nano particles in the matrix [35]. Clearly, great attention must be paid to the combination of polymeric plasticizer and micro/nano additives to control the eventual plasticizer migration and at the same time to reach an optimal combination of mechanical properties (good balance between stiffness and toughness).

Understanding the mechanism and the kinetics of plasticizer loss is fundamental also to evaluating the short- and long-term performance of the final plasticized material. For instance, if the correlation between mechanical properties and plasticizer concentration is known, it will be possible to predict the change of mechanical properties and, consequently, the lifetime of products. Furthermore, it will be also possible to develop new methods to eliminate/control the plasticizer migration.

In this paper, the attention is focused on the results of previous studies where suitable biobased skin compatible films were successfully prepared and investigated. Different additives were added: calcium carbonate, chitin nanofibrils as functional filler (dispersed by poly(ethylene glycol) (PEG)), melt strength enhancer (Plastistrength), and acetyl tributyl citrate (ATBC) plasticizer. The addition of all these additives has led to an improvement of the melt processability, of the mechanical properties (the films resulted flexible and high resistant) and of the antimicrobial and anti-inflammatory properties [44–47]. However, troubles were encountered during the storage of films where 4 wt % of micrometric calcium carbonate was present [44] due to the plasticizer migration. Interestingly, any data were reported about the effect on plasticizer migration of natural nanofibers obtained by sea food waste, chitin nanofibrils used in these bionanocomposites for their indirect anti-microbial activity [47].

In the present paper, the identification of the diffusion coefficients for the examined blends was carried out. Furthermore, in order to limit and/or to control the plasticizer migration, different types of commercial calcium carbonate particles (micro and nanometric) and chitin nanofibrils were used. The effect of their addition on the films processability and on the mechanical properties was carried out to ensure that no significant variations occurs in the already optimized rheological and mechanical properties. Finally, the effect of the introduction of $CaCO_3$ and chitin nanofibrils on the ATBC migration was evaluated through the diffusion coefficient calculation. The diffusion coefficients were obtained, applying analytical correlations based on the Fick's second law, by regression of the experimental migration data plotted as the square root of time.

2. Materials and Methods

2.1. Materials

The polymeric granules and additives used in this work for the blend's preparation are:

- Commercial poly(lactic) acid (PLA), trade name 2003D, produced by Nature Works LLC (Minnetonka, Minneapolis, MN, USA) was used. This is a commercial grade containing about 4% of D-lactic acid units that lower the melting point and the crystallization tendency, improving the processing ability. This PLA, according to the producer's data sheet has a density of 1.24 g/cm^3, a melt flow index (MFI) of 6 g/10 min (210 °C, 2.16 kg) and a nominal average molar mass of 200,000 g/mol.
- Poly(butylene succinate) (PBS), trade name BioPBS FD92PM, purchased from Mitsubishi Chemical Corporation (Tokyo, Japan). It is a copolymer of succinic acid, adipic acid and 1,4-butandiol with a melt flow index (MFI) of 4 g/10 min (190 °C, 2.16 kg) and a density of 1.24 g/cm^3.
- Acetyl tributyl citrate (ATBC), a product of Tecnosintesi S.p.A (Bergamo, Italy), was used as biobased and biodegradable plasticizer. It is a colorless and odorless liquid having a density of 1.05 g/cm^3 and a molecular weight of 402.5 g/mol.
- Plastistrength 550 (named PST for brevity), commercialized from Arkema (Paris, France), is a medium molecular-weight acrylic copolymer that appears as a white powder with a density of 1.17 g/cm^3. It is a commercial melt strength enhancer commonly added to improve the melt processability.
- Poly(ethylene glycol) (PEG6000) provided by Sigma-Aldrich (St. Louis, MO, USA) was used, for improving the dispersion of chitin nanofibrils [48]. It is a colorless solid, with a high molecular weight of 6000 g/mol and a solubility in water of 50 mg/mL at 20 °C.
- Chitin nanofibrils (CNs) water suspension (2 wt % of concentration) was supplied by MAVI SUD (Latina, Italy). CNs represent the pure and polysaccharidic molecular portion of α-chitin obtained after elimination of the protein portion. These fibrils have an average size of 240 × 7 × 5 nanometers (nm) and a shape like thin needles [48]. The production process of chitin nano-fibrils patented by MAVI results in the formation of a stable aqueous suspension of nanofibrils containing 300 billion nano crystals per milliliter with the addition of sodium benzoate. This substance is an anti-MLD, added to the suspension to avoid the possible attack of mold and bacteria on the chitin [46].
- Two different typologies of calcium carbonate, commercialized by Omya SpA (Avenza, Italy), with different particle size distributions were used: Omyacarb 2-AV (named 2AV), Omya Smartfill 55-OM (named Smartfill). Hakuenka CC-R (named CCR) is commercialized by Shiraishi. 2AV has a micrometric particle size with a diameter value (relative to the maximum distribution curve, d$_{98\%}$) of 15 μm, and 38% of particles of diameter less than 2 μm. The average statistical diameter (d$_{50\%}$) is 2.6 μm with a specific weight of 2.7 g/cm^3. Smartfill is fine ground and surface fatty acid treated calcium carbonate having 55% of particles with an average diameter <2 μm (bulk density: 1.1 g/mL). CCR is a precipitated nano-calcium carbonate coated with acids having an average particle size of 80 nm (specific weight of 2.6–2.7 g/cm^3).

2.2. Characterization of Fillers

A Brunauer-Emmett-Teller (BET) analysis of the chitin nano-fibrils and of the three calcium carbonate fillers was carried out, with a micromeritics instrument-Gemini V analyzer (Micromeritics, Atlanta, GA, USA)—in order to determine the surface area, the total porosity and the number of particles per gram of the fillers. The measurement was performed with the same procedure used in the previous works of Coltelli et al. [47].

2.3. Blends Preparation and Torque Characterization

PLA/PBS blends preparation was carried out with a Haake Minilab II (Thermo Scientific Haake GmbH, Karlsruhe, Germany) co-rotating conical twin-screw extruder. Before processing the materials were dried in air circulated oven at 60 °C for at least 24 h. The molten materials were recovered in filaments for the subsequent tests. For each extrusion cycle, 6 g of PLA/PBS pellets, manually mixed with the other additives were fed through a little hopper into the mini-extruder. After the feeding, the molten material flowed in a closed circuit for 1 min at the end of which it was recovered. During this time, the torque was measured as a function of time, at least ten experimental measurements were performed for each blend compositions to guarantee the reliability and the consistency of the test. The final torque value is taken once that the melt stabilizes. The extrusion was carried out at 190 °C with a screw rotating speed of 110 rpm.

The compositions of the PLA/PBS blends investigated in this work, chosen considering the results of previous studies [44,47], are reported in Table 1.

To this purpose, the PLA/PBS ratio was maintained equal to 0.8 for all the formulations and also the ratio between ATBC and the polymeric matrix of PLA/PBS was kept constant and equal to 0.2. Furthermore, a fixed quantity (2 wt %) of Plastistrength (PST) was used for all blend compositions (except for F1 blend). The F3 formulation, according to previous studies [44,47] contained also a few percent (4 wt %) of micro-calcium carbonate (2AV). However, as mentioned, plasticizer migration was observed in these samples. Consequently, it has been decided to add a greater amount of calcium carbonate (7 wt %) and two other types of calcium carbonate with different particle size distributions to the blends to evaluate their effect on ATBC migration. F5 and F6 blends Smartfill and CCR were respectively added. The addition of chitin nanofibrils alone (F7 formulation) and coupled with the calcium carbonate particles, that showed better capability in hindering the plasticizer diffusion, was also investigated (F8 formulation).

For the preparation of the blends containing chitin nanofibrils (F7 and F8) PEG 6000 was used to achieve a better dispersion of the fibrils. The procedure adopted and the quantity of chitin nanofibrils chosen are described in the previous works of Coltelli et al. [47,48].

Table 1. Blends name and composition.

Blends Name	PLA (wt %)	PBS (wt %)	ATBC (wt %)	PST (wt %)	$CaCO_3$ (wt %)	PEG6000 (wt %)	NC (wt %)
F1	63	17	20	-	-	-	-
F2	62	16	20	2	-	-	-
F3	59	15	20	2	4 (2AV)	-	-
F4	59	15	17	2	7 (2AV)	-	-
F5	59	15	17	2	7 (Smartfill)	-	-
F6	59	15	17	2	7 (CCR)	-	-
F7	61	15	18	2	-	2	2
F8	57.5	14.5	15	2	7 (2AV)	2	2

2.4. Melt Flow Rate

A CEAST Melt Flow Tester M20 (Instron, Canton, MA, USA) equipped with an encoder was used to investigate the melt flow behavior of the blends. The encoder, following the movement of the piston, acquires the melt volume rate (MVR) data. For each blend, three tests were carried out following the standard ISO 1133D [49]. According to the ISO procedure, the sample is simply preheated for 40 s at 190 °C after that, a weight of 2.160 kg is released on the piston and then, after 5 s, a blade cuts the spindle starting the real test. Every 3 s, MVR value is recorded by the encoder. The molten material, flowing through the capillary of specific diameter and length, is recovered and the MFR value is obtained. Before the test, the pelletized filaments of each polymer blend were dried in an air oven at 60 °C for one day.

2.5. Thermal Characterization by Differential Scanning Calorimetry (DSC)

Differential scanning calorimetry (DSC) measurements were performed with a DSC TA Instruments Q200 (TA Instruments, New Castle, UK), equipped with a RSC cooling system. Indium was used as the standard for calibration with aluminum hermetic pans. About 10 g of material was analyzed for each blend. Nitrogen was used as purge gas at a rate of 50 mL/min. The samples were heated at 10 °C/min from −40 °C to 220 °C. Only the first scan was considered to take into account the samples thermal history. Glass transition temperature (T_g), melting temperature (T_m), cold crystallization temperature (T_{cc}), melting and cold crystallization enthalpies (ΔH_m and ΔH_{cc}) were determined by using the TA Universal Analysis software. In particular, the enthalpies of melting (ΔH_m) and cold crystallization (ΔH_{cc}) were determined from the corresponding peak areas in the heating thermograms; while the melting temperature (T_m) and the cold crystallization temperature (T_{cc}) were recorded at the maximum of the melting peak and at the minimum of the cold crystallization peak, respectively.

The crystallinity percentage (Xcc) of PLA in the blends was calculated as [50]

$$X_c = \frac{\Delta H_{m,PLA} - \Delta H_{cc,PLA}}{\Delta H^{\circ}_{m,PLA} \cdot X_{PLA}} \quad (1)$$

where $\Delta H_{m,PLA}$ and $\Delta H_{c,PLA}$ are the melting enthalpy and the enthalpy of cold crystallization of PLA, X_{PLA} is the weight fraction of PLA in the sample and $\Delta H^{\circ}_{m,PLA}$ is the melting enthalpy of the 100% crystalline PLA, equal to 93 J/g [51].

2.6. Tensile Test

The mechanical properties of the blends were evaluated by tensile tests carried out at room temperature. An INSTON 5500R universal testing machine (Canton, MA, USA), equipped with a 100 N load cell, was used. The machine, interfaced with a computer running a MERLIN software (INSTRON version 4.42 S/N-014733H), was assembled with compressed air grips (initial grip separation: 25 mm). The crosshead speed was set at 100 mm/min. The preparation of tensile specimens was carried out using the pelletized strains come out from the micro-compounder. The pellets were dried in an oven at 60 °C for 24 h to avoid the water uptake; subsequently, they were used for the film preparation by compression molding. The pelletized materials were pressed between two Teflon sheets at 180 °C for 1 min with a pressure of 3 tons, using a NOSELAB ATS manual laboratory heat press.

The mechanical tests were performed on an ISO 527-2 type A [52] dumbbell specimens obtained from the films cut with a Manual Cutting Press EP 08 (Elastocon, Brahmult, Sweden). At least 10 specimens were tested for each sample and the average values were reported.

2.7. Scanning Electron Microscopy Analysis (SEM)

Samples morphologies were investigated by scanning electron microscopy (SEM) with a FEI Quanta 450 FEG instrument (Thermo Fisher Scientific, Waltham, MA, USA). The micrographs of samples fractured with liquid nitrogen and sputtered with a layer of gold were collected. The metallic layer makes the surface electrically conductive, allowing the backscattered electrons to generate the images.

2.8. Migration Tests

Films prepared by compression molding (adopting the same procedure of film preparation described in Section 2.5) were used for the migration tests. To evaluate the weight loss of films due to the ATBC migration, three pieces of film for each formulation were put between two paper sheets. In this way, thanks to the capillarity forces related to the ATBC absorption from the paper sheets, the plasticizer is removed from the surface of the film ensuring a migration kinetics controlled by diffusion. The samples were kept in an oven at 60 °C (above T_g) to make the test severe and to accelerate the migration process. Periodically, the films were weighed to estimate the weight loss as a function of time. The migration tests were stopped after 1500 h. For each formulation, the percentage weight loss of the film as a function of time was determined with the following relationship.

$$\%wt\ loss = \frac{wt_{t,film} - wt_{t0,film}}{wt_{t0,film}} \cdot 100 \qquad (2)$$

where $wt_{t,film}$ is the film weight at the time t and wt_{t0}, film is the film weight at the beginning of the test ($t = 0$). To separate the effect of calcium carbonate (used as an additive for hinders the plasticizer migration), from that of the loss of ATBC, was calculated the percentage of weight loss normalized respect to the amount of ATBC initially present in the film

$$\%wt\ lost\ ATBC = \frac{lost\ ATBC\ mass}{initial\ ATBC\ mass} \cdot 100 \qquad (3)$$

In this way, it is possible to evaluate the films' oiliness, thanks to the total weight loss of the film and appreciate the percentage by weight of ATBC that migrates from the film independently from the amount of plasticizer in the film.

3. Theoretical Analysis

For the prediction of the migration of monomers or additives, mathematical models are often used. In particular, the second Fick's law is generally applied.

For a 3D system, the second Fick's law is expressed as

$$\frac{\partial c}{\partial t} = D\left[\frac{\partial^2 c}{\partial x^2} + \frac{\partial^2 c}{\partial y^2} + \frac{\partial^2 c}{\partial z}\right] \qquad (4)$$

If the diffusion is one-dimensional, that means that there is a gradient concentration only along one axis (for example x-axis), Equation (4) can be written in a simpler form (Equation (5)).

$$\frac{\partial c}{\partial t} = D\frac{\partial^2 c}{\partial x^2} \qquad (5)$$

Commonly, the prediction of a substance migration through a polymer matrix above its glass transition temperature is described by the Fick's second law based on one directional transfer (Equation (5)) [53]. General solutions of the diffusion equation can be adopted for a variety of initial and boundary conditions, provided the diffusion coefficient is constant. Crank and Vergnaud [54,55] proposed and classified different types of solutions for different geometries and boundary conditions. Taking into account the sample geometry adopted for the migration experiments, it is possible to apply one of the Crank's solutions of the Fick's second law. The equation used is reported below (Equation (6)) and it has been adopted successfully for plasticized polymeric systems [56,57]

$$C_x = \frac{1}{2}C_0\left[\operatorname{erf}\left(\frac{h-x}{2\sqrt{D \cdot t}}\right) + \operatorname{erf}\left(\frac{h+x}{2\sqrt{D \cdot t}}\right)\right] \qquad (6)$$

The parameters involved in Equation (3) are:

- C_x (mg/cm^3) that is the concentration of the chemical species that diffuses at a distance x from the center of the sample at the time t
- C_0 (mg/cm^3) is the starting concentration of the chemical species that diffuses at $t = 0$; thus, it will coincide with the initial concentration of the plasticizer present in the sample
- D is the diffusion coefficient (cm^2/s)
- h (mm) is the sample thickness
- erf is the error function (where $erf z = \frac{2}{\sqrt{\pi}} \int_0^z \exp(-\eta^2)d\eta$)

It can be observed that Equation (6) is symmetrical about $x = 0$ this means that the system can be cut in half by a plane at $x = 0$ without affecting the concentration distribution. It must be pointed out that the equation was obtained assuming that the migration occurred from the plasticized matrix to the same pure matrix, whereas our experiments were performed using paper, thus by removing the migrated plasticizer on the surface. Furthermore, if it is assumed that the plasticizer is initially distributed with a known concentration in the film and, if it is also supposed that the diffusion coefficient of the plasticizer in the polymer can be treated as a constant, a simple form of the Crank's equation can be obtained by Equation (7) [54,58]

$$M_t = 2C_{P0}\sqrt{\frac{Dt}{\pi}} \leftrightarrow \frac{M_t}{C_{P0}} = \sqrt{D} \cdot 2\sqrt{\frac{t}{\pi}} \tag{7}$$

where M_t (mg/cm^2) is the total plasticizer lost from the film at time t (s), C_{p0} (mg/cm^3) is the initial migrant concentration in the polymer and D (cm^2/s) is the diffusion coefficient. Equation (7) also assumes that the film is sufficiently thick that the concentration of plasticizer at the mid-plane remains at its original value (C_{p0}) and this can be physically achieved if less than about of 15–20% of the plasticizer is lost [58]. If experimental migration data are available, the diffusion coefficient can be thus obtained by linear regression of the migration data as function of the square root of time (according to Equation (7)). Clearly, to adopt Equation (7), there must be sufficient data points available in the early part of the migration graph. Nevertheless, the principal criticism of this simplified equation is that the diffusion coefficient of plasticizer varies with the plasticizer concentration and thus this equation can be applied only for small amounts of plasticizer (low concentration). In fact, it is known in [59,60] that the diffusivity increases with the plasticizer concentration due to the increment of free volume and mobility of the polymer caused by the plasticizer addition. This dependence of diffusion with plasticizer concentration is well described by the exponential equation [54,61]

$$D(C) = D_{c0}e^{aC} \tag{8}$$

where D_{c0} is the zero-concentration diffusivity, and a is the plasticization coefficient related to the plasticizer efficiency. When the diffusivity is a function of the concentration, the differential equation (Equation (5)) is not linear. Normally for migration modeling the diffusion coefficient (D) is seen as concentration independent, which in most cases (if the system is not highly plasticized) can be acceptable [62].

However, it must be pointed out that two kinetics migration modes dominate the plasticizer loss: the diffusion mode (above mentioned) and the evaporation mode [23]. Considering only the one-dimensional problem in the x-direction (similarly to Equation (5)), the evaporation condition can be described by the mass balance [54,63]

$$-D(C)\left(\frac{\partial C}{\partial x}\right) = F(C - C_e) \tag{9}$$

In Equation (9), the mass transfer related to the diffusion process (left term) is equated to the mass transfer of the plasticizer from the surface (right term). Obviously, evaporation can occur if the plasticizer concentration at the surface is greater than the concentration corresponding to the environment saturated with plasticizer (C_e) [23].

Thus, the plasticizer migration will be affected by the coexistence of two phenomena: diffusion and evaporation; the overall rate of plasticizer loss will be determined by the slower process (diffusion- or evaporation-controlled). In the case of diffusion-controlled system, the evaporation rate will be faster than diffusion rate. On the other-hand, in the case of evaporation-controlled system the evaporation rate will be slower than diffusion rate and often this leads to a formation of a plasticizer film on the surface of the analyzed sample [64]. Generally, it is possible to understand if the system is diffusion- or evaporation-controlled, by the shape of the concentration profile obtained plotting the curves of mass loss versus the square root of the time (Figure 1).

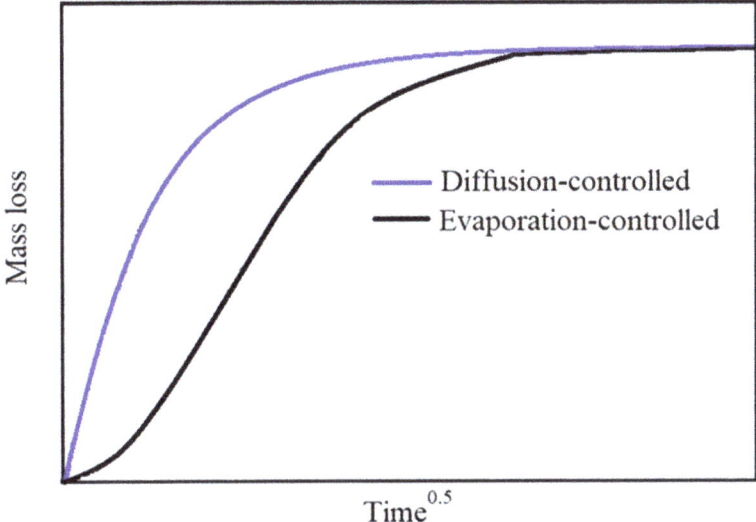

Figure 1. Examples of diffusion-controlled (blue line) and evaporation-controlled (black line) shape concentration profile.

The shape of concentration profile for an evaporation-controlled system, is flatter (S-shaped) if compared to the diffusion-controlled system [23].

It is evident that, for cases where the chosen plasticizer has a very high boiling point if compared to the temperature in which the material is currently used, the plasticizer will be accumulated on the external surfaces and it will form a thin film. The presence of this thin film of plasticizer that cannot easy evaporate due to the low temperature, limits the kinetic of plasticizer migration (the system will be evaporation-controlled). It is obvious that the concentration of plasticizer and the operative temperature will strongly influence the migration kinetics. Zhang et al. [65] demonstrated for a PLA system plasticized with acetyl triethyl citrate (ATC) that the ATC migration increases with the increasing of ambient temperature. The diffusion-controlled mechanism is activated for this system around 100–135 °C. Consequently, due to the similarity of the Zhang system with the polymeric system studied in this paper and also considering that the temperature adopted for the migration test (60 °C) is well below from the evaporation temperature of ATBC, it will be expected that the system will be evaporation-controlled.

The different film blends used for this work showed a plasticizer release phenomenon during the storage. From a practical point of view, it is difficult that the external film layer of plasticizer on the sample surface is not altered by external factors (presence of paper packaging which absorbs the plasticizer for example). Furthermore, there are some sectors (like cosmetics) in which a controlled release of these films is foreseen (for example in beauty masks). For these cases, it is evident how the diffusion process is the dominating mechanism.

In this work, the migration studies were carried out with the goal of investigating the diffusion-controlled mechanism in order to evaluate the effect of micro and nano calcium carbonate addition on the plasticizer diffusion coefficient. Consequently, the experimental migration tests were also carried out in order to make the diffusion the kinetically controlling mechanism. The shape of concentration profile obtained for all the compositions examined (Figure 2), being diffusion shaped, confirmed the diffusion-controlling mechanism.

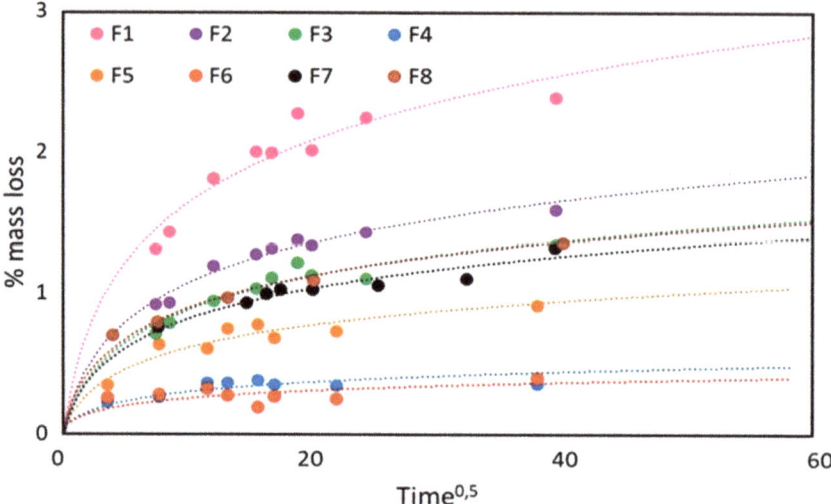

Figure 2. Concentration profile for all the compositions examined in this paper.

4. Results and Discussion

4.1. Migration Results and Determination of the Diffusion Coefficients

The trends of the weight loss percentage as function of the time are reported for all the blends in Figure 3. In Figure 3a, the loss percentages are reported for the reference formulations (F1, F2, and F3) compared to F4 where the quantity of micro-calcium carbonate (2AV) was incremented from 4 to 7 wt %. It can be noticed that F1 formulation has the higher weight loss if compared to F2. Both the formulations do not have micro-calcium carbonate; nevertheless, their plasticizer kinetic release is markedly different. This difference can be attributed to the addition of PST that, favoring the formation of intermolecular interaction [44], decreases the ATBC migration making more difficult the plasticizer diffusion. It can be observed that the addition of 4 wt % of micro-calcium carbonate (F3 blend), do not alter significantly the ATBC weight loss and only a slight slowing is observed if compered to F2. On the other hand, an increment from 4 wt % to 7 wt % significantly decreases the migration of ATBC and the weight loss percentage is reduced under the 0.5%. The major quantity of filler significantly slowed the plasticizer kinetic release. This can be ascribed to a major tortuosity path (generated by the increment of micro-calcium carbonate particles number) that the ATBC molecules must encounter. The plasticizer molecules have to follow a longer path in order to leave the polymeric structure [38].

The effect of the three different types of calcium carbonate particles (micro- and nano-, surface coated and not) are reported in Figure 3b. The best barrier properties to the ATBC migration are obtained with 2AV (F4). Moving from micro- to nano- calcium carbonate (F6 formulation), it would be expected a significant reduction of the plasticizer migration [35]. However, the data show a very similar trend with a slight decrease in the loss of ATBC. This result shows a poor efficiency of CCR (which also has a cost higher than Omycarb 2-AV). The explanation of this behavior can be ascribed to the presence of zones rich in agglomerates, as highlighted by the SEM analysis shown in Figure 4. It is likely that these surface coated nano-particles' agglomeration is related to the manufacturing process of CCR. CCR, in fact, is a precipitated calcium carbonate where a water coating process was used. Generally the 'wet' processes have a minor surface coated area if compared to the 'dry' processes [66]. The coating in aqueous medium is different from solvent or dry coating, the process is controlled by micelle absorption followed by the micelle collapse into double or multiple layers during the drying stage. It has been demonstrated that for this process the monolayer coating is incomplete [67]. This incomplete coverage is responsible of the partially particles agglomeration. However, the presence of some nano-agglomerates cannot be the only reason for which the efficiency of the nano-metric fillers is not good enough. In fact, it must be considered that these nano-fillers are surface coated with fatty acids to improve their dispersion, reducing the surface tension between a hydrophobic and non-polar polymer and inorganic polar hydrophilic particles [68,69]. These surface agents, containing a polar group and a long aliphatic chain, can alter the ATBC absorption on the particles' surface, worsening the hindering of ATBC migration. It is known in fact that the coating with fatty acids reduces the wettability of the particles to solvents such as water and n-decane [70]. Hence it is reasonable, on the basis of the results obtained, to hypothesize a negative chemical affinity (that reduces the ATBC absorption) between the ATBC and the aliphatic chains of fatty acids used for the surface coating. This conjecture can be reflected in the F5 formulation that contains surface-coated micro-calcium carbonate particles (Smartfill). The results achieved in the ATBC loss are worsened if compared to the not-surface-covered calcium carbonate particles, confirming the probable low absorption capacity of the calcium carbonate particles' surface treated with fatty acids.

The results are worse where the coating with fatty acids is better; in fact, Smartfill is a ground calcium carbonate surface coated with a 'dry' process, hence it has a better surface coating if compared to CCR. Consequently, the ATBC absorption capacity of the calcium carbonate appears to be greatly worsened by this coating. The best results on the other hand are obtained with a calcium carbonate having no surface coverage.

Finally, in Figure 3c the effect of chitin nanofibrils alone and coupled to 2AV can be observed. It can be observed that chitin nanofibrils alone are capable of limiting the plasticizer release, and the migration level of ATBC is comparable to the F4 formulation. The coupling between chitin nanofibrils and 2AV does not show any type of synergy. The presence of agglomerates (that can be observed in the SEM of Figure 4) leads to a worsening of the final material barrier properties.

The quantitative results of the maximum weight loss, calculated according to Equations (2) and (3), are reported in Table 2 and confirms the trends of Figure 3.

Table 2. Film weight loss and ATBC lost, calculated according to Equations (2) and (3), for all film formulations.

Blends Name	Weight Loss (wt %)	Lost ATBC (wt %)
F1	2.40 ± 0.31	12.01 ± 1.53
F2	1.61 ± 0.01	8.05 ± 0.08
F3	1.36 ± 0.19	6.78 ± 0.95
F4	0.39 ± 0.15	2.29 ± 0.08
F5	0.92 ± 0.14	5.44 ± 0.83
F6	0.22 ± 0.07	1.31 ± 0.43
F7	1.33 ± 0.43	7.40 ± 2.38
F8	1.37 ± 0.16	9.12 ± 1.06

The surficial area of the different fillers used in the present paper were determined by BET analysis and the results are reported in Table 3.

From Table 3, it can be noticed that CCR presents the higher Langmuir area among the calcium carbonates and a number of particles per gram two order of magnitude greater as a consequence of nanometric dimension. Moreover, chitin nanofibrils show a very high surficial area. The fillers with a higher surficial area should develop more interactions with the polymeric matrix.

However, it must be pointed out that a high surface area makes the aggregation of particles easier, confirming what has been observed by the SEM (Figure 4), where micrometric particles or agglomerates are revealed in all the samples. Chitin nanofibrils show the highest surficial area among the fillers, making them capable of agglomerating easily, resulting in micrometric bundles (Figure 4, F7 sample). This justifies the use of PEG to separate the fibrils as much as possible [47].

Table 3. Surficial areas, porosity, and number of particles per gram of the four fillers used in this paper.

Samples	Langmuir Area (m²/g)	BET (m²/g)	Total Pore Volume (cm³/g)	Number of Particles Per Gram
2AV	4.3094	2.7126	0.0013	1.3×10^{12}
Smartfill	6.1180	3.7659	0.0018	3.5×10^{12}
CCR	32.1068	18.4610	0.0085	4.0×10^{14}
NC	61.7235	39.1443	0.0194	4.1×10^{13}

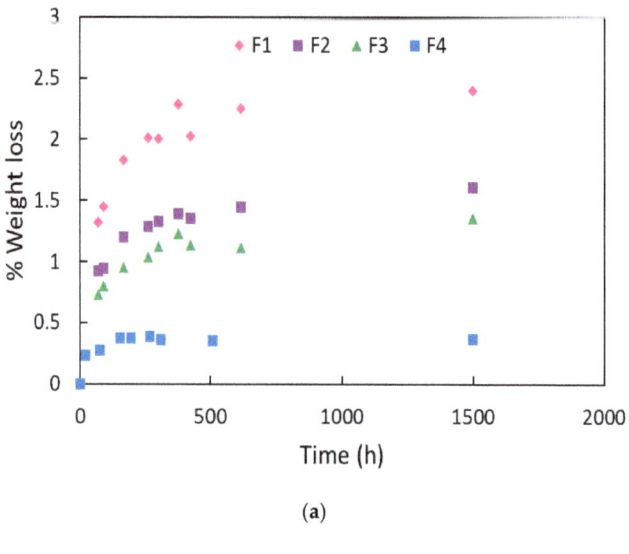

(a)

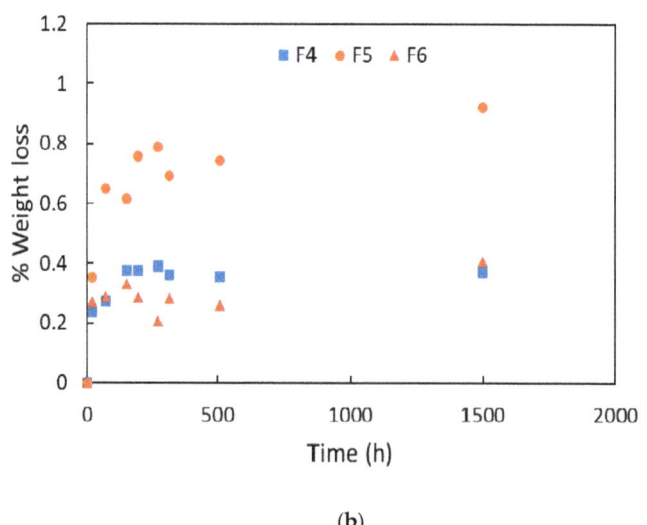

(b)

Figure 3. *Cont.*

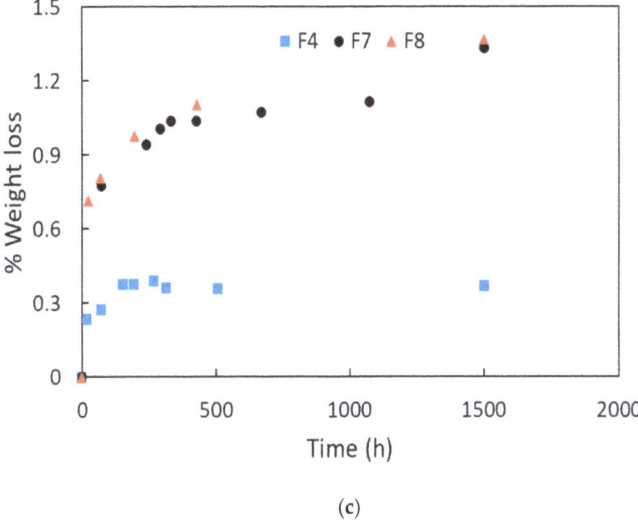

(c)

Figure 3. Weight loss percentage as function of the time for (**a**) F1, F2, F3 and F4 formulations, (**b**) F4 compared to F5, F6, (**c**) F4 compared to F7 and F8.

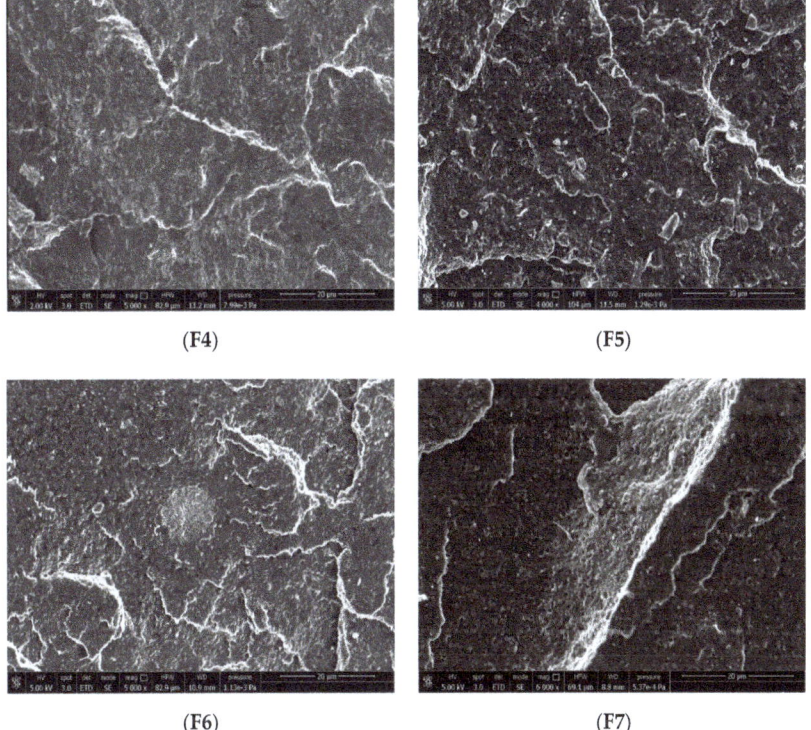

Figure 4. *Cont.*

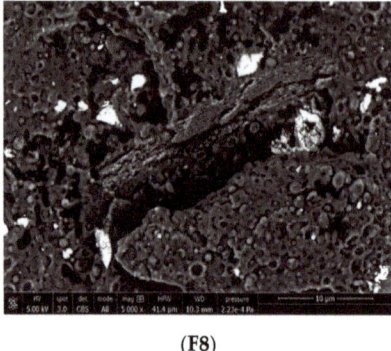

(F8)

Figure 4. SEM micrographs of **F4**, **F5**, **F6**, **F7**, and **F8** formulations. The **F8** micrographs was obtained by backscattered electrons to better evidence $CaCO_3$ particles (white).

From the weight loss trends reported in Figure 3, two migration regimes can be observed: a first linear regime where a high quantity of plasticizer is lost in a short time and a second regime where the mass loss is lower and almost constant. The diffusion mechanism, as reported previously, is clearly influenced by the addition of the fillers (by their typology and quantity). However, another factor must be considered. The films, in fact, were maintained at 60 °C, slightly above the PLA glass transition temperature. At this temperature, the presence of compatible plasticizer (like ATBC) enhances the free volume of chains which induces molecular mobility. This reduces the Tg, but it can also favor the crystallization process [71]. The plasticizer goes only in the amorphous regions and if these amorphous regions decrease due to the crystallinity increment, the plasticizer migration will be accelerated. Considering that fillers can also act as nucleating agents (especially $CaCO_3$) [40], film crystallization can become the main cause of ATBC loss. To better understand the role of the crystallization phenomena to the ATBC release, DSC analysis was performed considering only the first heating scan to take into account the real condition of the films produced by compression molding (i.e., considering their thermal history).

In Table 4 for all the formulations the values of glass transition temperature (T_g), cold crystallization temperature (T_{cc}), cold crystallization enthalpy (ΔH_{cc}), melting temperature (T_m), and melting enthalpy (ΔH_m) are reported. Furthermore, the PLA percentage of crystallinity (X_{cc}), calculated according to Equation (1), was reported. It can be observed that all the blends, which were held at 60 °C, are well above their T_g. Since these are plasticized films, their T_g, compared to pure PLA, are shifted towards lower values. A marked decrement of PLA crystallinity from F1 to F2 formulation containing the PST can be observed. In literature, other authors confirmed that the PST addition decreases the number of crystals [72], this capability combined to the intermolecular interactions created by PST, limits considerably the plasticizer migration. The addition of 4 wt % of micrometric calcium carbonate (2AV) does not alter the crystallinity percentage initially present in the film. By adding larger quantities of filler (4–7 wt %), the starting crystalline content is significantly reduced independently of the type of calcium carbonate used. A greater quantity of filler not only hinders the diffusion process by increasing the tortuosity diffusion paths, but it also worsens the mobility of the polymeric chains and slows down the crystallization kinetics. It is known, in fact, that rigid fillers (such as calcium carbonate) can act as heterogeneous nucleation sites (nucleating agents) only if added in small quantities and if enough time is provided to the system for crystallizing [6]. Initially, the crystallinity degree does not influence the plasticizer migration that it is only affected by the type of filler used. At this purpose, the data confirm the negative role of the surface coating on the migration of ATBC making the micrometric 2AV the best choice in limiting the migration process.

On the other hand, F8 formulation shows a different crystallization behavior due to the interactions occurring between calcium carbonate particles and chitin nanofibrils. The not uniform distribution related to the formation of agglomerates facilitates the plasticizer migration and, at the same time, the PLA crystallization. The combination of these factors explains the major loss of ATBC that was encountered.

Table 4. Results of differential scanning calorimetry analysis (first heating).

Blends	T_g (°C)	T_{cc} (°C)	ΔH_{CC} (J/g)	T_m (°C)	ΔH_m (J/g)	X_C (%)
F1	43.6	96.4	3.05	143.7	18.84	27
F2	31.7	86.7	12.97	145.3	19.94	12
F3	30.3	85.4	13.30	144.8	19.19	11
F4	44	92.9	17.56	145.9	20.44	5
F5	36.2	90	16.93	146.5	18.6	3
F6	44.1	93.3	18.37	147.9	19.92	3
F7	46.4	87.5	19.31	143.4	20.57	2
F8	43.5	85.7	11.97	145.5	18.82	13

To better understand the migration mechanism of the plasticizer and if this is influenced during the time by an eventual crystallinity increment (due to the permanence of the sample at 60 °C for a long period), DSC analysis were performed at the end of the migration tests. The comparison was made only between the reference formulation F1, the formulation F2 containing the PST, the best filler calcium carbonate added formulation (F4) and for the chitin nanofibrils added formulation (F7).

In Table 5, the DSC results at the end of the migration test for F1, F2, and F4 formulations are summarized. In this way, it is possible to verify how the crystallinity varies during the heat treatment at 60 °C. In Figure 5 the DSC thermograms before and after the migration test are also reported.

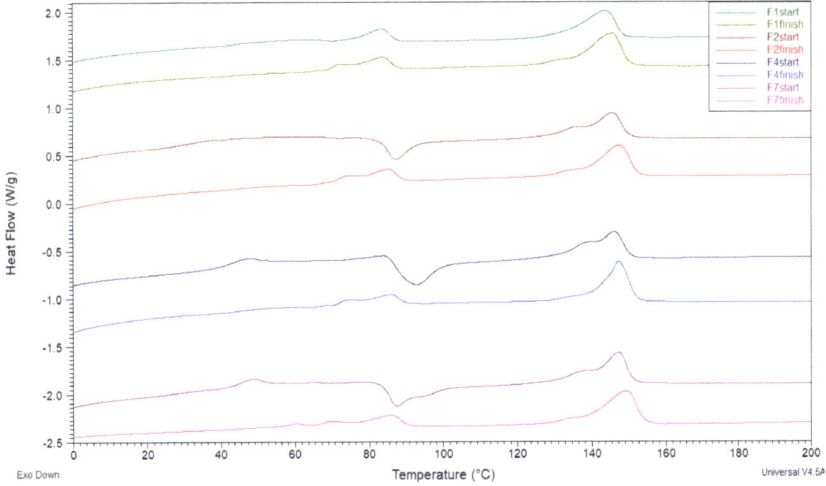

Figure 5. Comparison between thermograms before and after the migration test (first heating).

Table 5. DSC results for F1, F2, F4, and F7 formulations at the beginning and at the end of the migration test (first heating).

Blends	T_g (°C)	T_{cc} (°C)	ΔH_{CC} (J/g)	T_m (°C)	ΔH_m (J/g)	X_C (%)
F1 start	43.6	96.4	3	143.7	18.8	27
F1 finish	42.4	-	-	145.6	23.5	40
F2 start	42.7	86.7	12.9	145.3	19.9	12
F2 finish	43.5	-	-	146.8	21.8	38
F4 start	44	92.9	17.5	145.9	20.4	5
F4 finish	43.8	-	-	147.4	21.9	40
F7 start	46.4	87.5	19.31	143.4	20.57	2
F7 finish	48.8	-	-	149.1	21.84	38.5

It can be observed that, over time, the crystallinity of the samples increases independently if there is the presence of PST or filler. This means that the migration of plasticizer over the time is influenced by the crystallinity increase that reduces the amorphous regions in where the plasticizer is situated and favors its migration. Anyway, it seems that independently from the additives, the system reaches the same value of PLA crystallinity (about 40%) that can be assumed as the final value reached over the time.

At this point, the diffusion coefficients for all formulations were evaluated adopting the linear diffusion coefficient obtained by regression of the migration data as a function of the square root of the time according to the simple form of the Crank's equation (Equation (7)). According to Equation (7), by plotting the values of M_t/C_{p0} (where Mt is the weight variation of the samples at time t per film area (mg/cm^2) and C_{p0} is the initial density of the film (mg/cm^3)) as a function of the time square root, the diffusion coefficient corresponds to the slope of the linear part of the curve. Clearly, this simplified equation, considering only the early stages of the migration where the trend is linear, does not take into account the crystallization phenomenon that may occur in in long periods and can induce a further plasticizer migration over the time. Consequently, the diffusion coefficients were also calculated by applying the not-simplified Crank's solution of the second Fick's law (Equation (6)). Considering that the films are very thin (0.025 cm), the concentration variation of the ATBC along the entire film thickness can be neglected thus x can be imposed equal to zero in Equation (6). The results obtained with Equation (6) make the calculation of the diffusion coefficients at various instants possible until the end of the migration tests (after 1500 h), considering in this way also the crystallization processes. An average diffusion coefficient weighted as a function of the plasticizer concentration was thus considered and compared with the diffusion coefficient obtained from Equation (7); the results are reported in Table 6.

Table 6. Diffusion coefficient calculated according to Equation (7) and to Equation (6).

Blends	D (cm^2/s) Equation (7)	D (cm^2/s) Equation (6) Weight Averaged
F1	5.6×10^{-12}	1.3×10^{-10}
F2	4×10^{-12}	1.2×10^{-10}
F3	4×10^{-12}	1.1×10^{-10}
F4	4×10^{-13}	8.2×10^{-11}
F5	3×10^{-12}	9.8×10^{-11}
F6	1×10^{-12}	8.3×10^{-11}
F7	3×10^{-12}	9.7×10^{-11}
F8	9×10^{-12}	1.3×10^{-10}

The values of the diffusion coefficients calculated in the first stages by Equation (7) highlight the better efficiency of 2AV in limiting the plasticizer migration; in fact, for F4 formulation, the diffusion coefficient is lower of an order of magnitude. On the other hand, F8 shows the highest value of diffusion coefficients; the coexistence of the chitin nanofibrils and 2AV that forms agglomerates helps the ATBC diffusion leaving more routes for the plasticizer migration. Another good result is obtained by the diffusion coefficient of F6, containing the nano-metric CCR calcium carbonate. This result is coherent to what is observed from the ATBC mass loss. However, these diffusion coefficients consider only the initial diffusion mechanism where a very low value of crystallinity content was present for many formulations. As it was observed, the crystallinity increases up to 40% after 1500 h, consequently to consider the effects at longer times, the diffusion coefficients calculated by Equation (6) are more realistic.

First of all, it can be observed that considering the entire process, there is an increase of all diffusion coefficients and this confirms that the calculation only at the first stages lead to a D underestimation. A great worsening of the diffusion coefficients is registered for F1, F2, and F8 formulations coherently to what it was observed experimentally. In fact, these formulations lost a higher quantity of ATBC. The results confirm also that 4 wt % of 2AV is not a sufficient quantity to efficiently limit plasticizer migration; on the other hand, passing from 4 to 7 wt %, very good results are achieved. For longer times, the different behavior to the plasticizer migration for the three types of calcium carbonate emerges. The diffusion coefficients obtained confirm what was registered from the migration tests. In fact, it was observed that Smartfill was the worst filler in hindering the ATBC migration and, over long periods of time, its diffusion coefficient dramatically decreases. On the other hand, CCR seems to quite effectively limit plasticizer migration for long times with a slight decrement of its diffusion coefficient that passes from 1×10^{-12} cm^2/s to 8.3×10^{-11} cm^2/s. However, if we also compare these results with 2AV it can be observed that the improvement in the migration coefficients, and consequently in the ATBC loss, are not so evident to justify the use of this more expensive filler. Omycarb-2AV in fact, possess a D value of 8.2×10^{-11} cm^2/s perfectly comparable with F6 diffusion coefficient. Noteworthy is the greatest difference in the diffusion coefficient calculated with the Equations (6) and (7), registered for the F6 formulation. Additionally, considering the crystallinity values over time (reported in Table 5), it can be observed that the crystallization kinetics for this type of formulation is slower therefore the higher ATBC release will occur in longer times. Equation (7) can therefore give a fairly truthful estimation of the diffusion coefficients only for those systems that quickly lose the plasticizer and for which there are no great differences of crystallinity over time.

As far as concern the formulation containing only the chitin nanofibrils (F7), a decrease in plasticizer migration should be observed. In fact, it is known in [73] that their addition improves the barrier properties. Effectively, the diffusion coefficient obtained is lower and confirms the mass loss results obtained. However, to contrast the ATBC migration, 2AV is still the most efficient.

$$D = cost \cdot t^{-1} \qquad (10)$$

It can be observed from Figure 6 that the most significant variations of D, and therefore of the migration of the plasticizer, occur in the first 100 h—after the curves tend to flatten and the values of D tend to zero—meaning that the diffusion process finishes. The most flattened curve is that of the F4 formulation which will therefore have the lowest diffusion coefficient value over time and will therefore be able to release less plasticized content, further confirming the good efficiency of 2AV.

The rheological properties (torque, MFI, and MVR) properties were therefore evaluated and compared with those of the previous works [44,47].

The melt properties (MFR, MVR, and torque values) of each formulation are summarized in Table 7.

The addition of PST melt strength enhancer leads to an increment of torque (and in parallel to a decrement of MFR value) as it can be observed passing from F1 to F2 compositions. Clearly, the addition of a rigid filler (in this case micro calcium carbonate 2AV) leads to a marked decrement of the MFR value (F3 blend). As it can be expected, a further addition of micro-calcium carbonate (from 4 wt % to 7 wt %) further decreased the MFR from 9.4 to 7.6 (g/10 min) for F3 and F4 blends respectively.

The MFR (similarly to MVR), affected by the molecular weight and by the interactions occurring in the melt material, can be correlated to the degree of disentanglement of the fluid blend [74]. The more the melt material is disentangled, the more the diffusion coefficient increases as calculated by Equation (6) (so considering only the starting part of the mass loss versus time trend) The material above its glass transition consists of a net of entangled macromolecules. A high molecular weight as well as efficient interactions contribute to increasing the degree of entanglement making more rigid the net and more difficult the diffusion of ATBC molecules. On the contrary, a high degree of disentanglement, due to lack of interactions with additives or fillers, as well as a decrease of molecular weight (cutting macromolecules), make the net more suitable for diffusion increasing D. The data of MFR, measured in the melt, and D decrease from F1 to F4. In fact, the degree of disentanglement decreases and D decreases accordingly (Figure 7).

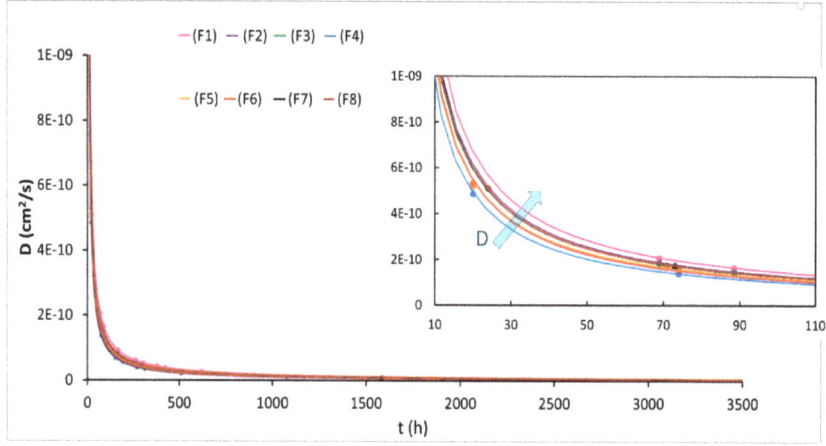

Figure 6. Trend of the diffusion coefficient over the time for all the formulation examined.

Table 7. Torque, MVR, and MFR values of the prepared blends.

Blends	Torque (N·cm)	MVR (cm³/10 min)	MFR (g/10 min)
F1	67.8 ± 5.4	22.5 ± 2.0	23.6 ± 2.1
F2	72.8 ± 6.0	11.8 ± 0.9	12.4 ± 0.9
F3	73.4 ± 9.8	8.7 ± 0.6	9.4 ± 0.6
F4	83.4 ± 4.0	8.4 ± 2.7	7.6 ± 2.4
F5	74.1 ± 5.4	12.0 ± 1.9	13.6 ± 2.2
F6	66.4 ± 5.9	14.1 ± 1.9	15.6 ± 2.1
F7	63.3 ± 3.6	11.8 ± 1.1	13.1 ± 1.3
F8	73.5 ± 5.2	10.0 ± 1.3	11.4 ± 1.5

However, in F5 using the same weight percent (7 wt %) of Smartfill with respect to F4 and CCR in F6, an increase in MFR was registered. It is likely that the surface coating of the calcium carbonate particles creates less friction in the molten matrix during the extrusion process. Moreover, as the removal of water during the drying of filler before the extrusion is more difficult for nanometric fillers, some chain scission due to water transesterification with polyester macromolecules can occur [75] resulting in shorter macromolecules. The latter effect influences the MFR confirming, this being the measurement done in the melt in dynamic conditions, so it is more dependent on the capability of macromolecules to flow. Both of these effects generate disentanglements that increase the value of D with respect to F4. In general, the observed variations in the Torque and MVR values are not so significant to alter the final processability independently of the filler used and also by the greater quantity of calcium carbonate added. Interestingly, F6 shows a very low value of D despite having the highest value of MFR. This is likely due to chain scission.

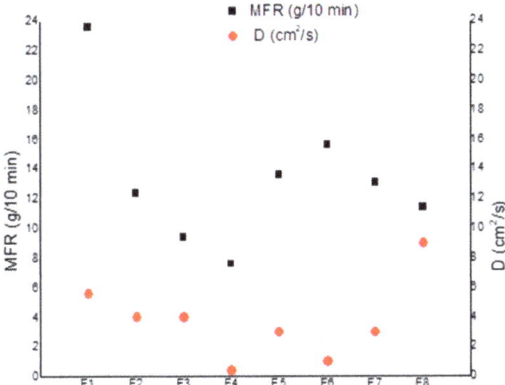

Figure 7. Comparison between MFR data and D calculated by Equation (6).

The occurring of chain scission was demonstrated by preparing a blend with the same composition of F6 but by previously drying better the nano-calcium carbonate, for two days at 60 °C at reduced pressure (50 KPa), and it was observed that the torque value was 77.0 ± 5.0 Ncm and the MFR value for this blend was 7.7 ± 0.6 cm^3/10 min, in agreement with a higher molecular weight of the biopolyesters with respect to F6.

It can be observed that, for F6, the correspondent low value of D evidenced very good interactions, counterbalancing the decrease in molecular weight and reasonably achieved thanks to the nano-dimension of the filler (extended surficial area) significantly enhancing interactions despite a sub-optimal chemical affinity due to fatty acids and despite partial agglomeration. Chitin nanofibrils (F7) and micro-nanometric uncoated carbonate (F5), having an intermediate dimension between that of micro-carbonate and nano-carbonate, showed a similar behavior in agreement with the similar dimensions shown in phase morphology analysis of blends. In general, the analysis of the melt fluidity of the blends was fundamental to show that molecular weight can be another important parameter affecting the diffusion behaviors of polymeric materials.

4.2. Mechanical Characterization of Blends

After the evaluation of the different types of filler added to the ATBC migration, a general screening concerning the mechanical properties was carried out in order to evaluate if the mechanical properties (that were optimized in previous works [44,47]) were significantly altered by the greater amount of calcium carbonate added and by the different types of fillers used.

The variations of the tensile properties (yield stress (σ_y), stress, and strain at break (σ_b and ε_b)) are reported in Table 8.

The increment up to 7 wt % of the different calcium carbonate fillers weakens the material. The greater quantity of rigid calcium carbonates added makes the final material more rigid, leading to a decrease of stress and elongation at break and to an increment of the yielding stress. However, comparing the final stress and strains values of the F3 formulation with those of F4, F5, and F6 it can be concluded that nevertheless this flexibility reduction the tensile properties of the plasticized PLA/PBS blends remains still good. The fillers introduced in these formulations, leads to a modification of the first part of the curve which assumes a typical trend of a more resistant system. In fact, it can be observed the capability of PST and fillers to increase the yield stress compared to the F1 formulation, which exhibits a behavior similar to an elastomer (Figure 8). The addition of an uncoated micrometric carbonate such as 2AV, capable of dispersing adequately during extrusion, ensures a yield value higher than other types of carbonates.

Table 8. Tensile properties of the films obtained from each blend.

Blends	σ_b (MPa)	ε_b (%)	σ_y (MPa)
F1	31.8 ± 1.4	572.7 ± 20.7	-
F2	33.0 ± 1.2	554.2 ± 12.3	10.2 ± 0.7
F3	32.5 ± 1.6	543.7 ± 29.8	23.3 ± 1.9
F4	29.3 ± 3.4	512.5 ± 13.8	31.1 ± 2.2
F5	29.0 ± 1.2	491.4 ± 25.9	20.7 ± 2.4
F6	28.4 ± 1.8	507.6 ± 13.9	24.5 ± 1.7
F7	25.5 ± 1.3	421.9 ± 25.1	11.6 ± 0.7
F8	25.5 ± 1.0	400.1 ± 21.9	10.8 ± 1.8

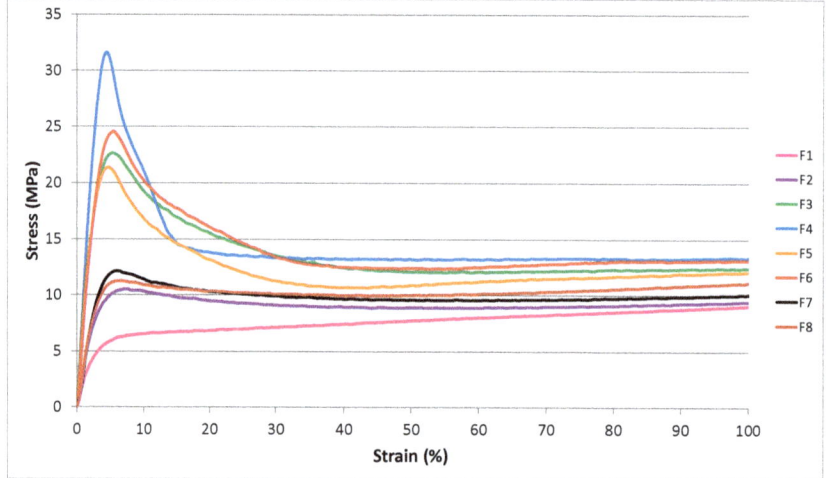

Figure 8. First part of the stress-strain curves for the formulations examined.

The combination of 2AV with chitin nanofibrils does not lead to an improvement of the tensile properties, as can be observed in F8 compared to F4, especially as regards the elongation at break. For F8 formulation, the yield stress is very similar to the result obtained with chitin nanofibrils alone. On the other hand, the comparison with the F4 formulation shows a marked decrease in the yield stress. The nanofibrils seem to not allow calcium carbonate to reinforce the formulation. Standing between the carbonate particles and the polymer matrix, they cause a reduction of the filler-matrix adhesion as it can be observed from the SEM image (Figure 4) where the micrometric calcium carbonate was found both in the matrix and near the areas containing nanofibrils.

5. Conclusions

On the basis of previous studies, the plasticized poly(lactic acid) (PLA)/poly(butylene succinate) (PBS)-based films containing chitin nanofibrils (CNs), calcium carbonate, and a small percentage of a commercial melt strength enhancer, have been investigated from the point of view of the plasticizer migration. It was observed that, during storage, these films lost a significant amount of plasticizer. The effect of the addition of different types of calcium carbonate (nanometric and micrometric, surface coated and not) and chitin nanofibrils was investigated with the purpose of reducing or controlling the plasticizer migration. The Crank's solution of the Fick's second law was adopted to obtain the quantative values of diffusion coefficients. In addition, the evolution of the crystallinity that can induce the plasticizer migration was considered.

The results showed that the not surface coated calcium carbonate (2AV) is the more effective in hindering the plasticizer migration. Micrometric calcium carbonate reduced significantly the ATBC migration but at the same time did not lead to a significant change in the processability and mechanical properties of the already optimized formulations.

Regarding the surface coated calcium carbonate, the surface coating (generally made with fatty acids) probably limits the ATBC absorption on the calcium carbonate particles surfaces, leading to worse results in ATBC migration. Moreover, the occurrence of chain scission due to difficulties in removing humidity from nanofiller before blending determined a decrease in molecular weight, resulting in macromolecular net disentanglement that allowed a better diffusion of ATBC, as concluded by studying the fluidity in the melt of blends. For these reasons, the migration control is not significantly improved passing from micro- to nano- calcium carbonate particles, despite the increased surficial area of fillers. From a point of view of cost, reduction of the plasticizer migration, and quantity added, the nanometric calcium carbonate particles are less efficient and this is ascribable to their agglomeration tendency and high hygroscopicity.

Chitin nanofibrils alone can also influence the plasticizer migration thanks to their capability of hindering the plasticizer diffusion, slowing but not efficiently limiting the migration, and thus suggesting new potentialities of such nano additives in films or products with a controlled release.

The coupling of chitin nanofibrils and 2AV did not lead to significant results. Negative effects have been encountered probably related to the synergistic agglomeration tendency of chitin nanofibrils and calcium carbonate.

The crystallinity reached from samples after migration test seems not affected by the type of filler added and about 40% of PLA crystallinity was obtained. Clearly, this marked crystallinity increment over time affects the plasticizer migration due to the decreasing portion of mobile amorphous fraction in which the plasticizer is located.

In general, the analysis of all the parameters affecting migration of plasticizers suggests promising strategies to better exploit the properties of inorganic micro and nano particles, as well as natural nanofibrils in biocomposites.

Author Contributions: Conceptualization, M.-B.C. and L.A.; Experiments, A.V., L.P., and L.A.; Calculations, V.G. and L.A.; Data curation, L.P. and A.V.; Data interpretation L.A. and M.-B.C.; Writing—original draft preparation, L.A. and A.V.; Writing—review and editing, M.-B.C.; Visualization, L.A.; Supervision, A.L.; Project administration, M.-B.C. and A.L.; Funding acquisition, M.-B.C. All authors have read and agreed to the published version of the manuscript.

Funding: This research was partially funded by the Bio-Based Industries Joint Undertaking under the European Union Horizon 2020 research program (BBI-H2020), PolyBioSkin project, grant number G.A. 745839.

Acknowledgments: We thank the Centre for Instrumentation Sharing-University of Pisa (CISUP) for their support in the use of FEI Quanta 450 FEG scanning electron microscope. We would also like to thank Pierfrancesco Morganti for interesting discussion.

Conflicts of Interest: The authors declare no conflict of interest.

References

1. Rao, M.G.; Bharathi, P.; Akila, R.M. A comprehensive review on biopolymers. *Sci. Rev. Chem. Commun.* **2014**, *4*, 61–68.
2. Mekonnen, T.; Mussone, P.; Khalil, H.; Bressler, D. Progress in bio-based plastics and plasticizing modifications. *J. Mater. Chem. A* **2013**, *1*, 13379–13398. [CrossRef]
3. Aliotta, L.; Gigante, V.; Coltelli, M.B.; Cinelli, P.; Lazzeri, A. Evaluation of Mechanical and Interfacial Properties of Bio-Composites Based on Poly(Lactic Acid) with Natural Cellulose Fibers. *Int. J. Mol. Sci.* **2019**, *20*, 960. [CrossRef] [PubMed]
4. Garrison, T.F.; Murawski, A.; Quirino, R.L. Bio-Based Polymers with Potential for Biodegradability. *Polymers* **2016**, *8*, 262. [CrossRef]

5. Scatto, M.; Salmini, E.; Castiello, S.; Coltelli, M.B.; Conzatti, L.; Stagnaro, P.; Andreotti, L.; Bronco, S. Plasticized and nanofilled poly(lactic acid)-based cast films: Effect of plasticizer and organoclay on processability and final properties. *J. Appl. Polym. Sci.* **2013**, *127*, 4947–4956. [CrossRef]
6. Aliotta, L.; Cinelli, P.; Coltelli, M.B.; Righetti, M.C.; Gazzano, M.; Lazzeri, A. Effect of nucleating agents on crystallinity and properties of poly(lactic acid) (PLA). *Eur. Polym. J.* **2017**, *93*, 822–832. [CrossRef]
7. Lim, L.T.; Auras, R.; Rubino, M. Processing technologies for poly(lactic acid). *Prog. Polym. Sci.* **2008**, *33*, 820–852. [CrossRef]
8. Auras, R.; Harte, B.; Selke, S. An overview of polylactides as packaging materials. *Macromol. Biosci.* **2004**, *4*, 835–864. [CrossRef]
9. Liu, H.; Zhang, J. Research progress in toughening modification of poly(lactic acid). *J. Polym. Sci. Part B Polym. Phys.* **2011**, *49*, 1051–1083. [CrossRef]
10. Krishnan, S.; Pandey, P.; Mohanty, S.; Nayak, S.K. Toughening of Polylactic Acid: An Overview of Research Progress. *Polym. Plast. Technol. Eng.* **2016**, *55*, 1623–1652. [CrossRef]
11. Urquijo, J.; Guerrica-Echevarría, G.; Eguiazábal, J.I. Melt processed PLA/PCL blends: Effect of processing method on phase structure, morphology, and mechanical properties. *J. Appl. Polym. Sci.* **2015**, *132*. [CrossRef]
12. Jawalkar, S.S.; Aminabhavi, T.M. Molecular modeling simulations and thermodynamic approaches to investigate compatibility/incompatibility of poly(l-lactide) and poly(vinyl alcohol) blends. *Polymer* **2006**, *47*, 8061–8071. [CrossRef]
13. Lipsa, R.; Tudorachi, N.; Vasile, C. Poly(vinyl alcohol)/poly(lactic acid) blends biodegradable films doped with colloidal silver. *Rev. Roum. Chim.* **2008**, *53*, 405–413.
14. Bhatia, A.; Gupta, R.K.; Bhattacharya, S.N.; Choi, H.J. Compatibility of biodegradable poly(lactic acid) (PLA) and poly (butylene succinate) (PBS) blends for packaging application. *Korea Aust. Rheol. J.* **2007**, *19*, 125–131. [CrossRef]
15. Zhao, P.; Liu, W.; Wu, Q.; Ren, J. Preparation, mechanical, and thermal properties of biodegradable polyesters/poly(lactic acid) blends. *J. Nanomater.* **2010**, *2010*. [CrossRef]
16. Aversa, C.; Barletta, M.; Puopolo, M.; Vesco, S. Cast extrusion of low gas permeability bioplastic sheets in PLA/PBS and PLA/PHB binary blends. *Polym. Technol. Mater.* **2020**, *59*, 231–240. [CrossRef]
17. Ostrowska, J.; Sadurski, W.; Paluch, M.; Tyński, P.; Bogusz, J. The effect of poly(butylene succinate) content on the structure and thermal and mechanical properties of its blends with polylactide. *Polym. Int.* **2019**, *68*, 1271–1279. [CrossRef]
18. Oguz, H.; Dogan, C.; Kara, D.; Ozen, Z.T.; Ovali, D.; Nofar, M. Development of PLA-PBAT and PLA-PBSA bio-blends: Effects of processing type and PLA crystallinity on morphology and mechanical properties. *AIP Conf. Proc.* **2019**, *2055*. [CrossRef]
19. Nofar, M.; Tabatabaei, A.; Sojoudiasli, H.; Park, C.B.; Carreau, P.J.; Heuzey, M.C.; Kamal, M.R. Mechanical and bead foaming behavior of PLA-PBAT and PLA-PBSA blends with different morphologies. *Eur. Polym. J.* **2017**, *90*, 231–244. [CrossRef]
20. Quero, E.; Müller, A.J.; Signori, F.; Coltelli, M.B.; Bronco, S. Isothermal cold-crystallization of PLA/PBAT blends with and without the addition of acetyl tributyl citrate. *Macromol. Chem. Phys.* **2012**, *213*, 36–48. [CrossRef]
21. Gigante, V.; Canesi, I.; Cinelli, P.; Coltelli, M.B.; Lazzeri, A. Rubber Toughening of Polylactic Acid (PLA) with Poly(butylene adipate-co-terephthalate) (PBAT): Mechanical Properties, Fracture Mechanics and Analysis of Ductile-to-Brittle Behavior while Varying Temperature and Test Speed. *Eur. Polym. J.* **2019**, *115*, 125–137. [CrossRef]
22. Pietrosanto, A.; Scarfato, P.; Di Maio, L.; Nobile, R.M.; Incarnato, L. Evaluation of the Suitability of Poly(Lactide)/Poly(Butylene-Adipate-co-Terephthalate) Blown Films for Chilled and Frozen Food Packaging Applications. *Polymers (Basel)* **2020**, *12*, 804. [CrossRef] [PubMed]
23. Wei, X.-F.; Linde, E.; Hedenqvist, M.S. Plasticiser loss from plastic or rubber products through diffusion and evaporation. *Mater. Degrad.* **2019**, *3*, 18. [CrossRef]
24. Zimmermann, L.; Dierkes, G.; Ternes, T.A.; Völker, C.; Wagner, M. Benchmarking the in Vitro Toxicity and Chemical Composition of Plastic Consumer Products. *Environ. Sci. Technol.* **2019**, *53*. [CrossRef]
25. Jamarani, R.; Erythropel, H.C.; Nicell, J.A.; Leask, R.L.; Marić, M. How green is your plasticizer? *Polymers* **2018**, *10*, 834. [CrossRef]

26. Arvanitoyannis, I.S.; Bosnea, L. Migration of substances from food packaging materials to foods. *Crit. Rev. Food Sci. Nutr.* **2004**, *44*, 63–76. [CrossRef]
27. Balestri, E.; Menicagli, V.; Ligorini, V.; Fulignati, S.; Raspolli Galletti, A.M.; Lardicci, C. Phytotoxicity assessment of conventional and biodegradable plastic bags using seed germination test. *Ecol. Indic.* **2019**, *102*, 569–580. [CrossRef]
28. Coltelli, M.B.; Danti, S.; Trombi, L.; Morganti, P.; Donnarumma, G.; Baroni, A.; Fusco, A.; Lazzeri, A. Preparation of innovative skin compatible films to release polysaccharides for biobased beauty masks. *Cosmetics* **2018**, *5*, 70. [CrossRef]
29. Hiremath, C.; Heggannavar, G.B.; Mitchell, G.R.; Kariduraganavar, M.Y. Biopolymers in Drug-Delivery Applications. In *Green Polymer Composites Technology: Properties and Applications*; Inamuddin, Ed.; CRC: Boca Raton, FL, USA; Taylor and Francis: Abingdon, UK, 2015; pp. 514–525.
30. Xiao, H.; Lu, W.; Yeh, J. Effect of plasticizer on the crystallization behavior of poly(lactic acid). *J. Appl. Polym. Sci.* **2009**, *113*, 112–121. [CrossRef]
31. Stark, T.D.; Choi, H.; Diebel, P.W. Influence of plasticizer molecular weight on plasticizer retention in PVC geomembranes. *Geosynth. Int.* **2005**, *12*, 99–110. [CrossRef]
32. Higa, C.M.; Tek, A.T.; Wojtecki, R.J.; Braslau, R. Nonmigratory internal plasticization of poly(vinyl chloride) via pendant triazoles bearing alkyl or polyether esters. *J. Polym. Sci. Part A Polym. Chem.* **2018**, *56*, 2397–2411. [CrossRef]
33. Reddy, N.N.; Mohan, Y.M.; Varaprasad, K.; Ravindra, S.; Vimala, K.; Raju, K.M. Surface treatment of plasticized poly(vinyl chloride) to prevent plasticizer migration. *J. Appl. Polym. Sci.* **2010**, *115*, 1589–1597. [CrossRef]
34. Dressler, S. Plasma treatments. In *Surface Modification Technologies: An Engineer's Guide*; Marcel Dekker: New York, NY, USA, 1989; pp. 317–419.
35. Li, X.; Xiao, Y.; Wang, B.; Tang, Y.; Lu, Y.; Wang, C. Effect of Poly(1,2-propylene glycol adipate) and Nano-CacCO$_3$ on DOP Migration and Mechanical Proprties of Flexible PVC. *J. Appl. Polym. Sci.* **2012**, *124*, 1737–1743. [CrossRef]
36. Scott, M.P.; Rahman, M.; Brazel, C.S. Application of ionic liquids as low-volatility plasticizers for PMMA. *Eur. Polym. J.* **2003**, *39*, 1947–1953. [CrossRef]
37. Ma, Y.; Liao, S.; Li, Q.; Guan, Q.; Jia, P.; Zhou, Y. Physical and chemical modifications of poly(vinyl chloride) materials to prevent plasticizer migration—Still on the run. *React. Funct. Polym.* **2020**, *147*, 104458. [CrossRef]
38. Scaffaro, R.; Maio, A.; Gulino, E.F.; Morreale, M.; La Mantia, F.P. The Effects of Nanoclay on the Mechanical Properties, Carvacrol Release and Degradation of a PLA/PBAT Blend. *Materials* **2020**, *13*, 983. [CrossRef]
39. Argon, A.S.; Cohen, R.E. Toughenability of polymers. *Polymer* **2003**, *44*, 6013–6032. [CrossRef]
40. Aliotta, L.; Cinelli, P.; Beatrice Coltelli, M.; Lazzeri, A. Rigid filler toughening in PLA-Calcium Carbonate composites: Effect of particle surface treatment and matrix plasticization. *Eur. Polym. J.* **2018**, *113*, 78–88. [CrossRef]
41. Cioni, B.; Lazzeri, A. The Role of Interfacial Interactions in the Toughening of Precipitated Calcium Carbonate–Polypropylene Nanocomposites. *Compos. Interfaces* **2010**, *17*, 533–549. [CrossRef]
42. Yang, B.; Bai, Y.; Cao, Y. Effects of inorganic nano-particles on plasticizers migration of flexible PVC. *J. Appl. Polym. Sci.* **2010**, *115*, 2178–2182. [CrossRef]
43. Zhou, Y.W.; Xiao, Y.; Li, X.H.; Wang, B. Effects of nano-particles on DOP migration and tensile properties of flexible PVC sheets. *Adv. Mater. Res.* **2012**, *548*, 119–122. [CrossRef]
44. Gigante, V.; Coltelli, M.; Vannozzi, A.; Panariello, L.; Fusco, A.; Trombi, L.; Donnarumma, G.; Danti, S.; Lazzeri, A. Flat Die Extruded Biocompatible Poly(Lactic Acid). *Polymers* **2019**, *11*, 1857. [CrossRef] [PubMed]
45. Coltelli, M.-B.; Gigante, V.; Vannozzi, A.; Aliotta, L.; Danti, S.; Neri, S.; Gagliardini, A.; Morganti, P.; Panariello, L.; Lazzeri, A. Poly(lactic) acid (PLA) based nano-structured functional films for personal care applications. In Proceedings of the AUTEX 2019—19th World Textile Confreence on Textiles at the Crossroads, Ghent, Belgium, 11–15 June 2019.
46. Coltelli, M.B.; Gigante, V.; Panariello, L.; Morganti, P.; Cinelli, P.; Danti, S.; Lazzeri, A. Chitin nanofibrils in renewable materials for packaging and personal care applications. *Adv. Mater. Lett.* **2018**, *10*, 425–430. [CrossRef]

47. Coltelli, M.; Aliotta, L.; Vannozzi, A.; Morganti, P.; Fusco, A.; Donnarumma, G.; Lazzeri, A. Properties and Skin Compatibility of Films Based on Poly(Lactic Acid) (PLA) Bionanocomposites Incorporating Chitin Nanofibrils (CN). *J. Funct. Biomater.* **2020**, *11*, 21. [CrossRef] [PubMed]
48. Coltelli, M.; Cinelli, P.; Gigante, V.; Aliotta, L.; Morganti, P.; Panariello, L.; Lazzeri, A. Chitin Nanofibrils in Poly(Lactic Acid) (PLA) Nanocomposites: Dispersion and Thermo-Mechanical Properties. *Int. J. Mol. Sci.* **2019**, *20*, 504. [CrossRef] [PubMed]
49. International Organization for Standardization. *ISO 1133-1:2011: Determination of the Melt Mass-Flow Rate (MFR) and Melt Volume-Flow Rate (MVR) of Thermoplastics—Part 1*; International Organization for Standardization: Geneva, Switzerland, 2011; Volume 2011, p. 24.
50. Aliotta, L.; Gigante, V.; Acucella, O.; Signori, F.; Lazzeri, A. Thermal, Mechanical and Micromechanical Analysis of PLA/PBAT/POE-g-GMA Extruded Ternary Blends. *Front. Mater.* **2020**, *7*, 1–14. [CrossRef]
51. Fischer, E.W.; Sterzel, H.J.; Wegner, G. Investigation of the structure of solution grown crystals of lactide copolymers by means of chemical reactions. *Kolloid Z. Z. Polym.* **1973**, *251*, 980–990. [CrossRef]
52. ISO. *ISO 527, Plastics, Determination of Tensile Properties. Part 2: Test Conditions for Moulding and Extrusion Plastics*; ISO: Geneva, Switzerland, 2012.
53. Hamdani, M.; Feigenbaum, A.; Vergnaud, J.M. Prediction of worst case migration from packaging to food using mathematical models. *Food Addit. Contam.* **1997**, *14*, 499–506. [CrossRef]
54. Crank, J. *The Mathematics of Diffusion*, 2nd ed.; Oxford University Press: Oxford, UK, 1975; ISBN 0-19-853411-6.
55. Vergnaud, J.M. General survey on the mass transfers taking place between a polymer and a liquid. *J. Polym. Eng.* **1995**, *15*, 57–78. [CrossRef]
56. Mercer, A.; Castle, L.; Comyn, J.; Gilbert, J. Evaluation of a predictive mathematical model of di-(2-ethylhexyl) adipate plasticizer migration from PVC film into foods. *Food Addit. Contam.* **1990**, *7*, 497–507. [CrossRef] [PubMed]
57. Lundsgaard, R.; Kontogeorgis, G.M.; Kristiansen, J.K.; Jensen, T.F. Modeling of the Migration of Glycerol Monoester Plasticizers in Highly Plasticized Poly(vinyl chloride). *J. Vinyl Addit. Technol.* **2009**, *15*, 147–158. [CrossRef]
58. Till, D.E.; Reid, R.C.; Schwartz, P.S.; Sidman, K.R.; Valentine, J.R.; Whelan, R.H. Plasticizer migration from polyvinyl chloride film to solvents and foods. *Food Chem. Toxicol.* **1982**, *20*, 95–104. [CrossRef]
59. Huang, J.-C.; Liu, H.; Liu, Y. Diffusion in polymers with concentration dependent diffusivity. *Int. J. Polym. Mater.* **2001**, *49*, 15–24. [CrossRef]
60. Storey, R.F.; Mauritz, K.A.; Cox, B.D. Diffusion of various dialkyl phthalate plasticizers in PVC. *Macromolecules* **1989**, *22*, 289–294. [CrossRef]
61. Philip, J.R. A very general class of exact solutions in concentration-dependent diffusion. *Nature* **1960**, *185*, 233. [CrossRef]
62. Lundsgaard, R. Migration of Plasticisers from PVC and Other Polymers. Ph.D. Thesis, Technical University of Denmark, Kgs. Lyngby, Denmark, 2010.
63. Ekelund, M.; Azhdar, B.; Gedde, U.W. Evaporative loss kinetics of di (2-ethylhexyl) phthalate (DEHP) from pristine DEHP and plasticized PVC. *Polym. Degrad. Stab.* **2010**, *95*, 1789–1793. [CrossRef]
64. Smith, G.S.; Skidmore, C.B.; Howe, P.M.; Majewski, J. Diffusion, evaporation, and surface enrichment of a plasticizing additive in an annealed polymer thin film. *J. Polym. Sci. Part B Polym. Phys.* **2004**, *42*, 3258–3266. [CrossRef]
65. Zhang, J.F.; Sun, X. Physical characterization of coupled poly(lactic acid)/starch/maleic anhydride blends plasticized by acetyl triethyl citrate. *Macromol. Biosci.* **2004**, *4*, 1053–1060. [CrossRef]
66. Cao, Z.; Daly, M.; Clémence, L.; Geever, L.M.; Major, I.; Higginbotham, C.L.; Devine, D.M. Chemical surface modification of calcium carbonate particles with stearic acid using different treating methods. *Appl. Surf. Sci.* **2016**, *378*, 320–329. [CrossRef]
67. Shi, X.; Rosa, R.; Lazzeri, A. On the coating of precipitated calcium carbonate with stearic acid in aqueous medium. *Langmuir* **2010**, *26*, 8474–8482. [CrossRef] [PubMed]
68. Osman, M.A.; Suter, U.W. Surface treatment of calcite with fatty acids: Structure and properties of the organic monolayer. *Chem. Mater.* **2002**, *14*, 4408–4415. [CrossRef]
69. Huang, H.; Tian, M.; Yang, J.; Li, H.; Liang, W.; Zhang, L.; Li, X. Stearic acid surface modifying Mg(OH)$_2$: Mechanism and its effect on properties of ethylene vinyl acetate/Mg(OH)$_2$ composites. *J. Appl. Polym. Sci.* **2008**, *107*, 3325–3331. [CrossRef]

70. Rezaei Gomari, K.A.; Denoyel, R.; Hamouda, A.A. Wettability of calcite and mica modified by different long-chain fatty acids (C18 acids). *J. Colloid Interface Sci.* **2006**, *297*, 470–479. [CrossRef] [PubMed]
71. Muller, J.; Jiménez, A.; González-Martínez, C.; Chiralt, A. Influence of plasticizers on thermal properties and crystallization behaviour of poly(lactic acid) films obtained by compression moulding. *Polym. Int.* **2016**, *65*, 970–978. [CrossRef]
72. Kaczmarek, H.; Nowicki, M.; Vuković-Kwiatkowska, I.; Nowakowska, S. Crosslinked blends of poly(lactic acid) and polyacrylates: AFM, DSC and XRD studies. *J. Polym. Res.* **2013**, *20*. [CrossRef]
73. Broers, L.; van Dongen, S.; de Goederen, V.; Ton, M.; Spaen, J.; Boeriu, C.; Schroën, K. Addition of Chitin Nanoparticles improves Polylactic Acid Film Properties. *Nanotechnol. Adv. Mater. Sci.* **2018**, *1*, 1–8.
74. Chremos, A.; Jeong, C.; Douglas, J.F. Influence of polymer architectures on diffusion in unentangled polymer melts. *Soft Matter* **2017**, *13*, 5778–5784. [CrossRef]
75. Signori, F.; Coltelli, M.B.; Bronco, S. Thermal degradation of poly(lactic acid) (PLA) and poly(butylene adipate-co-terephthalate) (PBAT) and their blends upon melt processing. *Polym. Degrad. Stab.* **2009**, *94*, 74–82. [CrossRef]

© 2020 by the authors. Licensee MDPI, Basel, Switzerland. This article is an open access article distributed under the terms and conditions of the Creative Commons Attribution (CC BY) license (http://creativecommons.org/licenses/by/4.0/).

Article

Dog Wool Microparticles/Polyurethane Composite for Thermal Insulation

Francisco Claudivan da Silva [1,2], Helena P. Felgueiras [3], Rasiah Ladchumananandasivam [1,4], José Ubiragi L. Mendes [1], Késia Karina de O. Souto Silva [4] and Andrea Zille [3,*]

1. Post-graduate Program in Mechanical Engineering (PPGEM), Department of Mechanical Engineering, Federal University of Rio Grande do Norte, Natal 59078-970, Brazil; clauditextil@yahoo.com.br (F.C.d.S.); rlsivam@gmail.com (R.L.); ubiragi@ct.ufrn.br (J.U.L.M.)
2. Foundation for the Promotion of Research of the State of Rio Grande do Norte–FAPERN, Natal 59064-901, Brazil
3. 2C2T-Centro de Ciência e Tecnologia Têxtil, Universidade do Minho, Campus of Azurém, 4804-533 Guimarães, Portugal; helena.felgueiras@2c2t.uminho.pt
4. Textile Engineering Department, Federal University of Rio Grande do Norte, Natal 59078-970, Brazil; kesiasouto@hotmail.com
* Correspondence: azille@2c2t.uminho.pt

Received: 10 April 2020; Accepted: 9 May 2020; Published: 11 May 2020

Abstract: A polyurethane (PU)-based eco-composite foam was prepared using dog wool fibers as a filler. Fibers were acquired from pet shops and alkaline treated prior to use. The influence of their incorporation on the PU foams' morphological, thermal, and mechanical properties was investigated. The random and disorganized presence of the microfibers along the foam influence their mechanical performance. Tensile and compression strengths were improved with the increased amount of dog wool microparticles on the eco-composites. The same occurred with the foams' hydration capacity. The thermal capacity was also slightly enhanced with the incorporation of the fillers. The fillers also increased the thermal stability of the foams, reducing their dilatation with heating. The best structural stability was obtained using up to 120 °C with a maximum of 15% of filler. In the end, the dog wool waste was rationally valorized as a filler in PU foams, demonstrating its potential for insulation applications, with a low cost and minimal environmental impact.

Keywords: dog wool fibers; fillers; polyurethane; eco-composites; renewable resources

---

1. Introduction

Global energy consumption is estimated to increase by 53% within the next 10 years [1]. One of the simplest and most cost-effective ways to reduce the energy demands and the greenhouse gas emissions is through building insulation. If properly selected, an effective insulation may save energy by requiring less for space cooling in the summer and heating in the winter, and thus reducing the use of natural resources (e.g., petroleum and gas) [2]. Thermal insulation is achieved by means of a material or composite materials endowed with high thermal resistance. Over the years, many options have been proposed and tested, including fiberglass, mineral wool, and foams (e.g., polyurethane, PU, and polyvinyl chloride, PVC) [3].

PU is formed of stiff and flexible segments, endowing PU foams with versatile properties and light weight and making them particularly desirable for insulation. They are obtained by a reaction between polyfunctional alcohols (polyol polyether or polyol polyester) and polyisocyanate [4]. Their foaming appearance is possible due to the production of a blowing agent (e.g., carbon dioxide) during exothermal polymerization, which remains enclosed within the material and ensures the foam insulating performance [5,6]. Depending on the amount, proportions, and characteristics of the components, three categories of PU foams can be defined: Flexible, semi-rigid, and rigid, the last being the preferred for insulation purposes due to its highly cross-linked and closed-cell structure, good mechanical and chemical resistance, low density, and low water adsorption [7]. Further, the R-value (measure of how well a two-dimensional barrier resists the conductive flow of heat) of rigid PU foams is among the highest of any insulating material, thus ensuring efficient heat retention and/or consistent temperature control of refrigerated environments [8].

In recent years, the development of PU-based composite foams with interdisciplinary functions has expanded considerably with the goal of increasing their mechanical performance, broadening their application, and preserving the environment by using lower amounts of PU [9,10]. Natural fibers have attracted much attention as potential reinforcements for composites due to their availability, biodegradability, and low cost [11]. These fibers have been involved in a growing type of polymer composites, the eco-composites, which describe combinations of materials with environmental and ecological potential and/or produced using materials from renewable resources [12–14]. So far, vegetable fibers such as flax, hemp, jute, and kenaf have been the most explored due to their low density, variable mechanical properties, and intrinsic biodegradability [15–17]. However, animal-based fibers are starting to demonstrate their potential as well. Feather keratin fibers have been shown to possess a hollow structure, filled with air, responsible for their low density and low dielectric constant, properties highly desirable in composites for electronic or automotive applications [18,19]. Silk fibers have been investigated to produce composites for tissue engineering due to their increased oxidation resistance and improved antibacterial and UV-light protection properties [20]. Animal-derived wastes, such as wool fibers, have also been successfully embedded in a polymeric film-forming matrix of cellulose acetate, with potential applications in the packaging and agricultural industries [21].

Even though this is an environmentally friendly solution to animal waste disposal, very few reports have been published on the subject.

In the present work, we explore the use of discarded dog wool fibers as a reinforcement agent in the production of PU-based eco-composites for thermal insulation. According to the Statistical Institute of Brazil (IBGE), there are in the country 52 million dog pets. The goal was to determine the efficiency of this mixture and the potentialities of animal wastes for industrial applications. To the authors' knowledge, this is the first report on the use of dog wool fibers as reinforcement in PU-based eco-composites. Various fiber percentages were combined with PU castor oil. The resulting eco-composite foams were characterized in terms of their physical, thermal, and mechanical properties in light of the desirable application.

2. Materials and Methods

2.1. Materials

Respan, a semi-flexible and biodegradable polyurethane from castor oil (PU) resin, acquired from Resichem Chemicals LTDA (São Paulo, Brazil), was used as matrix. Castor oil is a vegetable oil pressed from castor beans. Dog wool fibers were used as reinforcement and were collected in pet shops in the metropolitan area of the city of Natal (Natal, Brazil). PU was used as control. All remainder chemicals were acquired from VWR International and used without further purification.

2.2. Treatment of Dog Wool Fibers

Dog wool fibers were initially washed in 0.05 M sodium hydroxide (NaOH) solution, to remove impurities present along the surface, and dried at 50 °C for 24 h. After, they were ground in a micro-slicer (Urschel, Chesterton, IN, USA) to obtain microparticles of ≈30 mesh screen.

2.3. Preparation of Eco-Composites

Dog wool microparticles were thoroughly mixed with the semi-flexible PU resin using a commercial mixer, to guarantee the homogeneity of the composite structure. The mixture was then poured onto a steel mold, which was tightly closed, and submitted to a controlled expansion process to induce strong interactions between matrix and reinforcement (Figure 1). Eco-composite plates were produced with dimensions of 30 × 30 × 1 cm³ and different ratios of fiber in their composition (Table 1). Then, 100% PU plates were also produced and used as control.

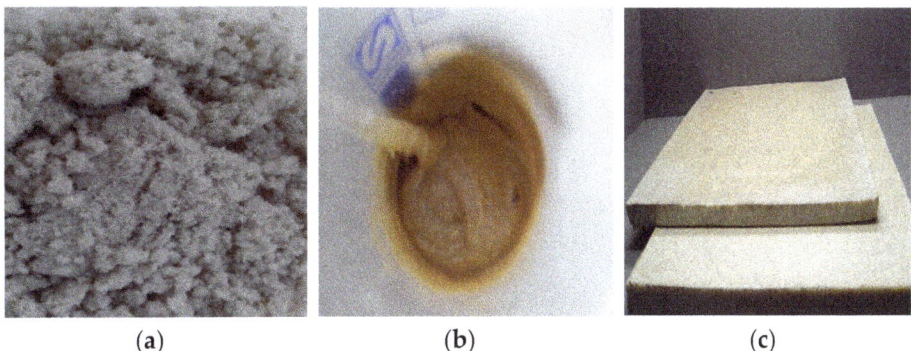

Figure 1. (a) Fiber microparticles, (b) mixture of PU resin and the fiber microparticles, and (c) the eco-composite.

Table 1. Eco-composites' composition.

Eco-Composite (%)	Dog Wool (g)	PU (g)
5	12.5	237.5
10	25.0	225.0
15	37.5	212.5
20	50.0	200.0

2.4. Scanning Electron Microscopy (SEM)

Morphological analyses of the fibers and the eco-composites were carried out using a SEM TM 3000 HITACHI (Hitachi, Chiyoda, Tokyo, Japan). Backscattering electron images were realized with an acceleration voltage of 15 kV that enabled the visualization of the distribution of the fiber reinforcement along the polymeric matrix.

2.5. Particle Size Distribution

The particle size distribution was performed in a laser diffraction particle size analyzer model CILAS 1180 (Cilas, Orléans, France) at the laser light wavelength of 635 nm. The equipment is able to measure particles ranging from 0.04 to 2500 µm. The size distributions of the samples were determined based on Fraunhofer diffraction theory and expressed as frequency (%) vs. particle diameter (µm). The measurement was carried out with samples of 0.2 g in accordance to the standard BS ISO 13320:2009.

2.6. Fourier-Transformed Infrared (FTIR)

FTIR spectra of the eco-composites with various reinforcement percentages were collected using a Shimadzu spectrometer, model FTIR-8400S, IRAffinity-1 (Shimadzu, Kyoto, Japan), coupled with an attenuated total reflectance (ATR) accessory, the PIKE MIRacle™ single reflection with a ZnSe crystal (PIKE Technologies, Madison, WI, USA). Spectra were obtained in the range of 4000–500 cm^{-1}, from 30 scans at a resolution of 4 cm^{-1}. All measurements were performed in triplicate.

2.7. Thermal Properties

The thermal properties in the castor polyurethane samples were determined using the KD2 Pro (Decagon Devices, Pullman, WA, USA) equipment coupled with a thermal sensor twin needle SH1, which uses the transient line heat source method to measure thermal diffusivity, specific heat (heat capacity), thermal conductivity, and thermal resistivity. All analyses were performed at room temperature following the standards ISO EN 31092-1994. An average of 10 readings was taken for each sample and the data were reported as mean ± standard deviation. TGA was performed on a DTG-60H model (Shimadzu, Kyoto, Japan) using a platinum pan. The TGA trace was obtained in the range of 30–300 °C under nitrogen atmosphere, flow rate of 50 mL/min, and temperature rise of 10 °C/min. Results were plotted as percentage of mass loss vs. temperature. DSC was carried on a Power Compensation Diamond DSC (Perkin Elmer, Waltham, MA, USA) with an Intracooler ILP, based on the standards ISO 11357-1:2016, ISO 11357-2:1999, and ISO 11357-3:1999. Samples were dried at 60 °C for 1 h and placed in an aluminum sample pan before testing. The analysis was carried out in nitrogen atmosphere with a flow rate of 50 mL/min. The DSC analysis was carried out at three stages: The first heating, cooling, and second heating, all at the heating rate of 10 °C/min, in order to eliminate the thermal history of the samples. The thermogram was obtained in the range of 20 to 500 °C.

2.8. Mechanical Properties

The eco-composites' tensile strength and compression capacities were examined using an X 300KN Universal Testing Machine (Shimadzu, Kyoto, Japan). The tensile strength of the eco-composites was determined following the ASTM D3039 with a specimen of 3 mm of thickness and 25 mm of width (75 mm^2 of cross-section) and the compression test according to NBR 8082. In the compression test, deformation was measured when the machine was activated to reduce the thickness of the specimen in 10% at speed of 0.25 cm/min. It was calculated by the formula $R_c = F/A$, where R_c is the compression strength at 10% deformation (Pa), F is the force (N), and A is the test area of the sample (m^2).

2.9. Hydration Capacity

The eco-composites' water absorption capacity was measured following the ASTM D2842. Three replicates were used of each eco-composite. Samples were initially dried at 50 °C for 24 h and then transferred to a desiccator and left for 15 min until they reached room temperature. Samples were weighed in their dry state (mdry). After, they were immersed in distilled water (dH$_2$O) and measured continuously (24 random intervals) until saturation was reached. The saturation point was determined when the sample weight reached a constant value (mwet). The samples' hydration capacity was determined using the Equation (1):

$$\% \text{ Water Absorbed} = \frac{m_{wet} - m_{dry}}{m_{dry}} \times 100 \tag{1}$$

2.10. Dilatometry

The thermal expansion coefficient of the samples was determined on the device NETZSCH model DIL 402 PC (Netzsch, Selb, Germany). The samples were made with dimensions of 25 mm in length and 8 mm in diameter. The tests were carried out under an argon gas flow of 5 mL/min at the heating gradient from room temperature to 170 °C. The heating rate was 5 °C/min.

3. Results and Discussion

3.1. Particle Size and Eco-Composites' Morphology

SEM micrographs of the dog wool fibers, in their natural state (untreated), treated with NaOH, and combined with the PU resin as reinforcement to form eco-composites were taken (Figure 2). As expected, there were substantial differences between the fibers before and after treatment with NaOH.

The impurities present along the fibers (Figure 2a) were eliminated after NaOH washing, revealing the efficiency of this alkali treatment and leaving the surface clean and unspoiled (no evidences of degradation, Figure 2b), with a desirable open structure capable of interacting with the polymeric matrix. The incorporation of the fibers within the PU matrix was evidenced in Figure 2d. As can be observed, the porosity and morphology of the composite up to 15% of fiber content (Figure 2e) was not significantly different from the pure PU resin (Figure 2c). A very porous structure is characteristic of the PU foam. The average pore size (mean of 50 measures) of the composite was 84 ± 50 µm and the pore size of the pure PU was 83 ± 40 µm. At the fiber content of 20%, the PU structure became instable with large (~0.5 mm) collapsed structures and holes of ~0.2 mm around the fibers. Even though fibers' distribution and orientation were random, they were preferentially detected in compact areas, both in the interior and the borders of the foam cells. Similar outcomes were obtained with composites of rigid PU foam reinforced with cellulose fiber residues [5].

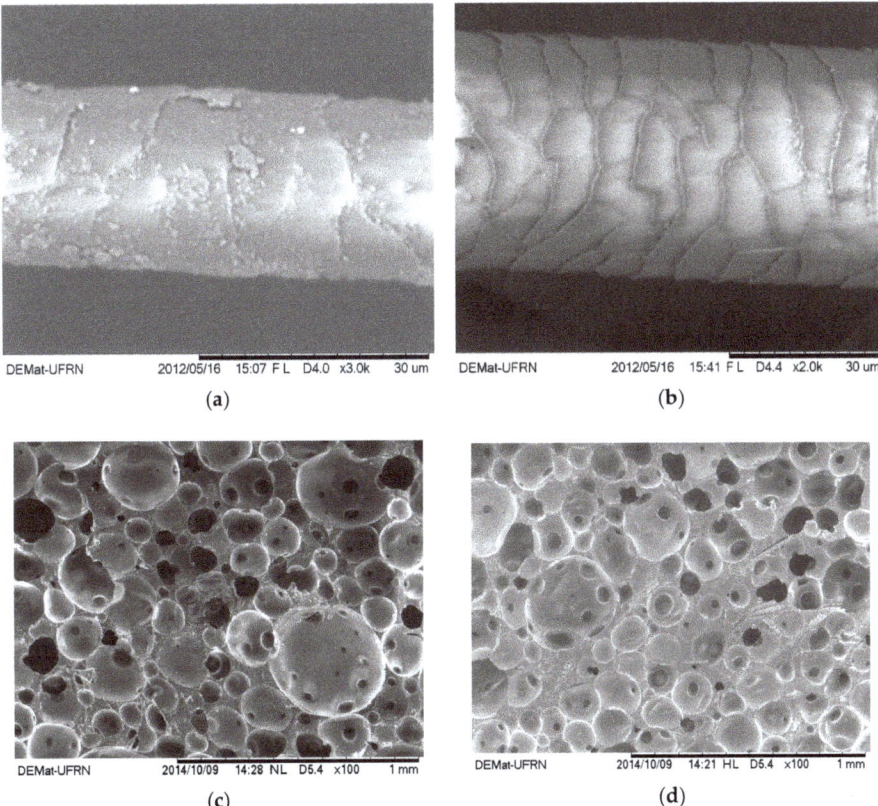

Figure 2. Cont.

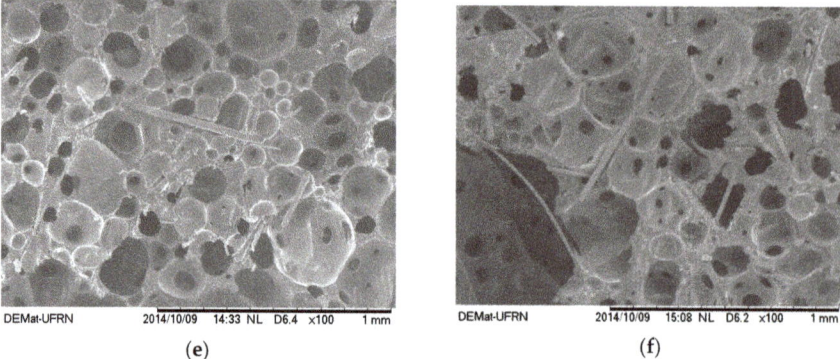

Figure 2. SEM micrographs of the dog wool fibers (**a**) in their natural state and (**b**) after treatment with 0.05 M of NaOH. Micrographs of the pure PU (**c**), PU + 10% of fibers (**d**), PU + 15% of fibers (**e**), and PU + 20% of fibers (**f**) at 100× magnification.

Regarding their size, the dog wool fibers were considered microparticles. According to data from Figure 3, they were very heterogeneous in size, varying from 1 to 700 µm, with the largest amount measuring between 30 and 40 µm. This heterogeneity is to be expected since the dog wool wastes were collected from various pet shops that use different wool treatments and cutting tools. These factors can then condition the micro-slicer precision and, consequently, the grounding process.

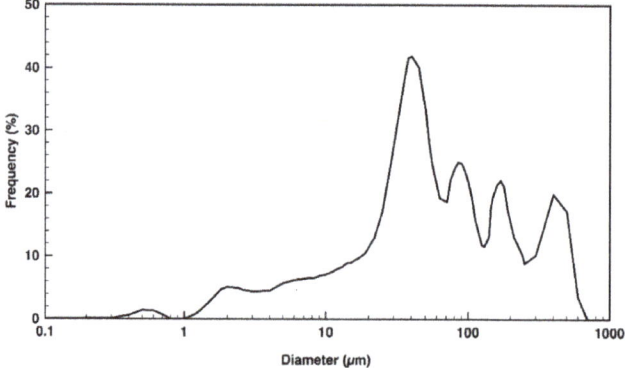

Figure 3. Dog wool microparticle size distribution.

3.2. ATR-FTIR Spectra

The spectra profiles of the eco-composites formed of PU and different amounts of dog wool microparticles are shown in Figure 4. Between 3200–3450 cm^{-1} was located one of the most important PU bands. This was attributed to the symmetric and asymmetric stretching vibrations of the N-H groups from the urethane and urea, which result from the reaction between water and isocyanate [22]. However, as observed by the spectrum of the dog wool fibers, a large peak at 3300 cm^{-1} is typical of the stretching vibrations of –NH groups in keratin [23]. As the amount of dog wool fibers increased in the composite, this region became broader, which indicates a larger number of intermolecular hydrogen being promoted by these microparticles [24]. A very small peak corresponding to C–H stretching of the aliphatic CH=CH was identified at 3008 cm^{-1}, while at 2950 and 2850 cm^{-1} the asymmetric and symmetric stretching vibrations of C–H were observed, respectively. A peak around 2270 cm^{-1}, associated with the stretching vibrations of the NCO group of the isocyanate, was detected in all formulations. However, it was more important in those eco-composites containing higher amounts of dog wool microparticles. This is indicative of the presence of unreacted isocyanate [25]. The peaks at 1710, 1240, and 1070 cm^{-1} relate to the stretching vibrations of the C=O and C–O of the ester groups, while the overlapping bands between 1540 and 1517 cm^{-1} can be attributed to the stretching and bending vibrations of the C–N and N–H of the urethane moieties, respectively. These two peaks could also be assigned to the C–N stretching and N–H bending vibrations of amide II in wool fibers. This explains the increasing definition and clarity of these two peaks as the percentage of dog wool fibers rose in the eco-composite [26]. A very small increase in the composite of the peak at 1650 cm^{-1} can be observed. This peak in the dog wool spectrum is attributed to the α-helix of the keratin structure [27]. This peak can be considered as a direct measure of the presence of the fiber in the composite.

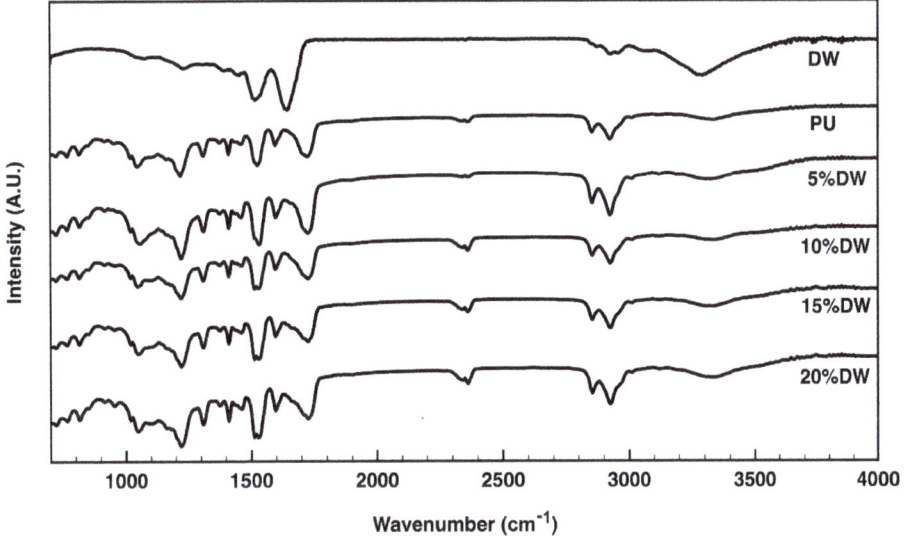

Figure 4. ATR-FTIR spectra of the pristine PU and Dog Wool (DW) and of the eco-composites containing 5% to 20% of dog wool microparticles.

3.3. Thermal Properties

Degradation steps associated with temperature rising were identified on PU and PU-based eco-composites via TGA (Figure 5). In the pure dog wool, a weight loss between 25 and 100 °C was observed due to the evaporation of the incorporated water. The second decomposition starting at around 200 °C could be attributed to the denaturation and degradation of the keratin molecules. According to literature, the disulfide bonds are cleaved between 230 and 250 °C [28]. In the composite, but not in the pure PU, a first very small step of degradation was detected between 25 and 100 °C (Figure 5 inset) and refers to the initial volatilization of moisture from the foams due to the evaporation or dehydration of hydrated cations [29,30]. This step was more important on the fiber-reinforced composites because of the wool fibers' affinity towards water molecules, which tends to increase moisture retention [24,31]. The first degradation step for the pristine PU was detected at ≈260 °C and was attributed to the cleavage of the PU polymeric backbone, initiating with the polyol component degradation (urethane chains) and, then, progressing to the isocyanate component degradation (ester bonds) [32]. At 300 °C, 12% of the original mass was already lost with the remaining 88% being further decomposed into amines, small transition components, and CO_2 [33]. Because of the wool fibers' incorporation, the eco-composites were more quickly prone to degradation. This occurred because keratin wool fibers, such as dog hair fibers, start decomposing at temperatures superior to 200 °C. In fact, from this point, denaturation of the helix structure and the destruction of chain linkages, peptide bridges, and the skeletal degradation occurs. At temperatures closer to 300 °C, several chemical reactions take place with the fibers being decomposed into lighter products and volatile compounds such as CO_2, H_2S, H_2O, and HCN [34]. From all formulations, the eco-composites containing 5% of dog wool microparticles were capable of retaining more of their original mass, ≈ 91%, at 300 °C.

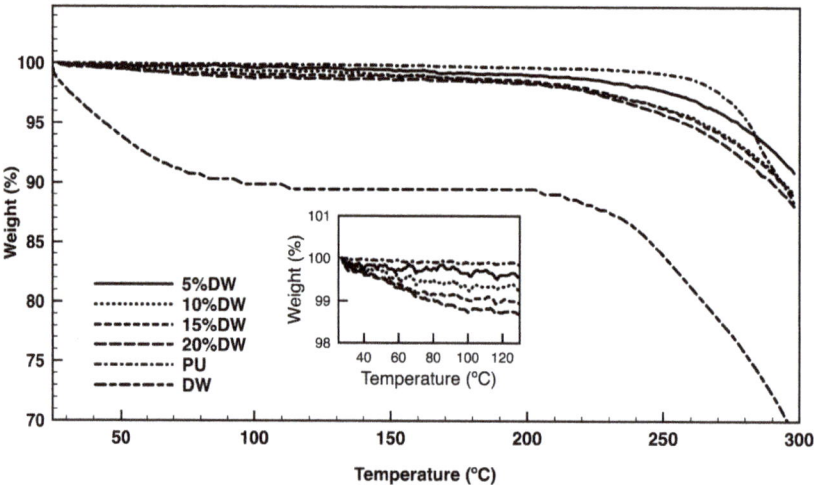

Figure 5. TGA of pristine PU and dog wool (DW) and the eco-composites containing 5, 10, 15, and 20% of dog wool microparticles measured between 25 and 300 °C, performed at a heating rate of 10 °C/min in a nitrogen atmosphere. The inset represents the initial part of the PU and PU composites between 25 and 130 °C.

DSC thermograms of the PU and PU-based composites prepared with different percentages of dog wool microparticles were acquired between 20 and 500 °C (Figure 6). The first heating cycle between 20 and 120 °C (Figure 6a) and the cooling cycle between 120 and 20 °C (Figure 6b) did not shown any significant event. In the second heating cycle (Figure 6c), the first endothermic peak for PU was detected at ≈300 °C, which, as seen earlier, is associated with the cleavage of the PU polymeric backbone, initiating with urethane chains and continuing to the ester bonds. In Figure 6d, it can be observed a detail of the second heat cycle between 100 and 180 °C. In this region a T_g is observed in all the thermograms relative to the hard urethane segments [35]. However, the T_g of the composites starting from 10% of dog wool content were lower (~150 °C) than corresponding PU control (~160 °C) and the 5% composite. It seems that the presence of the fibers affected the state of crystallinity in the PU matrix by reducing the T_g towards lower temperatures. These T_g are very small since polyurethane is mostly amorphous and suggest that the fibers improve the mobility of soft segment in PU, reducing the hydrogen bonding interactions [36]. For the eco-composite foams, the first endothermic peak occurred earlier, at ≈220 °C, with the initial denaturation of the wool fiber helix structure and the destruction of chain linkages. At temperatures ranging from 300 to 340 °C, the main polymeric chains in the eco-composite started degrading together with the remaining components of the wool fibers. These data are consistent with the TGA observations. The last endotherm peak registered for all foams was detected around 460 °C and can be attributed to the final degradation of the remaining residual polymeric chains and dog wool fibers into carbon char, small transition components, and volatile species [33,34].

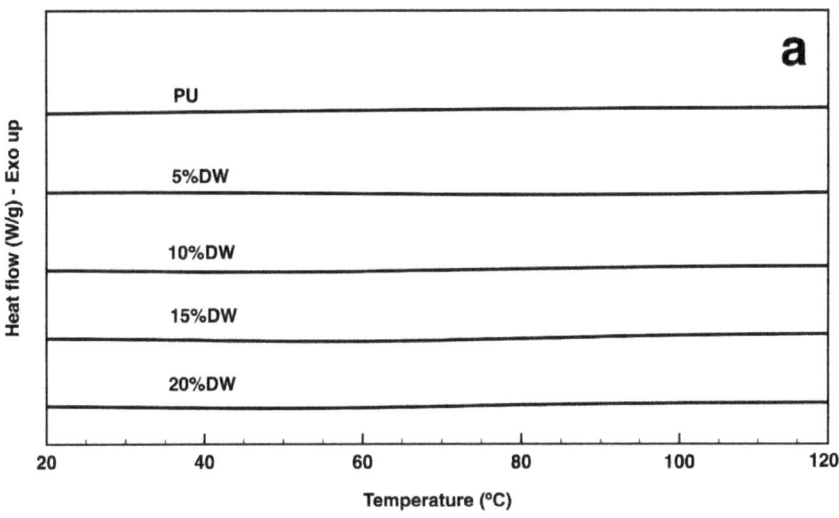

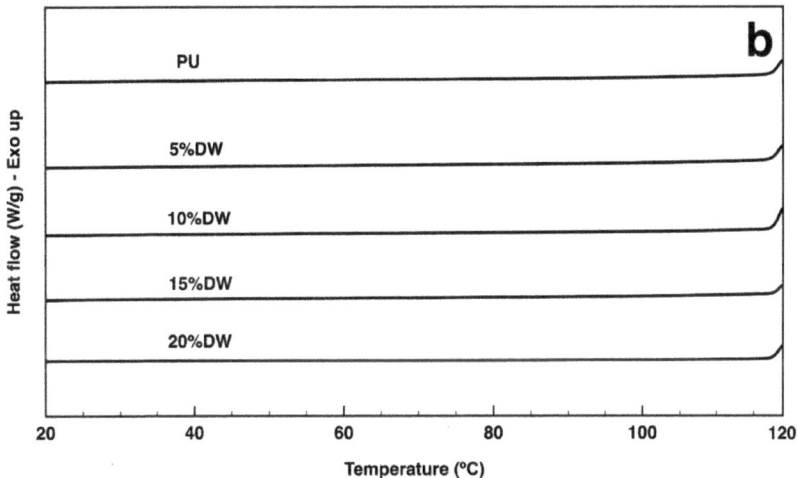

Figure 6. Cont.

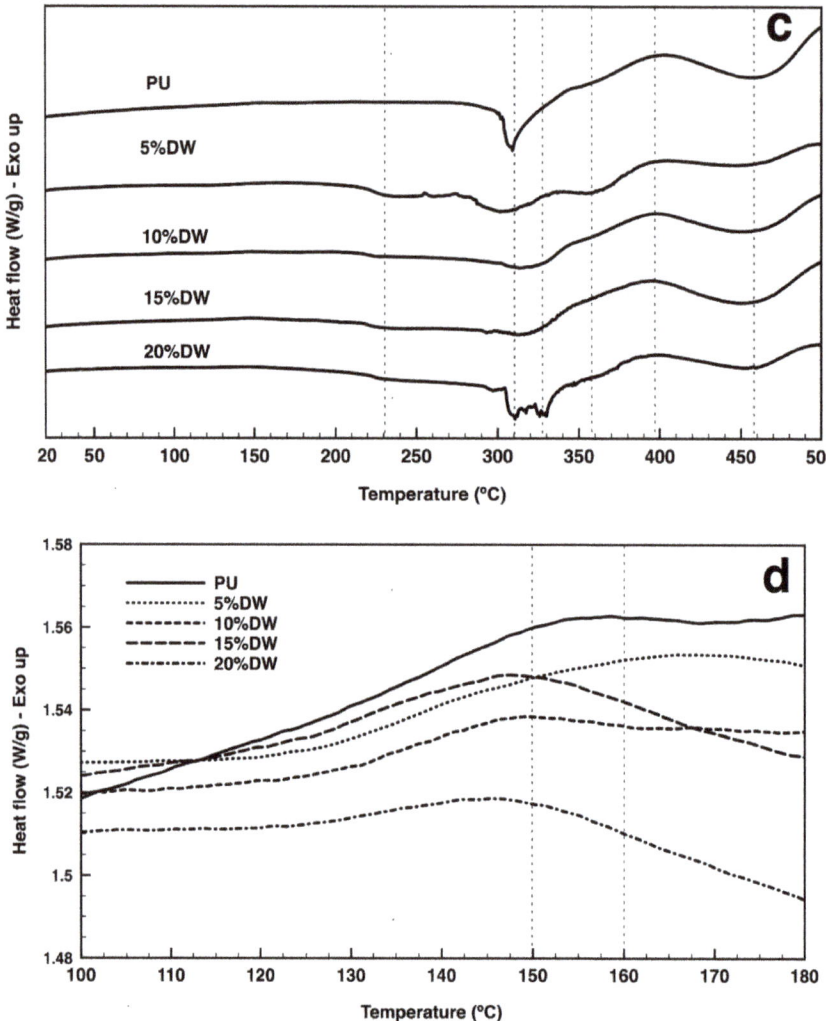

Figure 6. DSC thermogram of the pristine PU and the eco-composites containing 5, 10, 15, and 20% of dog wool microparticles collected at a heating rate of 10 °C/min in a nitrogen atmosphere. (**a**) The first heating cycle between 20 and 120 °C, (**b**) the first cooling cycle between 120 and 20 °C, (**c**) the second heating cycle between 20 and 500 °C. (**d**) Detail of the second heating cycle between 100 and 180 °C showing the T_g.

In foamed systems, the dominant heat transfer modes are thermal radiation and gas-gas and solid-solid conduction. In PU foams, the total conductivity ranges about two-thirds of the conductivity of stagnant air because there is low conductivity gas, or foaming agent, inside the foam [33]. Here, the addition of the wool fibers to the eco-composites had little influence on the foams' thermal conductivity (Table 2), maintaining the values within the expected ranges, desirable for thermal insulation, and approximated to those of polystyrene (one of the most common materials applied in thermal insulation) [37–39]. The thermal or heat capacity measures the amount of energy required to raise the temperature of a material one degree. Data from Table 2 demonstrates, again, that the PU and the eco-composites presented very similar values. However, the addition of 20% dog wool microparticles increased the composite thermal capacity above the pristine PU. Hence, this formulation requires more heat for temperature variations to occur, thus maintaining insulation more effectively. Thermal diffusivity describes the rate of temperature spread through a material and is a function of the thermal conductivity and the heat thermal capacity. As such, since thermal conductivity was the lowest in the eco-composites containing 20% of dog wool fibers, the same happened with the thermal diffusivity. It has been shown that thermal diffusivity is dependent on the organization of the foaming cells, their dimension, and the type of blowing agent applied [40]. Here, it is likely that the random disposition of the microparticles along the polymeric matrix may have compromised these specific thermal properties. Finally, in order to be classified as an insulating material, the foam must be endowed with a high thermal resistance. Data shows that thermal resistance decreased slightly with the addition of dog wool fibers. Even though these values are acceptable for thermal insulation, it seems that by increasing the percentage of fibers within the eco-composite this property is also enhanced. Thus, future studies will be conducted to confirm this premise.

Table 2. Main thermal properties of pristine PU and the eco-composites ($n = 3$, S.D. ± 3).

Samples	Thermal Conductivity (W/mk)	Thermal Capacity (MJ/m^{3}k)	Thermal Diffusivity (m^2/s)	Thermal Resistance (°C cm/W)
PU	0.053 ± 0.004	0.561 ± 0.045	0.091 ± 0.003	1878.5 ± 153.3
5% DW	0.064 ± 0.006	0.454 ± 0.015	0.141 ± 0.016	1576.0 ± 153.3
10% DW	0.070 ± 0.002	0.603 ± 0.048	0.122 ± 0.006	1411.5 ± 61.7
15% DW	0.063 ± 0.002	0.530 ± 0.046	0.120 ± 0.009	1590.0 ± 38.2
20% DW	0.061 ± 0.002	0.615 ± 0.053	0.098 ± 0.012	1647.5 ± 45.4

3.4. Mechanical Properties

The foams' tensile stress and compression performance were measured with and without the addition of the dog wool microparticles (Figures 7 and 8, respectively). PU achieved the highest percentage of elongation from the tested formulations ($\approx 50\%$), although requiring less stress (≈ 1.25 MPa) to break than the reminding eco-composites. In fact, with the addition of only 5% dog wool microparticles, the stress necessary to reach a similar elongation state ($\approx 46\%$) was almost double, ≈ 2 MPa (Figure 7). This behavior is explained by the interactions established between the polymer matrix and the fiber arrays, which led to the disorganization of the PU original structure. At this percentage, small changes were induced in the foam's morphology. It is possible the microparticles migrated and filled existing defects, thus increasing the force necessary to break the material. A stiffness enhancement was registered with superior percentages of dog wool fibers. This is related to the higher rigidity of the foam solid phase in consequence of the fiber's contribution [5]. Because of the heterogeneous and disorganized orientation and distribution of the fibers along the composite, there was no proportion between the force applied-elongation capacity and the percentage of fiber reinforcement.

Even though the 5% eco-composites registered the most balanced performance between stress applied and elongation capacity (Figure 7), their resistance to compression was the lowest from the group (Figure 8). It is likely the rearrangements the polymeric foam underwent, to accommodate the fibers, promoted the development of an anisotropic-like material in which the mechanical resistance was more important in one direction than in the other [41]. Also, the presence of gaps along the foam in response to the addition of the microparticles and the alterations in the PU original structure may have contributed to this phenomenon. Irregularities in the foams' organization are more likely to occur in composites containing smaller amounts of reinforcement fibers than higher [5]. In turn, the increased percentage of filler can induce a decrease in the reactivity of the components in the system, affecting the foam expansion and increasing its density and rigidity, consequently, improving the compression strength [12]. As such, it was expected the maximum compressive stress to strain to be endured by the composite with the largest percentage of filler (20%).

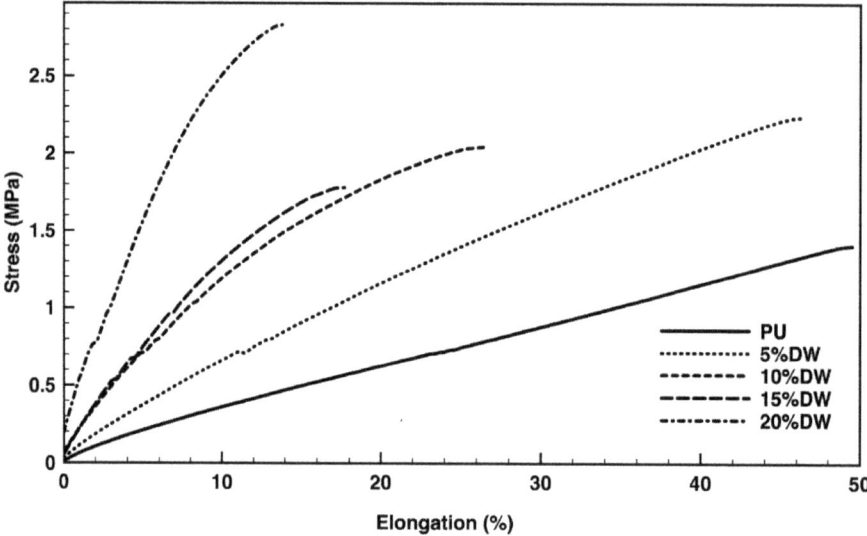

Figure 7. Stress (MPa) versus elongation at break (%) of the pristine PU and the dog wool-reinforced eco-composites.

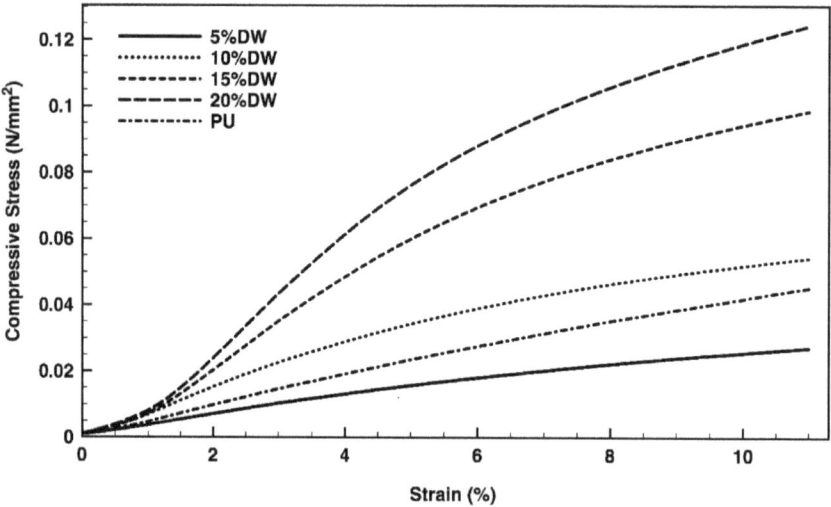

Figure 8. Compressive stress versus strain of the pristine PU and the dog wool-reinforced eco-composites.

3.5. Hydration Capacity

The water adsorption capacity of the pristine PU and the dog wool-reinforced eco-composites was followed up to six days in dH_2O, with samples being weighed every 24 h until water saturation was reached. Data from Figure 9 revealed pristine PU as the foaming material with the least hydration capacity, reaching a saturation state with only 4% of water in its composition. The eco-composites were found more attractive to water molecules with their hygroscopic capacity augmenting as the percentage of microparticles increased, that is, from 5% water content in the 5% dog wool eco-composites to 11%

water content registered for the 20% dog wool-reinforced eco-composite. These results are explained by the ability of wool fibers to bind and absorb large amounts of water [31]. Water permeability in wool fibers is dictated mainly by cell membrane lipids. However, much remains to be understood on this front. The interaction between fibers and water is quite complicated; at low relative humidity, a water molecule monolayer can be formed by the interaction with specific fiber polar side chains, while at high relative humidity water associates with the peptide backbone of the fiber, generating a multilayer absorption. Fiber swelling also occurs, as a result of the breaking of hydrogen bonds between and within protein chains, due to water molecules rising over the surface and within the intercellular spaces; thus, generating even more interaction sites for water molecules [43,44].

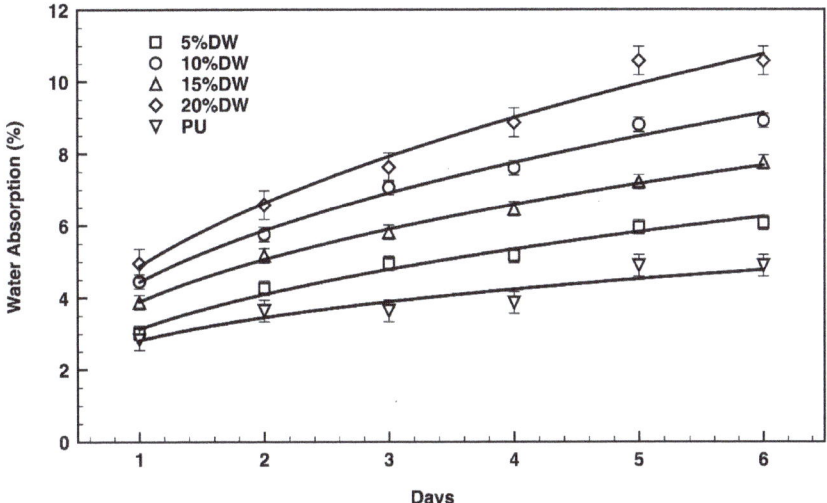

Figure 9. Water adsorption capacity of the pristine PU and the dog wool-reinforced eco-composites over time.

3.6. Dilatometry

Thermal expansion is defined as the increase in a material's volume in response to temperature rising. As the temperature rises, molecular agitation increases thereby growing the distance between the molecules. Figure 10 shows evidences of PU dilation with the increase in temperature from 50 to 160 °C. The addition of the dog wool fibers reduced significantly the polymer expansion with temperature variations. In fact, with a 5% addition of microparticles, the foam's volume did not even alter. By incorporating a higher content of wool fibers, the foam's dilatation reached negative values, indicative of the loss in material stability and capacity to maintain its structural integrity with heating. The best results showing structural stability up to 120 °C were obtained using a maximum of 15% of dog wool fibers.

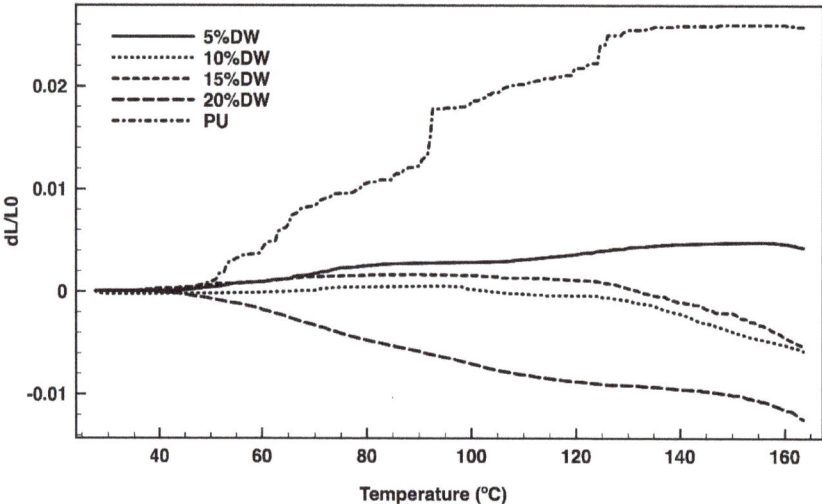

Figure 10. PU and eco-composites' dilatation with increasing temperature, from 30 to 160 °C.

4. Conclusions

PU eco-composites reinforced with dog wool fibers were successfully produced at different percentages. Alkaline treatment was effective in removing impurities from the fibers without compromising their integrity. Fibers were incorporated along the polymeric matrix in a random and disorganized manner. Still, they were effective in increasing the foams' mechanical resistance, namely tensile and compression strengths. The thermodynamic behavior suffered little changes with the incorporation of the dog wool fillers, being the most important the improvement in thermal capacity. Additionally, the hydration capacity was significantly improved in response to the wool fibers' water permeability and increased capacity to bind with water molecules. Dilatometry studies revealed the capacity of the fillers to increase the thermal stability of the foam, reducing their expansion with heating. Data demonstrated the potential of this combination to produce new alternative solutions for insulation using low-cost, sustainable resources and with minimal environmental impact.

Author Contributions: F.C.d.S., K.K.d.O.S.S., and J.U.L.M. performed the main analysis and data collection, finalized the paper, and performed data interpretation. R.L. supervised and performed data interpretation. A.Z. and H.P.F. performed data analysis and finalized the manuscript. All authors have read and agreed to the published version of the manuscript.

Funding: Authors acknowledge the *Fundação de Apoio a Pesquisa do Estado do Rio Grande do Norte* (FAPERN) for financing this work. They thank the pet shops from Natal city for donating the dog wool fibers used in the experiments. H.P. Felgueiras and A. Zille also acknowledge project UID/CTM/00264/2019 of Centre for Textile Science and Technology (2C2T), funded by national funds through FCT/MCTES.

Conflicts of Interest: The authors declare no conflict of interest.

References

1. Aditya, L.; Mahlia, T.; Rismanchi, B.; Ng, H.; Hasan, M.; Metselaar, H.; Muraza, O.; Aditiya, H. A review on insulation materials for energy conservation in buildings. *Renew. Sust. Energy. Rev.* **2017**, *73*, 1352–1365. [CrossRef]
2. Bozsaky, D. The historical development of thermal insulation materials. *Period. Polytech. Archit.* **2010**, *41*, 49–56. [CrossRef]
3. Willoughby, J. Insulation. In *Plant Engineer's Reference Book (Second Edition)*; Snow, D.A., Ed.; Butterworth-Heinemann: Oxford, UK, 2002; pp. 301–3018.

4. Xue, B.-L.; Wen, J.-L.; Sun, R.-C. Lignin-based rigid polyurethane foam reinforced with pulp fiber: Synthesis and characterization. *ACS Sustain. Chem. Eng.* **2014**, *2*, 1474–1480. [CrossRef]
5. Silva, M.; Takahashi, J.; Chaussy, D.; Belgacem, M.; Silva, G. Composites of rigid polyurethane foam and cellulose fiber residue. *J. Appl. Polym. Sci.* **2010**, *117*, 3665–3672. [CrossRef]
6. Ciecierska, E.; Jurczyk-Kowalska, M.; Bazarnik, P.; Kowalski, M.; Krauze, S.; Lewandowska, M. The influence of carbon fillers on the thermal properties of polyurethane foam. *J. Therm. Anal. Calorim.* **2016**, *123*, 283–291. [CrossRef]
7. Rasheed, A.K.; Marhoon, I.I. Mechanical and physical properties of glass wool-rigid polyurethane foam composites. *Al-Nahrain J. Eng. Sci.* **2015**, *18*, 41–49.
8. Schuetz, M.; Glicksman, L.R. A basic study of heat transfer through foam insulation. *J. Cell. Plast.* **1984**, *20*, 114–121. [CrossRef]
9. Ashida, K. *Polyurethane and Related Foams: Chemistry and Technology*; CRC Press: Boca Raton, FL, USA, 2006.
10. Członka, S.; Strąkowska, A.; Strzelec, K.; Adamus-Włodarczyk, A.; Kairytė, A.; Vaitkus, S. Composites of Rigid Polyurethane Foams Reinforced with POSS. *Polymers* **2019**, *11*, 336. [CrossRef]
11. Rajkumar, G.; Srinivasan, J.; Suvitha, L. Natural protein fiber hybrid composites: Effects of fiber content and fiber orientation on mechanical, thermal conductivity and water absorption properties. *J. Ind. Text.* **2015**, *44*, 709–724. [CrossRef]
12. Gutiérrez, M.C.; De Paoli, M.-A.; Felisberti, M.I. Cellulose acetate and short curauá fibers biocomposites prepared by large scale processing: Reinforcing and thermal insulating properties. *Ind. Crop. Prod.* **2014**, *52*, 363–372. [CrossRef]
13. Zhao, Q.; Yam, R.C.; Zhang, B.; Yang, Y.; Cheng, X.; Li, R.K. Novel all-cellulose ecocomposites prepared in ionic liquids. *Cellulose* **2009**, *16*, 217–226. [CrossRef]
14. Dicker, M.P.; Duckworth, P.F.; Baker, A.B.; Francois, G.; Hazzard, M.K.; Weaver, P.M. Green composites: A review of material attributes and complementary applications. *Compos. Part A Appl. Sci. Manuf.* **2014**, *56*, 280–289. [CrossRef]
15. Zini, E.; Scandola, M. Green composites: An overview. *Polym. Compos* **2011**, *32*, 1905–1915. [CrossRef]
16. John, M.J.; Thomas, S. Biofibres and biocomposites. *Carbohydr. Polym.* **2008**, *71*, 343–364. [CrossRef]
17. Briga-Sa, A.; Nascimento, D.; Teixeira, N.; Pinto, J.; Caldeira, F.; Varum, H.; Paiva, A. Textile waste as an alternative thermal insulation building material solution. *Constr. Build. Mater.* **2013**, *38*, 155–160. [CrossRef]
18. Hong, C.K.; Wool, R.P. Development of a bio-based composite material from soybean oil and keratin fibers. *J. Appl. Polym. Sci.* **2005**, *95*, 1524–1538. [CrossRef]
19. Cheng, S.; Lau, K.-T.; Liu, T.; Zhao, Y.; Lam, P.-M.; Yin, Y. Mechanical and thermal properties of chicken feather fiber/PLA green composites. *Compos. Part B-Eng.* **2009**, *40*, 650–654. [CrossRef]
20. Cheung, H.-Y.; Ho, M.-P.; Lau, K.-T.; Cardona, F.; Hui, D. Natural fibre-reinforced composites for bioengineering and environmental engineering applications. *Compos. Part B Eng.* **2009**, *40*, 655–663. [CrossRef]
21. Aluigi, A.; Vineis, C.; Ceria, A.; Tonin, C. Composite biomaterials from fibre wastes: Characterization of wool–cellulose acetate blends. *Compos. Part A Appl. Sci. Manuf.* **2008**, *39*, 126–132. [CrossRef]
22. Gama, N.V.; Soares, B.; Freire, C.S.R.; Silva, R.; Neto, C.P.; Barros-Timmons, A.; Ferreira, A. Bio-based polyurethane foams toward applications beyond thermal insulation. *Mater. Des.* **2015**, *76*, 77–85. [CrossRef]
23. Pielesz, A.; Freeman, H.S.; Wesełucha-Birczyńska, A.; Wysocki, M.; Włochowicz, A. Assessing secondary structure of a dyed wool fibre by means of FTIR and FTR spectroscopies. *J. Mol. Struct.* **2003**, *651–653*, 405–418. [CrossRef]
24. Moody, V.; Needles, H.L. Major Fibers and Their Properties. In *Tufted Carpet*; William Andrew Publishing: Norwich, NY, USA, 2004; pp. 35–59.
25. Cruz-Lopes, L.P.; Rodrigues, L.; Domingos, I.; Ferreira, J.; Lemos, L.T.d.; Esteves, B. Production of Polyurethane Foams from Bark Wastes. *Int. J. Chem. Mol. Nucl. Mater. Metall. Eng.* **2016**, *10*, 1056–1059.
26. Idris, A.; Vijayaraghavan, R.; Rana, U.A.; Patti, A.F.; MacFarlane, D.R. Dissolution and regeneration of wool keratin in ionic liquids. *Green. Chem.* **2014**, *16*, 2857–2864. [CrossRef]
27. Aluigi, A.; Zoccola, M.; Vineis, C.; Tonin, C.; Ferrero, F.; Canetti, M. Study on the structure and properties of wool keratin regenerated from formic acid. *Int. J. Biol. Macromol.* **2007**, *41*, 266–273. [CrossRef] [PubMed]
28. Liu, X.; Nie, Y.; Meng, X.; Zhang, Z.; Zhang, X.; Zhang, S. DBN-based ionic liquids with high capability for the dissolution of wool keratin. *RSC. Adv.* **2017**, *7*, 1981–1988. [CrossRef]

29. Yavuz, G.; Felgueiras, H.P.; Ribeiro, A.I.; Seventekin, N.; Zille, A.; Souto, A.P. Dyed poly (styrene-methyl methacrylate-acrylic acid) photonic nanocrystals for enhanced structural color. *ACS Appl. Mater. Interfaces* **2018**, *10*, 23285–23294. [CrossRef]
30. Felgueiras, H.P.; Teixeira, M.A.; Tavares, T.D.; Homem, N.C.; Zille, A.; Amorim, M.T.P. Antimicrobial action and clotting time of thin, hydrated poly (vinyl alcohol)/cellulose acetate films functionalized with LL37 for prospective wound-healing applications. *J. Appl. Polym. Sci.* **2019**. [CrossRef]
31. Sakabe, H.; Ito, H.; Miyamoto, T.; Inagaki, H. States of water sorbed on wool as studied by differential scanning calorimetry. *Text. Res. J.* **1987**, *57*, 66–72. [CrossRef]
32. Zhang, Y.; Shang, S.; Zhang, X.; Wang, D.; Hourston, D. Influence of structure of hydroxyl-terminated maleopimaric acid ester on thermal stability of rigid polyurethane foams. *J. Appl. Polym. Sci.* **1995**, *58*, 1803–1809. [CrossRef]
33. Mosiewicki, M.A.; Dell'Arciprete, G.; Aranguren, M.I.; Marcovich, N.E. Polyurethane foams obtained from castor oil-based polyol and filled with wood flour. *J. Compos. Mater.* **2009**, *43*, 3057–3072. [CrossRef]
34. Ma, B.; Qiao, X.; Hou, X.; Yang, Y. Pure keratin membrane and fibers from chicken feather. *Int. J. Biol. Macromol.* **2016**, *89*, 614–621. [CrossRef] [PubMed]
35. Cuvé, L.; Pascault, J.P.; Boiteux, G.; Seytre, G. Synthesis and properties of polyurethanes based on polyolefine: 1. Rigid polyurethanes and amorphous segmented polyurethanes prepared in polar solvents under homogeneous conditions. *Polymer* **1991**, *32*, 343–352. [CrossRef]
36. Zia, F.; Zia, K.M.; Zuber, M.; Kamal, S.; Aslam, N. Starch based polyurethanes: A critical review updating recent literature. *Carbohydr. Polym.* **2015**, *134*, 784–798. [CrossRef] [PubMed]
37. Zhang, H.; Fang, W.-Z.; Li, Y.-M.; Tao, W.-Q. Experimental study of the thermal conductivity of polyurethane foams. *Appl. Therm. Eng.* **2017**, *115*, 528–538. [CrossRef]
38. Wu, G.; Wang, Y.; Wang, K.; Feng, A. The effect of modified AlN on the thermal conductivity, mechanical and thermal properties of AlN/polystyrene composites. *RSC Advanc.* **2016**, *6*, 102542–102548. [CrossRef]
39. Pau, D.S.W.; Fleischmann, C.M.; Spearpoint, M.J.; Li, K.Y. Thermophysical properties of polyurethane foams and their melts. *Fire. Mater.* **2014**, *38*, 433–450. [CrossRef]
40. Prociak, A.; Pielichowski, J.; Sterzynski, T. Thermal diffusivity of rigid polyurethane foams blown with different hydrocarbons. *Polym. Test.* **2000**, *19*, 705–712. [CrossRef]
41. Hawkins, M.C.; O'Toole, B.; Jackovich, D. Cell morphology and mechanical properties of rigid polyurethane foam. *J. Cell. Plast.* **2005**, *41*, 267–285. [CrossRef]
42. Banik, I.; Sain, M.M. Role of refined paper fiber on structure of water blown soy polyol based polyurethane foams. *J. Reinf. Plast. Comp.* **2008**, *27*, 1515–1524. [CrossRef]
43. Lo Nostro, P.; Fratoni, L.; Ninham, B.W.; Baglioni, P. Water absorbency by wool fibers: Hofmeister effect. *Biomacromolecules* **2002**, *3*, 1217–1224. [CrossRef]
44. Ormondroyd, G.A.; Curling, S.F.; Mansour, E.; Hill, C.A. The water vapour sorption characteristics and kinetics of different wool types. *J. Text. Inst.* **2017**, *108*, 1198–1210. [CrossRef]

© 2020 by the authors. Licensee MDPI, Basel, Switzerland. This article is an open access article distributed under the terms and conditions of the Creative Commons Attribution (CC BY) license (http://creativecommons.org/licenses/by/4.0/).

Article

Polypropylene/Basalt Fabric Laminates: Flexural Properties and Impact Damage Behavior

Pietro Russo [1], Ilaria Papa [2], Vito Pagliarulo [3] and Valentina Lopresto [2,*]

[1] Institute for Polymers, Composites and Biomaterials, National Research Council, Via Campi Flegrei 34, 80078 Pozzuoli-Naples, Italy; pietro.russo@unina.it
[2] Department of Chemical, Materials and Production Engineering, University of Naples Federico II, P.le Vincenzo Tecchio 80, 80125 Naples, Italy; ilaria.papa@unina.it
[3] Institute of Applied Sciences & Intelligent Systems "E. Caianiello", Via Campi Flegrei 34, 80078 Pozzuoli-Naples, Italy; v.pagliarulo@isasi.cnr.it
* Correspondence: lopresto@unina.it

Received: 27 April 2020; Accepted: 7 May 2020; Published: 8 May 2020

Abstract: Recently, the growing interests into the environmental matter are driving the research interest to the development of new eco-sustainable composite materials toward the replacement of synthetic reinforcing fibers with natural ones and exploiting the intrinsic recyclability of thermoplastic resins even for uses in which thermosetting matrices are well consolidated (e.g., naval and aeronautical fields). In this work, polypropylene/basalt fabric composite samples were prepared by film stacking and compression molding procedures. They have been studied in terms of flexural and low-velocity impact behavior. The influence related to the matrix modification with a pre-optimized amount of maleic anhydride grafted PP as coupling agent was studied. The mechanical performances of the composite systems were compared with those of laminates consisting of the pure matrix and obtained by hot-pressing of PP pellets and PP films used in the stacking procedure. Results, on one side, demonstrated a slight reduction of both static and dynamic parameters at the break for specimens from superimposed films to ones prepared from PP pellets. Moreover, an outstanding improvement of mechanical performances was shown in the presence of basalt layers, especially for compatibilized samples.

Keywords: polypropylene; basalt fibers; composite laminate; flexural; impact damage

---

1. Introduction

The growing diffusion of plastics in all industrial fields and concerns related to their environmental impact and secure disposal at the end of the useful life has been the main reason motivating academic and industrial research towards the study and development of new eco-sustainable materials. So, an increasing interest in natural fibers as polymers reinforcement (conventional and bio-sources derived) has been detected. In this context, basalt fibers have gained outstanding attention as reinforcing fibers concerning traditional glass and carbon-based ones [1,2]. Specifically, the interest of the research toward the use of basalt fibers was driven by their remarkable properties such as relatively low cost, sustainability, enhanced mechanical properties, non-flammability, high chemical stability, excellent sound and thermal insulation [3]. Basalt fibers, derived from natural rocks in volcanic regions with technologies similar to that of glass fibers' forming, are mainly composed of silica (SiO_2) and alumina (Al_2O_3) [4] and have a melting temperature ranging from 1500 °C and 1700 °C with an average diameter between 9 and 13 micrometers. In the last decades, basalt fibers, included in thermosetting and thermoplastic resins for uses in several fields as transportation, defense and constructions, have largely demonstrated their potentiality as eco-sustainable substitutes of glass fibers, especially if an adequate adhesion at the interface with the surrounding polymer phase is ensured.

Concerning thermosetting matrices, it is possible to obtain great benefits, primarily in terms of mechanical and thermal properties, including basalt fibers in epoxy, polyester and vinyl ester resins [5–9]. However—with the awareness that the sustainability of new products can be encouraged by the recyclability of the involved matrices—recently, the focus of the research has been increasingly devoted to thermoplastic systems mainly based on polypropylene [10–12], polyethene [13,14], polyamides [15–18], polyesters [19,20] as well as polymer blends [21] and hybrids [22,23].

In particular, polypropylene resins, primarily used for mass productions due to their low cost and full versatility in terms of processability and properties, still receive an extraordinary interest with research efforts mainly aimed to improve their performances further and extend their field of application more and more.

In this regard, Szabo et al. [11] demonstrated that the mechanical properties of polypropylene composites packed with different short fibers are strongly affected by the content and direction of the same. As witnessed by scanning electron micrographs, samples showed damage mechanisms as fiber pull out in the perpendicular direction and debonding along the longitudinal direction depending on the manufacturing process used.

Guo et al. [12] considered basalt fabric modified with the aid of a silane coupling agent KH550 to reinforce a commercial polypropylene resin. Hydrophilicity and lipophilicity tests of modified basalt fibers allowed a preliminary optimization of the pre-treatment parameters. Time and coupling agent content at values of 1.5 h and 6% by weight have been varied, respectively. The surface modification of basalt fibers significantly improved tensile, impact resistance, bending properties and thermal stability of the PP matrix giving rise to composites with better dynamic viscoelasticity compared to ones containing the same content of unmodified basalt fibers.

Greco et al. [24] highlighted the crucial role of structural features as the microcrystal size and the amorphous content of basalt fibers in determining their mechanical properties as well as the expected enhancement of mechanical performances of polypropylene composites by increasing the fiber-matrix interface goodness.

In the present paper, polypropylene/basalt fabric composite laminates, prepared by film stacking and hot-pressing procedures, were analyzed. Flexural and low-velocity impact properties were studied, taking pure PP specimens as reference materials and considering influences related to the matrix modification with a pre-optimized amount of a maleic anhydride grafted PP as a coupling agent. Mechanical results, also interpreted in light of indentation measurements, demonstrated that, although filming of the matrix does not significantly compromise mechanical performances of the PP matrix, the presence of basalt layers induces a marked improvement of composite performances. Moreover, the preliminary modification of the matrix to enhance the interfacial adhesion further improves the flexural performances, especially in terms of strength allowing to withstand higher impact loading concerning neat PP based composite laminates. Premised that for both investigated composites, no penetration seems to occur and the more significant matrix-reinforcement interaction for compatibilized composite laminamakesmake them less prone to plastic deformations as also evidenced by the indentation measurements.

2. Experimental

2.1. Materials

In this paper a polypropylene matrix (Hyosung Topilene PP J640, MFI@230 °C, 2.16 kg: 10 g/10 min) provided by Songhan Plastic Technology Co. Ltd. (Shangai, China) and a plain wave basalt fabric (areal weight: 210 g/m^2) from Incotelogy GmbH (Pulheim, Germany). The resin was used as received (PP) or pre-modified (PPC) by the inclusion of 2 wt.% of a coupling agent Polybond 3000 (MFI@190 °C, 2.16 kg: 400 g/10 min) from Chemtura (Philadelphia, Pennsylvania, USA). This latter is a polypropylene grafted with maleic anhydride (PP-g-MA) with 1.2 wt.% of MA.

In more details, the modification of the commercial PP was performed with the aid of a co-rotating twin-screw extruder Collin Teach-Line ZK25T (Ebersberg, Germany), operating with the following temperature profile: 180–190–205–195–85 °C, from the hopper to the die, and at a screw speed of 60 rpm. The neat PP and the extruded pellets of modified PP were transformed in flat films with a thickness approximately equal to 35–40 µm using a Collin flat die extruder Teach-Line E20T equipped with a calendar CR72T (Ebersberg, Germany). For this stage, the processing was conducted at a screw speed of 55 rpm, setting the temperature profile along the screw at 180, 190, 200, 190 and 185 °C.

2.2. Laminates Preparation

The conventional film stacking technique was used to obtain laminates. Layers of plastic films and basalt fabric, alternatively overlapped, are subjected to a pre-optimized pressure and temperature cycle (see Figure 1a) by press Collin GmbH (Edersberg, Germany) Mod. P400E.

In this way, 380 mm × 380 mm plaques constituted by 16 plies, symmetrically settled with respect to the medium plane, with an average thickness of 2.6 mm and a volumetric content of reinforcement of about 50% (ASTM D 3171-04, Test Method II) have been produced.

As a reference, plates of only PP matrix, with a thickness approximately equal to 2.5 mm, were prepared from PP granules and by superposition of plastic films mentioned above, according to the conditions shown in Figure 1b.

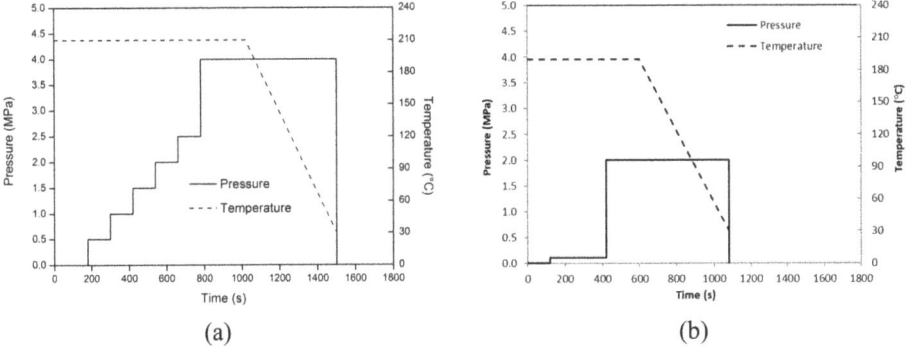

Figure 1. Hot-pressing conditions for (a) composite and reference (b) plates.

2.3. Experimental Techniques

2.3.1. Static Mechanical Properties

Flexural tests have been carried out by a universal Instron dynamometer Mod. 5564 equipped with a load cell of 1 kN. Specimens 12.7 mm wide and 100 mm long were cut from each sample laminate and loaded at room temperature in the three-point bending mode, according to the ASTM D790 standard, using a cross-head speed set at 2.5 mm/min and a span equal to 70 mm.

Flexural parameters evaluated by processing typical stress–strain curves were averaged on at least 5 determination for each investigated sample.

2.3.2. Impact Properties

Falling weight machine (Ceast Fractovis, Torino, Italy) was used for the impact tests at complete penetration, to obtain and study the full load–displacement curves to get useful information about the response of the laminates and for investigating the effect of the varied parameters. The penetration energy, U_p, set to obtain the complete penetration of all composite systems studied is equal to 100 J. Then, different energy levels (U = 3, 8, 15 J) were chosen to carry out the so-called indentation tests, useful to study the damage start and evolution. The rectangular samples, 100 × 150 mm, cut by a diamond saw from the original panels, were supported by the clamping device suggested by the ASTM D7137 Standard and were loaded in the center by an instrumented cylindrical impactor with a hemispherical nose, 19.8 mm in diameter. Tests were carried out using an impactor (m = 3.640 kg) placed at specific heights to obtain the selected impact energies. After all the impact tests, the samples were observed by a visual 2D stylus profilometer (Dektak XT) to derive quantitative information about step heights, which is indentation depth along the impacted areas.

3. Results and Discussion

The flexural stress–strain curves of composite specimens based on the neat and pre-modified film of polypropylene (coded as PP and PPC, respectively) are reported in Figure 2. At the same time, the evaluated mechanical parameters (modulus and strength) are summarized in Table 1. Flexural parameters compare quite favorably with those of other studies [5].

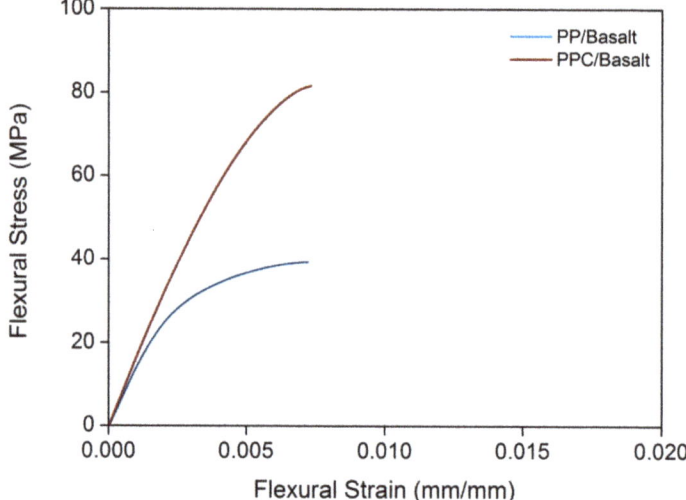

Figure 2. Flexural stress–strain curves.

Table 1. Flexural stress–strain results.

Sample	Flexural Modulus (MPa)	Flexural Strength (MPa)
PP/Basalt	13790 ± 820	39.9 ± 2.0
PPC/Basalt	15940 ± 788	81.1 ± 2.2

With regard to composite systems, previous morphological analyses conducted by electron scanning microscopy (SEM) on the same laminates have shown that for PP/Basalt composites, as expected given the nature of the phases involved, a poor interfacial adhesion is evident with images showing reinforcing fibers predominantly smooth and clean. On the contrary, for the PPC/basalt system, the authors reported that the presence of PP-g-MA improves the fiber wetting [25].

In light of the foregoing consideration, the significant improvement of the flexural parameters such as modulus and strength detected for the specimens containing the coupling agent is reasonable. In particular, the flexural strength of PPC/Basalt specimens was about twice the value estimated for composites based on neat PP.

Figure 3 compares reference materials in terms of load–displacement curves at penetration. From the curves, it is possible to note the higher maximum load, F_{max} and the corresponding displacement value, d, for the PP film-based specimens (see Table 2).

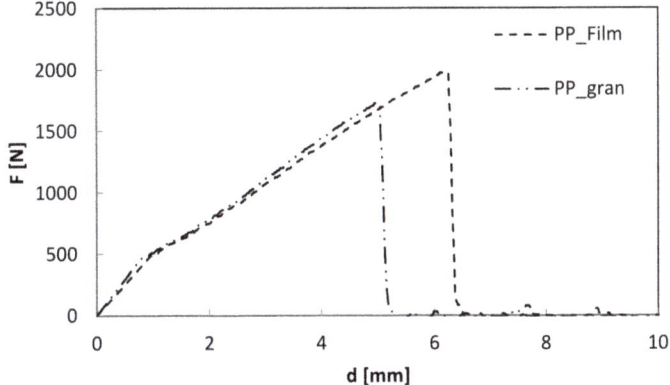

Figure 3. Load–displacement curves at penetration of PP granule and PP film-based samples.

Table 2. Impact parameters at penetration for matrix.

Matrix	F_{max} (N)	d (mm)
PP_gran	1689	4.99
PP_film	1964	6.25

In Figure 4, the impact curves at the penetration of composite materials are reported.

Remembering that the slope of the first linear part of the curve is an index of the impact rigidity of the system tested, the compatibilized system showed a slightly lower impact rigidity compared to the PP/Basalt one.

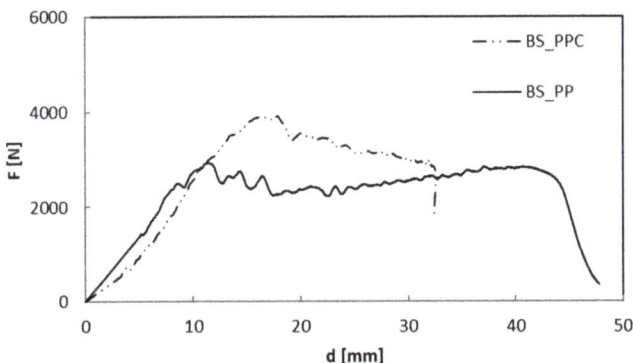

Figure 4. Load–displacement curves up to penetration: comparison between BS_PP and BS_PPC.

However, as also highlighted by the average data summarized in Table 3, it is capable of supporting higher loads (F_{max}) before the reinforcement breaks recording a higher deflection, d, in correspondence of the maximum load.

Furthermore, interestingly, in both cases, it was not possible to penetrate the panels even if a higher value of the maximum load, U_{max}, is recorded for the PP/Basalt composite sample (Table 3). This behavior is highlighted in Figures 5 and 6. It is evident that both the non-compatibilized and the compatibilized specimens battle to the penetration presenting only a hunching due to the impact event.

Table 3. Impact at penetration parameters for both composite systems.

Type	F_{max} (N)	d (mm)	U_{max} (J)
BS/PP	2919.58	14.88	102.88
BS/PPC	3918.91	17.12	84.34

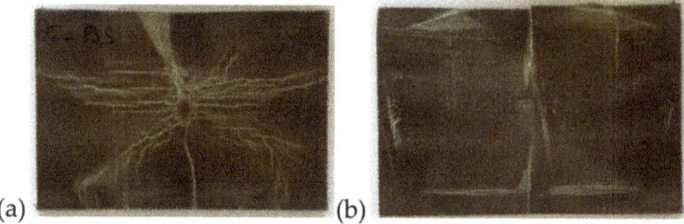

Figure 5. Impacted PPC basalt samples at penetration: (a) Front; (b) back.

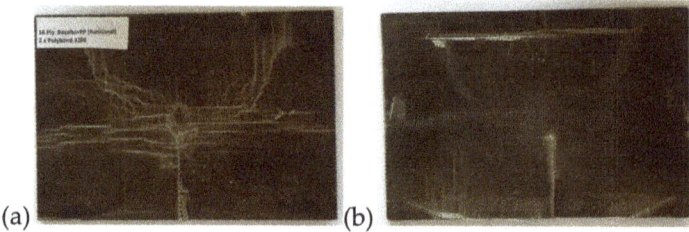

Figure 6. Impacted PP basalt samples at penetration: (a) Front; (b) back.

As far as the indentation measurements are concerned, in Figures 7–9 load–displacement curves of PP and PPC samples for three impact energy levels, U, are reported.

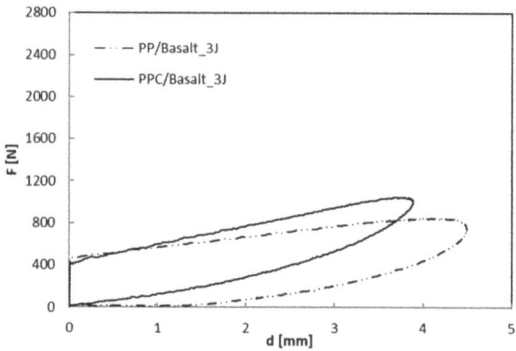

Figure 7. Load–displacement curves at indentation: comparison between BS_PP and BS_PPC; U = 3 J.

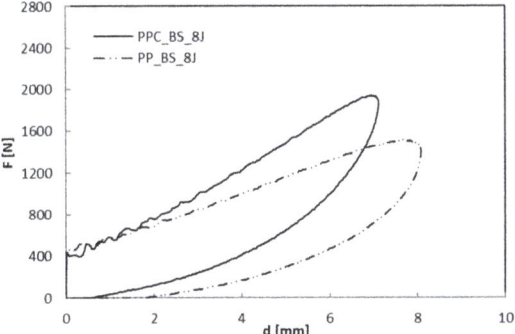

Figure 8. Load–displacement curves at indentation: comparison between BS_PP and BS_PPC; U = 8 J.

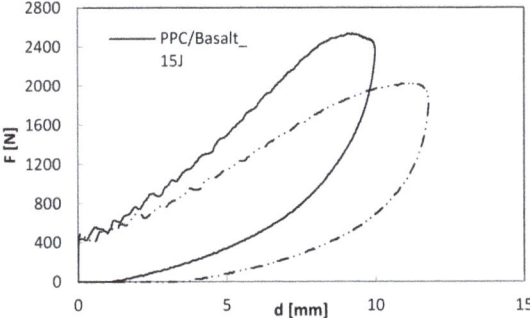

Figure 9. Load–displacement curves at indentation: comparison between BS_PP and BS_PPC; U = 15 J.

Taking into account that the area enclosed in the indentation curves represents the energy absorbed by the laminate to create damage, U_a, this parameter and the maximum impact load, F_{max}, increase as the impact energy, U, increases as shown in Figures 10 and 11, respectively. In particular, the U_a values, reflecting the extent of induced internal damage, are more significant for the PP/Basalt samples than for the PPC/Basalt ones for every tested energy indicating the occurrence of minor damage for the latter, under similar impact conditions.

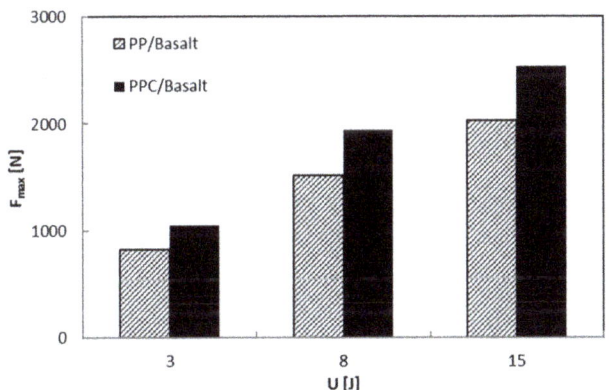

Figure 10. Maximum load, F_{max}, versus impact energy, U.

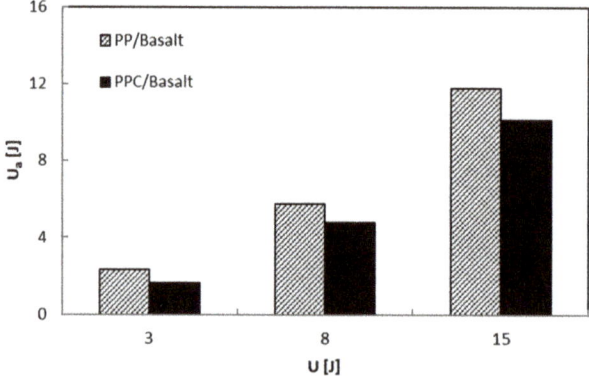

Figure 11. Absorbed energy, U_a, versus impact energy, U.

The increasing load usually results in a greater maximum deflection (Figure 12). The presence of the compatibilizer does not change this effect even if it gives rise to minor deflections to indicate a small amount of energy spent on bending. For the PP/Basalt system, the trend of the deflection has a stronger rise after U = 8 J, lowering the bending. The latter results in a greater amount of energy absorbed (see Figure 11), indicating global higher damage in the PP/Basalt system.

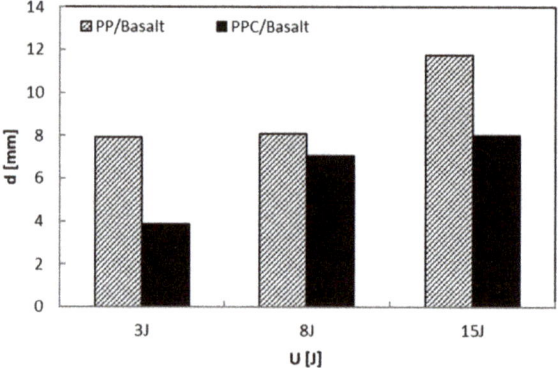

Figure 12. Maximum deflection, d, versus impact energy, U.

The last assertion is confirmed by measurements of the indentation depth representing the footprint impress by the impactor on the impacted side of the sample (residual plastic deformation) shown in Figure 13. The results demonstrated that this parameter increases at the increasing of the impact energy, U, for both types of basalt composite samples.

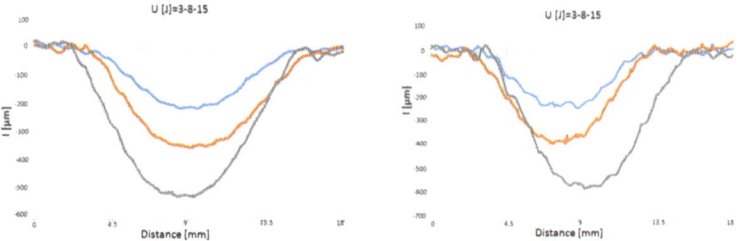

Figure 13. BS-PP and BS-PPC Indentation profiles.

Figure 14 shows the trend of the indentation depth, I, as a function of the impact energy, U. It is possible to note that the indentation, I, measured on the basalt PP samples is higher than that shown by the PPC ones on the entire range of impact energies examined: it was an expected effect since the best interfacial adhesion prevents plastic deformation of the matrix.

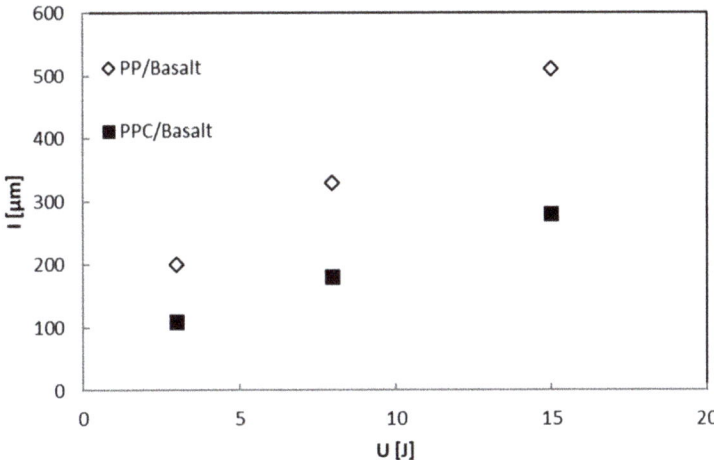

Figure 14. PP/Basalt and PPC/Basalt Indentation depth, I, versus the impact energy, U.

4. Conclusions

Polypropylene/basalt fabric composite laminates prepared by film piling and compression molding techniques were analyzed. Flexural and low-velocity impact properties were evaluated, taking neat PP plates as references and considering effects related to matrix modification with a pre-optimized amount of a maleic anhydride grafted PP as a coupling agent. Mechanical results demonstrated the preliminary change of the matrix to enhance the interfacial adhesion leads to samples with improved flexural performances especially the strength and ability to withstand higher impact loading with respect to neat PP based composite laminates. Premised that for both investigated systems, no penetration seems to occur, the more significant matrix-reinforcement interaction for compatibilized composite laminates make them less prone to plastic deformations as also evidenced by the indentation measurements. The indentation, I, recorded for the basalt PP samples is higher than the PPC ones confirming the higher absorbed energy, U_a that denotes greater damage due to the impact tests.

Author Contributions: Conceptualization, I.P, V.L. and P.R.; methodology, I.P, V.L. and P.R.; validation, I.P, V.L., V.P. and P.R.; investigation, I.P; resources, P.R.; data curation, I.P and V.P. writing—original draft preparation, I.P. and P.R. writing—review and editing, V.L.; supervision, V.L.; funding acquisition, V.L. All authors have read and agreed to the published version of the manuscript.

Funding: This research was funded by ONR Solid Mechanics Program, Survivability of marine composites and structures under impact and blast in extreme environments N00014-14-1-0380 ONR.

Acknowledgments: The authors gratefully acknowledge the ONR Solid Mechanics Program, in the person of Yapa D.S. Rajapakse, Program Manager, for the financial support provided to this research.

Conflicts of Interest: The authors declare no conflicts of interest. The funders had no role in the design of the study; in the collection, analyses, or interpretation of data; in the writing of the manuscript, or in the decision to publish the results.

References

1. Fiore, V.; Scalici, T.; di Bella, G.; Valenza, A. A review on basalt fibre and its composites. *Compos. Part B* **2015**, *74*, 74–94. [CrossRef]

2. Dhand, V.; Mittal, G.; Rhee, K.Y.; Park, S.-J.; Hui, D. A short review on basalt fiber reinforced polymer composites. *Compos. Part B* **2015**, *73*, 166–180. [CrossRef]
3. Jamshaid, H.; Mishra, R. A green material from rock: basalt fiber—A review. *J. Text. Inst.* **2016**, *107*, 923–937. [CrossRef]
4. Volkan, A.; Cakir, F.; Alyamac, E.; Seydibeyoglu, M.O. *Fiber Technology for Fiber-Reinforced Composites*; Woodhead Publishing: Sawston, UK, 2017; Chapter 8; pp. 169–185.
5. Lopresto, V.; Leone, C.; de Iorio, I. Mechanical characterization of basalt fiber reinforced plastic. *Compos. Part B* **2011**, *42*, 717–723. [CrossRef]
6. Shokrieh, M.M.; Memar, M. Stress corrosion cracking of basalt/epoxy composites under bending load. *Appl. Compos. Mater.* **2010**, *17*, 121–135. [CrossRef]
7. Varley, R.J.; Tian, W.; Leong, K.H.; Leong, A.Y.; Fredo, F.; Quaresimin, M. The effect of surface treatments on the mechanical properties of basalt-reinforced epoxy composites. *Polym. Compos.* **2013**, *34*, 320–329. [CrossRef]
8. Manikandan, V.; Jappes, J.T.W.; Kumar, S.M.S.; Amuthakkannan, P. Investigation of the effect of surface modifications on the mechanical properties of basalt fibre reinforced polymer composites. *Compos. Part B Eng.* **2012**, *43*, 812–818. [CrossRef]
9. Russo, P.; Simeoli, G.; Cimino, F.; Papa, I.; Lopresto, V. Impact damage behaviour of vinyl ester, epoxy and nylon 6 based basalt fiber composites. *J. Mater. Eng. Perform.* **2019**, *28*, 3256–3266. [CrossRef]
10. Botev, M.; Betchev, H.; Bikiaris, D.; Panayiotou, C. Mechanical properties and viscoelastic behavior of basalt fiber-reinforced polypropylene. *J. Appl. Polym. Sci.* **1999**, *74*, 523–531. [CrossRef]
11. Szabo, J.S.; Czigany, T. Static fracture and failure behavior of aligned discontinuous mineral fiber reinforced polypropylene composites. *Polym. Test.* **2003**, *22*, 711–719. [CrossRef]
12. Guo, J.; Mu, S.; Yu, C.; Hu, C.; Guan, F.; Zhang, H.; Gong, Y. Mechanical and thermal properties of polypropylene/modified basalt fabric composites. *J. Appl. Polym. Sci.* **2015**, *132*, 42504. [CrossRef]
13. Bashtannik, P.I.; Kabak, A.I.; Yakovchuk, Y.Y. Eff. Adhes. Interact. Mech. Prop. Thermoplast. Basalt Plast. *Mech. Compos. Mater.* **2003**, *39*, 85–88.
14. Akinci, A. Mechanical and morphological properties of basalt filled polymer matrix composites. *Arch. Mater. Sci. Eng.* **2009**, *35*, 29–32.
15. Song, J.; Liu, J.; Zhang, Y.; Chen, L.; Zhong, Y.; Yang, W. Basalt fiber-reinforced PA1012 composites: Morphology, mechanical properties, crystallization behaviors, structure and water contact angle. *J. Compos. Mater.* **2015**, *49*, 415–424. [CrossRef]
16. Deak, T.; Czigany, T.; Marsalkova, M.; Militky, J. Manufacturing and testing of long basalt fiber reinforced thermoplastic matrix composites. *Polym. Eng. Sci.* **2010**, *50*. [CrossRef]
17. Deák, T.; Czigány, T.; Tamás, P.; Németh, C. Nemeth Enhancement of interfacial properties of basalt fiber reinforced nylon 6 matrix composites with silane coupling agents. *Express Polym. Lett.* **2010**, *4*, 590–598. [CrossRef]
18. Prajapati, R.S.; Jain, S.; Shit, S.C. Development of basalt fiber-reinforced thermoplastic composites and effect of PE-g-MA on composites. *Polym. Compos.* **2017**, *38*. [CrossRef]
19. Ying, Z.; Wu, D.; Zhang, M.; Qiu, Y. Polylactide/basalt fiber composites with tailorable mechanical properties: Effect of surface treatment of fibers and annealing. *Compos. Struct.* **2017**, *176*, 1020–1027. [CrossRef]
20. Sorrentino, L.; de Vasconcellos, D.S.; D'Auria, M.; Sarasini, F.; Iannace, S. Thermoplastic composites based on poly(ethylene 2,6-naphthalate) and basalt woven fabrics: Static and dynamic-mechanical properties. *Polym. Compos.* **2016**, *37*, 2549–2556. [CrossRef]
21. Zhang, Z.F.; Xin, Y. Mechanical properties of basalt-fiber-reinforced polyamide 6/polypropylene composites. *Mech. Compos. Mater.* **2014**, *50*, 509–514. [CrossRef]
22. Bandaru, A.K.; Patel, S.; Sachan, Y.; Ahmad, S.; Alagirusamy, R.; Bhatnagar, N. Mechanical behavior of Kevlar/basalt reinforced polypropylene composites. *Compos. Part A Appl. Sci. Manuf.* **2016**, *90*, 642–652. [CrossRef]
23. Dehkordi, M.T.; Nosraty, H.; Shokrieh, M.M.; Minak, G.; Ghelli, D. The influence of hybridization on impact damage behaviour and residual compression strength of intraply basalt/nylon hybrid composites. *Mater. Des.* **2013**, *43*, 283–290. [CrossRef]

24. Greco, A.; Maffezzoli, A.; Casciaro, G.; Caretto, F. Mechanical properties of basalt fibers and their adhesion to polypropylene matrices. *Compos. Part B* **2014**, *67*, 233–238. [CrossRef]
25. Simeoli, G.; Acierno, D.; Sorrentino, L.; Iannace, S.; Sarasini, F.; Tirillò, J.; Russo, P. Comparison of Low-Velocity Impact Behavior of Thermoplastic Composites Reinforced with Glass and Basalt Woven Fabrics. In Proceedings of the 16th ECCM European Conference on Composite Materials, Seville, Spain, 22–26 June 2014.

© 2020 by the authors. Licensee MDPI, Basel, Switzerland. This article is an open access article distributed under the terms and conditions of the Creative Commons Attribution (CC BY) license (http://creativecommons.org/licenses/by/4.0/).

Article

Falling Weight Impact Damage Characterisation of Flax and Flax Basalt Vinyl Ester Hybrid Composites

Hom Nath Dhakal *, Elwan Le Méner, Marc Feldner, Chulin Jiang and Zhongyi Zhang

Advanced Materials and Manufacturing (AMM) Research Group, School of Mechanical and Design Engineering, University of Portsmouth, Anglesea Road, Anglesea Building, Portsmouth, Hampshire PO1 3DJ, UK; elwan.le-mener@port.ac.uk (E.L.M.); feldner.marc@gmail.com (M.F.); chulin.jiang@port.ac.uk (C.J.); zhongyi.zhang@port.ac.uk (Z.Z.)
* Correspondence: hom.dhakal@port.ac.uk; Tel.: +44-(0)23-9284-2582; Fax: +44-(0)23-9284-2351

Received: 28 February 2020; Accepted: 1 April 2020; Published: 3 April 2020

Abstract: Understanding the damage mechanisms of composite materials requires detailed mapping of the failure behaviour using reliable techniques. This research focuses on an evaluation of the low-velocity falling weight impact damage behaviour of flax-basalt/vinyl ester (VE) hybrid composites. Incident impact energies under three different energy levels (50, 60, and 70 Joules) were employed to cause complete perforation in order to characterise different impact damage parameters, such as energy absorption characteristics, and damage modes and mechanisms. In addition, the water absorption behaviour of flax and flax basalt hybrid composites and its effects on the impact damage performance were also investigated. All the samples subjected to different incident energies were characterised using non-destructive techniques, such as scanning electron microscopy (SEM) and X-ray computed micro-tomography (πCT), to assess the damage mechanisms of studied flax/VE and flax/basalt/VE hybrid composites. The experimental results showed that the basalt hybrid system had a high impact energy and peak load compared to the flax/VE composite without hybridisation, indicating that a hybrid approach is a promising strategy for enhancing the toughness properties of natural fibre composites. The πCT and SEM images revealed that the failure modes observed for flax and flax basalt hybrid composites were a combination of matrix cracking, delamination, fibre breakage, and fibre pull out.

Keywords: polymer-matrix composites (PMCs); composite laminates; low-velocity impact; delamination; X-ray micro CT

1. Introduction

Over the past decade, consumers' increased awareness of and expectations towards environmental sustainability, government legislation, an increased sense of corporate social responsibility (CSR) from industry sectors for achieving sustainable development aspirations through a triple bottom line performance (environment, economic, and social), have inspired research on materials which are renewable and recyclable [1,2].

Natural fibre-reinforced polymeric composites have been used in a wide range of engineering applications in recent years due to their abundant availability, lower density, and much higher specific strength and modulus than conventional glass and carbon fibre-reinforced composites [3–5]. Moreover, these reinforced materials possess a low embodied energy to process and use compared to energy-intensive conventional fibre-reinforced composites. Despite several benefits, there are still some issues which limit the use of natural fibre-reinforced composites in semi-structural and structural applications [6,7]. One of the key issues facing these composites is their hydrophilic nature, which leads to poor fibre matrix interfacial adhesion and lower mechanical properties [8].

Among the natural fibres used in polymeric composites, bast fibres (flax, hemp, jute, and kenaf) stand out as the most promising reinforcements [9,10]. Due to their unique hollow structure, these fibres provide a good damping property, which is very important when it comes to dealing with impact damage and vibration damping behaviours. The good damping properties of bast fibre-reinforced composites make them an attractive alternative to be used in automotive components where impact and damping properties are very important [11,12]. However, their high natural variability, strong affinity to water until saturation, limited processing temperature range, relatively low impact resistance, and low thermal stability negatively influence their long-term durability [13,14]. Moreover, their low impact resistance behaviour under different operation conditions is another concern when these materials are used in automotive and marine sectors.

Mitrevski et al. [15] studied the influence of impactor shape on the impact damage of composite laminates. The results demonstrated that the impactor shape plays a large role in the damage response of composite materials. Composite laminates undergo various impact-induced damage modes under impact loadings. A review carried out by Cantwell and Morton [16] has reported the various impact-induced failure modes of composite laminates. Wisheart and Richardson [17] investigated the impact damage response of complex geometry pultruded glass/polyester composites.

During the past two decades, there have been many reported works, especially those covering the mechanical and thermal properties of natural bast fibre-reinforced composites. This highlights the significant increase in the demand for these fibres. The automotive sector is leading the way towards using natural bast fibres due to the drives to produce lightweight and sustainable materials and reduce health risks during manufacturing and recycling. Many leading original equipment manufacturers (OEMs) in the automotive sector have been using natural fibre-reinforced composites in various parts, such as door linings, seat cushions, door cladding, and mainly non-structural applications. Nevertheless, due to the lack of sufficient mechanical properties for structural applications, natural fibre-reinforced biocomposites are not fully utilised in semi-structural and structural applications [18,19].

In recent years, widespread research has been focused on utilising a hybrid approach which consists of combining two or more reinforcements, in which the synergic effects of both reinforcements are utilised. In order to compensate for the shortcomings of the natural fibre-reinforced composites, glass and carbon fibres, as well as nano particulates, have been used as hybrid constituents [20–23]. There are many reported works where basalt fibres have been introduced as hybrid reinforcements on natural fibre composites owing to their high thermal stability, good mechanical properties, good corrosion resistance, and natural origin (coming from volcano rock). The work carried out by Dhakal et al. [24] investigated the influence of basalt fibre hybridisation on the post-impact mechanical behaviour of hemp fibre-reinforced composites, and their report suggests that basalt fibre hybrid systems significantly improved the post-impact mechanical properties. Similarly, carbon fibre hybridised flax fibre composites were investigated and it was observed that the hybrid system offered excellent mechanical properties compared to flax fibre non-hybrid composites [25]. Paturel and Dhakal [26] investigated the water absorption and low velocity impact damage characteristics of flax/glass fibre hybrid vinyl ester composites. Their findings suggest that glass fibre hybridised composites significantly reduced the water uptake percentage compared to flax fibre vinyl ester composites without hybridisation. It is evident from the various literature that the impact damage characteristics of natural fibre composites have been well-documented. However, not many studies have been focused on investigating the influence of basalt fibre hybrid flax composites subjected to low-velocity impact loading at different incident energy levels. Additionally, there has been limited work on the influence of hybridisation on the moisture absorption and its effects on the low-velocity impact damage mechanisms.

This study aimed to investigate the effect of basalt fibre hybridisation on the water absorption and low-velocity falling weight impact behaviour of flax fibre-reinforced vinyl ester hybrid composites with varying incident impact energies. For this, the flax/VE composites were impacted at low impact energies ranging from 50 to 70 Joules, which was sufficient to create impact damage up to penetration.

The impact performance of flax and flax/basalt/VE hybrid composites was evaluated in terms of the load bearing capability, energy absorption, and damage modes. In addition, the water absorption behaviour of flax and flax basalt hybrid composites and its effects on the impact damage were also investigated. The damage mechanisms of impacted composite specimens were characterised using non-destructive evaluation techniques, such as X-ray computed micro-tomography (μCT) and scanning electron microscopy (SEM).

2. Materials and Methods

2.1. Materials and Laminate Fabrication

The matrix material used was vinyl ester, Scott-Bader Crystic VE676-03, obtained from Scott-Bader. Woven flax and woven basalt fibres were used as the reinforcements (±45) as biaxial stitched non-crimp fabrics of 600 g/m². Figure 1 shows the flax and basalt fabric used to make flax and flax/basalt hybrid composites. The chemical and mechanical properties of key bast fibres (flax, kenaf, hemp, and jute), along with basalt fibre included for comparison purposes, are presented in Table 1.

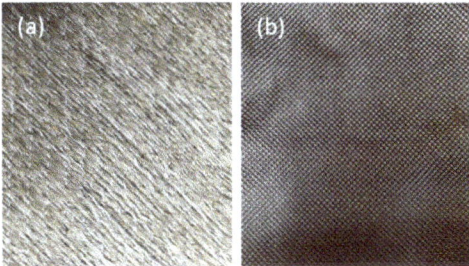

Figure 1. Reinforcements used, (a) flax woven fabric, and (b) basalt woven fabric.

Table 1. Chemical composition, and physical and mechanical properties of commonly used natural bast fibres [27–30].

Fibres	Cellulose	Hemi-Cellulose	Lignin	Pectin	Density (g/cm³)	Tensile Strength (MPa)	Young's Modulus (GPa)	Failure Strain (%)
Flax	70.5	16.5	2.5	0.9	1.45	700	60	2.3
Kenaf	78.5	8–13	21.5	0.6	1.40	350–600	21–60	1.6–3.5
Hemp	81	20	4	0.9	1.48	530	45	3
Jute	67	16	9	0.2	1.40	325	37.5	2.5
Basalt *	-	-	-	-	2.70	4800	90	3.15
E-glass *	-	-	-	-	2.55	3400	72	3.4
VE matrix	-	-	-	-		70	3.5	0.02

* For comparison purposes.

2.2. Composite Laminate Fabrication

The flax and flax/basalt hybrid laminates were fabricated by the vacuum infusion technique. Two types of samples were fabricated to investigate the influence of hybridisation. In the first set of samples, six layers of flax fibres were used. A second set of samples, including one layer of basalt fibres on the top face and five layers of flax fabric on the rear side, were employed. The main reason for investigating the fibre contents and orientation is to optimise the hybrid materials in the loading direction. The average of the fibre volume fraction of Flax/VE and Flax/Basalt/VE was approximately 31% and 33%, respectively. The void content was approximately 3%. The sample size of a 70 mm by

70 mm square was cut using water jet cutting of the composite panel. Schematics the of flax and flax basalt hybrid composite laminates are shown in Figure 2.

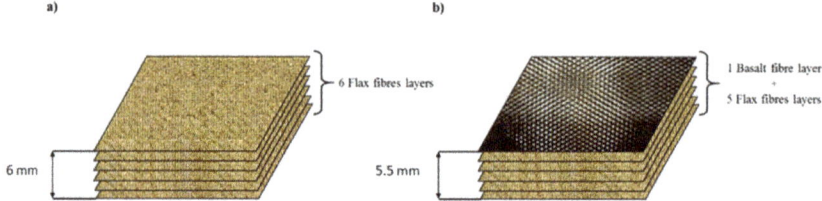

Figure 2. Formulation of laminates: (**a**) Flax/VE laminate with a thickness of 6 mm, and (**b**) flax/basalt/VE hybrid laminates with a thickness of 5.5 mm.

2.3. Moisture Absorption Measurement

The moisture uptake behaviour of flax/VE and flax/VE/basalt hybrid laminates was investigated in accordance with BS EN ISO:1999 [31]. Five specimens, consisting of 70 mm by 70 mm squares of flax/VE and flax/VE/basalt hybrid composites, were placed in a desiccator for 48 h and weighted near to 0.1 mg. Then, the specimens were immersed in de-ionised water at room temperature. After 24 h of immersion, the specimens were taken out and the surface was dried with absorbent paper. The process was repeated until the saturation moisture was reached. The percentage of moisture uptake was calculated using Equation (1):

$$M(\%) = \frac{M_t - M_0}{M_0} \times 100 \qquad (1)$$

where M (%) is the moisture uptake in percentage, M_t is the weight of the water-immersed specimen at a given time, and M_0 is the initial mass of the specimen in a dry condition.

2.4. Low-Velocity Drop Weight Impact Testing

Low-velocity instrumented falling weight impact testing was conducted by using an instrumented falling weight impact testing (IFWIT) machine. A Zwick/Roell impact test machine (IFW 413) was used for the testing in accordance with the British Standard BSEN ISO 6603-2 recommendations [32]. The hemispherical impact tup used was made of steel and had a 20 mm diameter. The incident energy was tailored by adjusting the release height of the impact mass, i.e., changing the impact velocities (while keeping the other parameters constant). To analyse the damage behaviour of flax and flax/basalt hybrid vinyl ester hybrid composites, the three different incident energies employed were 50, 60, and 70 Joules, respectively (corresponding impact velocities of 2.08, 2.28, and 2.46 m/s). The test specimens were 70 mm by 70 mm squares.

2.5. X-Ray Computed Micro-Tomography (πCT)

πCT, XT H 225 was used to assess the barely visible impact damage failure. The samples were subjected to the three different incident energies of 50, 60, and 70 Joules and examined using X-ray (πCT) to effectively evaluate the extent of damage due to impact loadings.

2.6. Scanning Electron Microscopy (SEM)

In order to investigate the damage mechanisms of the impacted samples, the surfaces of dried specimens and room temperature immersed specimens were examined using SEM Zeiss EVO LS10. Before examination, the samples were placed in a desiccator to remove all of the water of the samples, in order to avoid evaporation during characterisation. The specimens were also surface prepared and the damaged area was imaged.

3. Results and Discussion

3.1. Flax/Basalt/VE Hybrid Composite Lay Up

One of the main aims of this study was to find out if hybridising one side of the composite plate would provide optimal hybrid effects. Figure 3 shows the hybrid sample with a rear basalt layer. It is clear from the figure that the basalt layer exhibits push-out delamination. As the incident energy increases, the delamination also increases. Delamination is one of the most prevalent failure mechanisms in composite laminates. This phenomenon becomes even greater when two different types of fibres are hybridised. It can be seen that the impactor has perforated and all the flax layers have fractured, but the basalt layer has not fractured; these impacted images show delamination of the basalt layer.

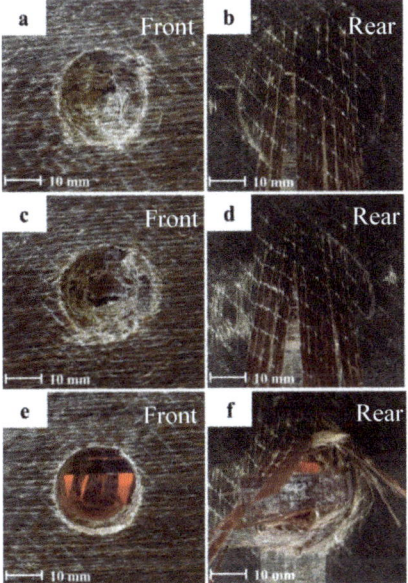

Figure 3. Images of damage progression on the front and rear faces of a flax/VE/basalt hybrid composite where basalt was placed on the rear side of the composite panels impacted in the range of (**a,b**) 50 J, (**c,d**) 60 J, and (**e,f**) 70 J.

The influence of the basalt fibre layer on the front and rear side of the laminates is further explain in Figure 4. It is evident from the figure that flax composites with a basalt layer on the top impact face provided optimal properties in comparison to the basalt fibre on the rear side under the three different incident impact energies of 50, 60, and 70 Joules. The main reason for this phenomenon is that when basalt fibre was placed on the rear side of the laminates, a significant amount of delamination was observed, which is shown in Figure 4. Employing this evidence, the remaining investigation was carried out on samples where basalt fibre was placed on the top side of the flax/VE composites. Just placing one layer of basalt fibre on the top of the flax/VE composite laminate provides a good design choice to fabricate high-performance composite laminates using a simple and cost-effective method.

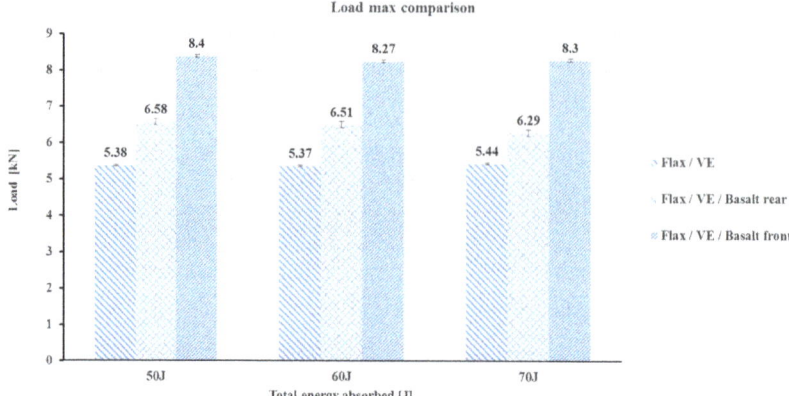

Figure 4. Maximum load comparison of different impact configurations of flax/VE, flax/VE/basalt rear, and flax/VE/basalt front for each incident energy employed.

3.2. Impact Damage Characteristics

Load and Energy Absorption Capabilities

Important impact parameters and corresponding values obtained from the low-velocity testing for flax and flax basalt hybrid samples at three different incident impact energies (50, 60, and 70 Joules) are presented in Table 2.

Table 2. Important impact parameters and corresponding values obtained from impact testing.

Specimen	Peak Load, F_m (kN)	Incipient Damage Load, F_i (kN)	Maximum Energy, E_m (J)	Incipient Energy, E_i (J)	Total Energy, E_t (J)
Energy at 50 J					
Flax/VE	5.39	4.09	16.81	2.22	43.03
Flax/VE/basalt	8.40	2.36	31.67	1.67	59.88
Energy at 60 J					
Flax/VE	5.42	4.11	14.67	2.70	40.44
Flax/VE/basalt	8.27	2.36	31.67	1.67	59.88
Energy at 70 J					
Flax/VE	5.51	4.21	15.73	2.82	41.44
Flax/VE/basalt	8.30	3.90	28.58	2.34	62.89

Load–deformation–energy traces obtained from the impact testing for flax/VE composites are shown in Figure 5.

Two of the most used parameters to assess damage resistance in composites after an impact are the impact energy and absorbed energy. The impact energy represents the maximum energy that the specimen can transform (it is equal to the kinetic energy of the impactor right before dart contact with the sample when the impact takes place), whereas the absorbed energy is the unrecoverable energy dissipated by the system (including energy dissipated by friction and, most importantly, by mechanisms which are peculiar to the material).

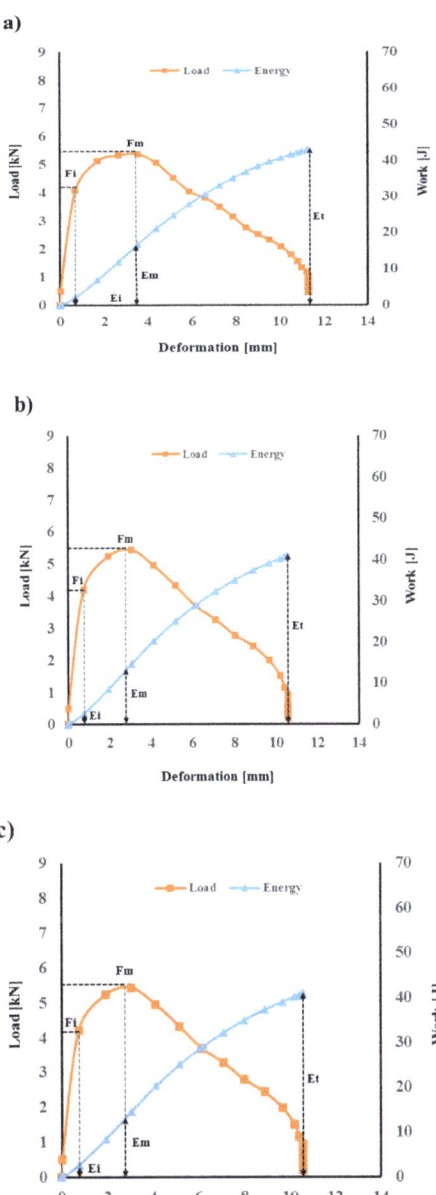

Figure 5. Load and work vs. deformation for flax/VE composites: (**a**) flax 50 J, (**b**) flax 60 J, and (**c**) flax 70 J. FM, load maximum; FI, incipient damage load; EM, energy maximum; EI, incipient damage energy; ET, energy total.

The absorbed energy can be calculated from load vs. deformation curves. In order to evaluate the laminate's performances, the transient response of each laminate was recorded in terms of the load, energy, and displacement. It can be observed from Figures 5 and 6 that the peak contact force is higher for the flax/basalt hybrid composite than that of flax/VE without hybridisation, which indicates that the hybrid specimens offer a higher impact resistance during impact events. A similar positive hybrid effect can be observed in load–time traces (Figure 7).

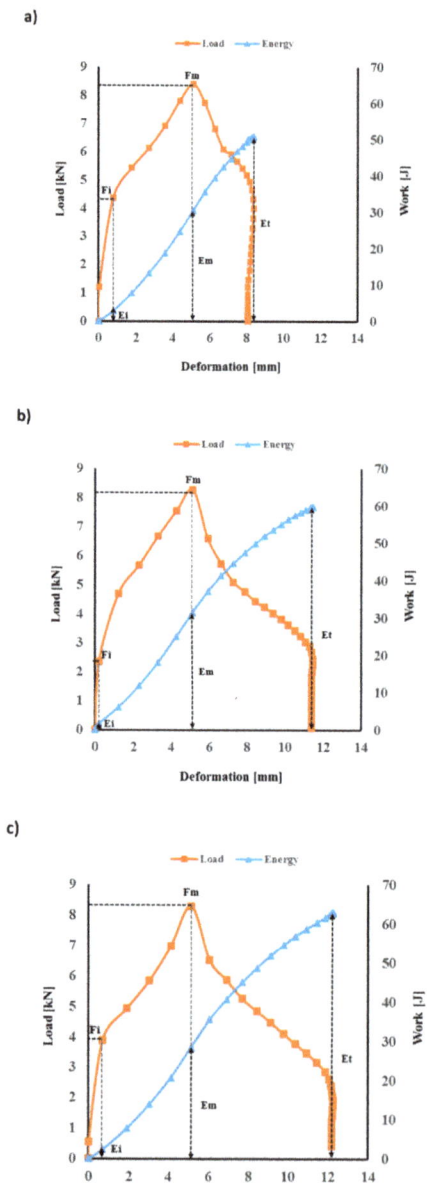

Figure 6. Load and work vs. deformation for flax/VE/basalt hybrid composites: (**a**) flax 50 J, (**b**) flax 60 J, and (**c**) flax 70 J. FM, load maximum; FI, incipient damage load; EM, energy maximum; EI, incipient damage energy; ET, energy total.

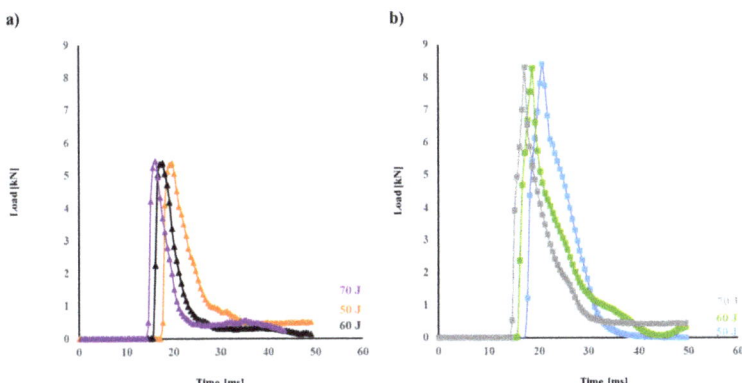

Figure 7. Typical load vs. time traces for 50 J, 60 J, and 70 J impacted (**a**) flax and (**b**) flax/basalt hybrid composites.

The load–deformation–energy traces for flax/basalt/VE hybrid composites are depicted in Figure 6. Flax/VE/basalt hybrid composites absorb more energy than flax/VE, as illustrated in Figure 6. We can observe similar curves for all three energy levels for flax/basalt hybrid composites. It is evident from Figure 6 that the applied incident energies (50, 60, and 70 Joules) were not high enough to penetrate or perforate (showing rebound energy) the hybrid samples, indicating their superior mechanical behaviour and higher energy dissipation potential compared to flax/VE samples without hybridisation. It is clear that flax/basalt hybrid composites exhibited a significantly improved impact performance compared to flax/VE composites with hybridisation.

Figure 6 illustrates the performance of flax basalt hybrid composites in terms of representing different impact parameters. It is clear from the results that flax/basalt hybrid composites exhibited significantly improved impact performances compared to flax/VE composites without hybridisation (Figure 5). It can also be observed that basalt hybridisation contributes to increasing the deformation of composites. This improvement could be attributed to the higher failure strain of basalt fibres compared to commonly used natural fibres such as flax and hemp. This phenomenon provides a balanced property, as one would expect for hybrid systems. These results indicate that basalt fibre hybridisation into natural fibre composites provides a promising strategy for enhancing the overall impact toughness. Such hybrid effects have been reported for improved mechanical properties, such as tensile, flexural, and fracture toughness behaviours [33,34].

Figure 7 shows the load–time traces of impacted flax and flax/basalt hybrid composite specimens. In this case, the peak load is the same as previously shown (Figures 5 and 6). However, the test time required to complete the impact event is important to consider. The time the striker was in contact with the impacted specimens is approximately the same for each composite. However, the time taken to complete the impact event is longer for flax/basalt hybrid composites compared to flax/VE composites. This is an indication that hybrid composites have better impact resistance behaviour as a result of the hybrid effect.

3.3. Moisture Absorption Behaviour

Figure 8 depicts the moisture absorption curves of flax/VE and flax/basalt/VE hybrid composites. It is evident from the curves that the moisture uptake at the beginning is linear and rapidly increases for flax/VE composites compared to flax/basalt hybrid composites. After the initial rapid rise, the moisture uptake slows down and reaches saturation at 768 h (30 days) for flax composites, whereas it takes longer—1008 h (42 days)—for flax/basalt hybrid composites. The longer time taken for hybrid composites to reach saturation moisture absorption can be attributed to the influence of basalt fibres restricting the flow of water molecules, as basalt fibres have better water repellence behaviour compared to flax/VE composites. Nonetheless, for the side where only flax is exposed, there would still be moisture ingress taking place at a higher rate than where basalt fabric was placed as a hybrid layer. This can be observed by the moisture uptake percentage difference between flax and flax/basalt hybrid composites, which is only 0.5%. If the basalt fabric was placed on both sides of the flax samples, the moisture uptake percentage of hybrid composites would have been far lower. Moreover, the sides of the both types of composites were not sealed, which is another reason for the higher moisture absorption displayed by both composites.

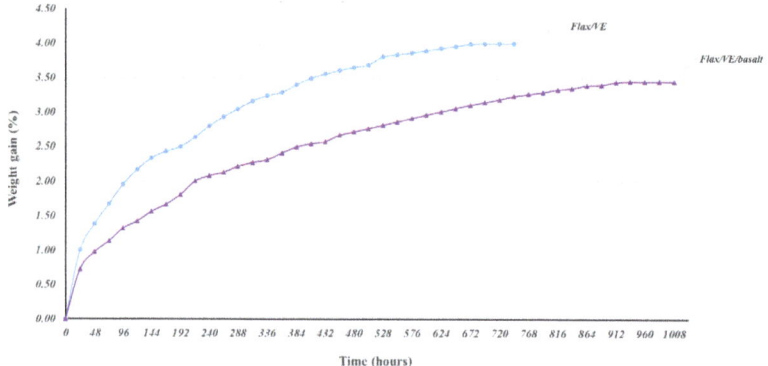

Figure 8. Water absorption comparison of flax/VE and flax/VE/basalt hybrid composites.

The maximum weight gain reported for vinyl ester matrix is 1.07% at room temperature [30]. The maximum weight gain percentages for flax/VE and flax/basalt/VE hybrid composites were approximately 4% and 3.5%, respectively. The lower moisture uptake percentage for flax/basalt hybrid composites is attributed to the barrier effects of top-layer basalt fibre on flax/VE composites. The moisture absorption behaviour for both composites indicates Fickian behaviour, which is rapid in the beginning and slowly reaches saturation.

By comparing the flax-basalt hybrid specimens with those made entirely of flax, it can be seen that the addition of basalt fibre improves the moisture absorption resistance of the hybrid composite. Since the five layers of flax fibre were sandwiched by one ply of basalt fibre, the total area of flax exposed to the water was decreased for flax/VE composites. The reason for the difference in moisture absorption between the flax and basalt hybrid specimens can be further explained by considering the chemical composition of flax fibres. The cellulose in the flax fibre is what provides the majority of the stiffness and strength; however, the semi-crystalline structure contains a large amount of hydroxyl groups, which give the fibre hydrophilic characteristics. By covering flax fibres with basalt fibres in a hybrid composite, the surface area exposed to water is reduced and therefore absorbs less moisture.

Influence of Moisture Absorption on the Impact Resistance Behaviour

The influence of moisture absorption on the flax composite in dry and wet conditions was investigated. The effects of moisture absorption on the load bearing capability of flax/VE composites impacted at two different incident energy levels are shown in Figure 9. The wet flax/VE specimens displayed a higher peak load compared to wet specimens, slightly outperforming the dry sample. This could be attributed to water absorption-induced plasticisation of the vinyl ester matrix leading to an increase of deformation and impact energy absorption [5,23].

Generally, when natural fibres absorb moisture, they swell, which promotes the development of adverse effects on the mechanical properties, such as tensile, flexural, and fatigue properties, due to the weak fibre matrix interface. However, as far as the impact performance is concerned, the results from this study suggest that there was no negative influence of moisture absorption on the load. Instead, the wet samples withstood a slightly higher peak load than the dry ones. This could be attributed to engrossed amounts of water causing swelling of the flax fibres, which could fill the gaps between the fibre and vinyl ester matrix and could have eventually led to an increase of impact load [5].

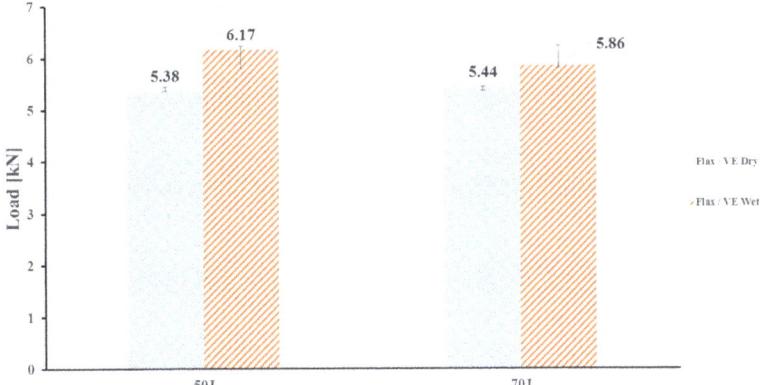

Figure 9. Impact load at different energy levels for flax/VE composites in dry and wet conditions.

3.4. Damage Characterisation

3.4.1. Damage Behaviour in Dry Conditions

Figure 10 depicts the damage of front and rear faces of impacted flax/VE composite specimens. As can be clearly observed, all the samples impacted at 50, 60, and 70 Joules were fully penetrated. The incident energy of 50 Joules was enough to cause damage to these groups of samples.

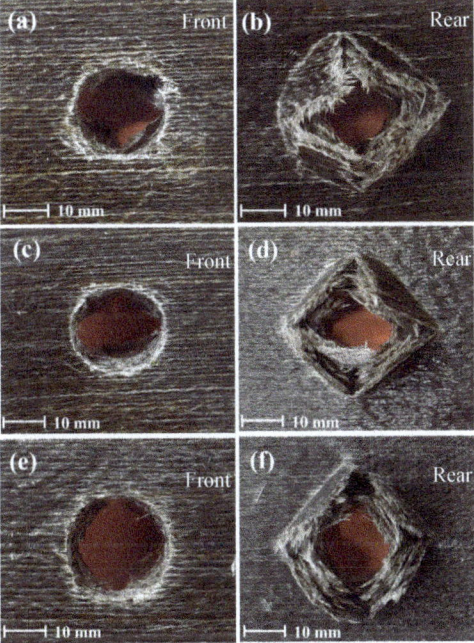

Figure 10. Images of damage progression on front and rear faces of flax/VE composite samples impacted in the range of (**a,b**) 50 J, (**c,d**) 60 J, and (**e,f**) 70 J.

Figure 11 shows the damage of the hybrid sample with one basalt layer on the top. In the pictures (a) and (b), a 50 J incident energy was not high enough to perforate the sample thanks to the basalt layer on the top. However, in the other images (Figure 11e–f) for 60 and 70 Joules of incident energy, the maximum energy absorbed is exceeded and all the layers of fibres are broken. As a result, the samples are fully penetrated.

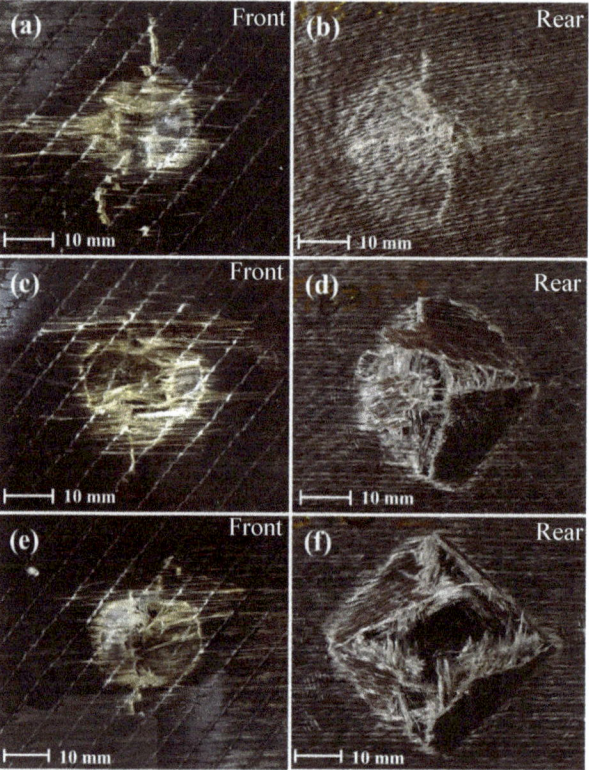

Figure 11. Images of damage progression on front and rear faces of flax/VE/basalt hybrid composite panels impacted in the range of (**a,b**) 50 J, (**c,d**) 60 J, and (**e,f**) 70 J.

3.4.2. Visual Observations of Damage Behaviour in Wet Conditions

Figure 12 shows impacted front and rear specimens for water-immersed flax/VE samples impacted at 50 and 70 Joules of incident energies. It can be observed that the wet samples do not have as clear holes as those of dry specimens (Figure 10), indicating the increased ductility of water-immersed samples.

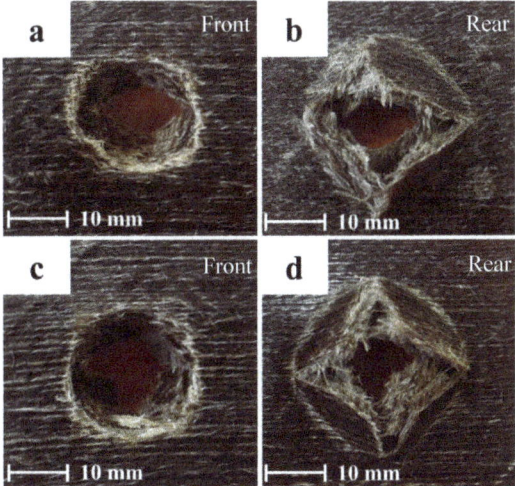

Figure 12. Images of damage progression on front and rear faces of flax/VE composite panels impacted in the range of (**a**,**b**) 50 J and (**c**,**d**) 70 J after water absorption.

3.4.3. Damage Behaviour from SEM Observations

SEM images illustrated in Figures 13 and 14 reveal matrix cracking, fibre fracture, and delamination on the fractured surfaces of impacted flax/VE and flax/basalt/VE hybrid samples. These damages are in close agreement with those observed in previous studies on flax/carbon epoxy and flax/glass vinyl ester matrix hybrid composites [25,26]. The SEM images further reveal that the extent of damage increased with the increase of the incident energy level. These images further suggest that at a higher energy level, the composites undergo severe damage, with evidence of fibre breakage and pull out, especially in the case of flax/VE composites (Figure 13). For flax/basalt/VE hybrid composites, delamination and fibre bending can be observed (Figure 14). It is also worth mentioning that with basalt fibre on the top layer of hybrid composites, the energy dissipation is increased, which allows an enhancement in impact and fracture toughness behaviours [35,36]. Moreover, it can be observed that for higher energy-impacted samples, more severe fibre damage and breakage can be observed. Different failure modes for flax/VE composites are further explained by the Micro-CT scan illustrated in Figure 15, which compliments the observation made via SEM images. From the annotation provided for the SEM images of flax/VE composites, as shown in Figure 13a–d, the following can be observed:

(1) Fibre breakage; (2) delamination; (3) fibre debonding; and (4 and 5) fibre pull out.

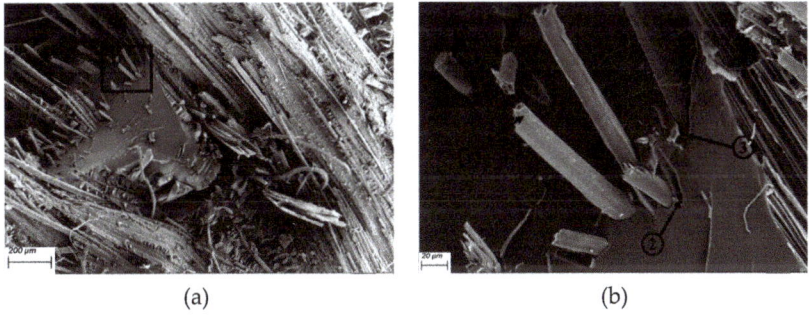

(a)   (b)

Figure 13. *Cont.*

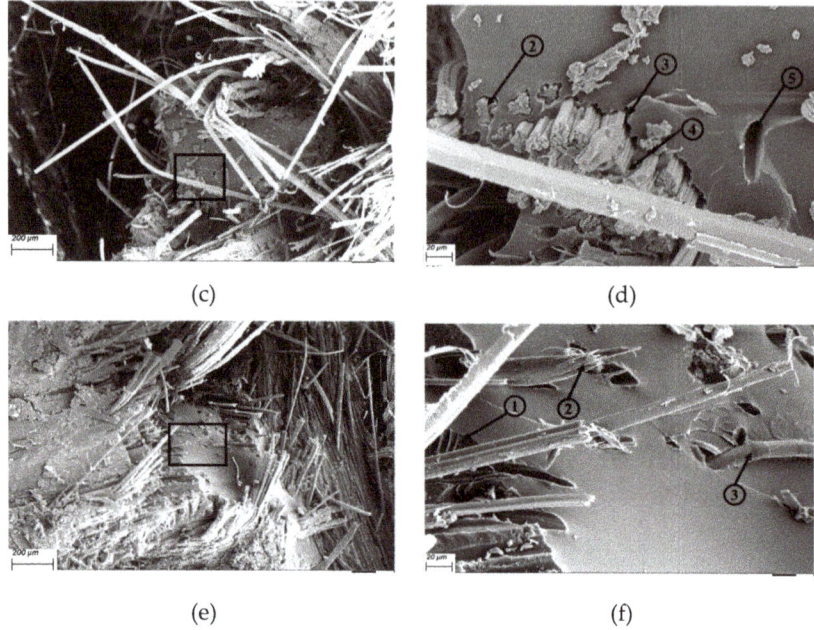

Figure 13. SEM images of flax/VE composites: (**a**) 150 (**b**) 1000× magnification with 50 J, (**c**) 150× magnification with 60 J, magnification with 50 J, (**b**) 1000× magnification with 50 J, (**c**) 150× magnification with 60 J, (**d**) 1000× magnification with 60 J, (**e**) 150× magnification with 70 J, and (**f**) 1000× magnification with 70 J.

From the annotation provided for the SEM images of flax/basalt/VE hybrid composites, as shown in Figure 14a–d, the following can also be observed:

(1) Matrix cracking and fibre bending and (2, 3, and 4) basalt fibre breakage and fracture.

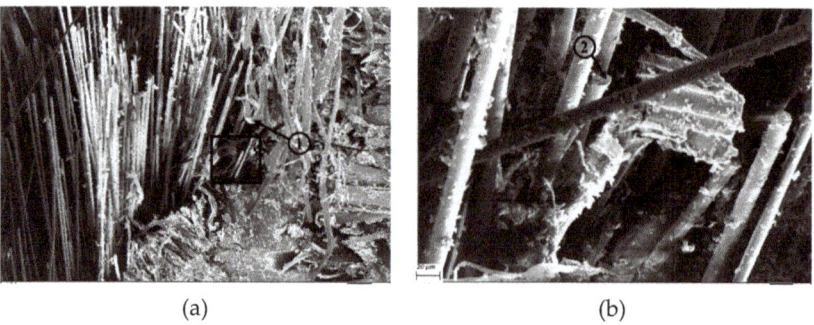

Figure 14. *Cont.*

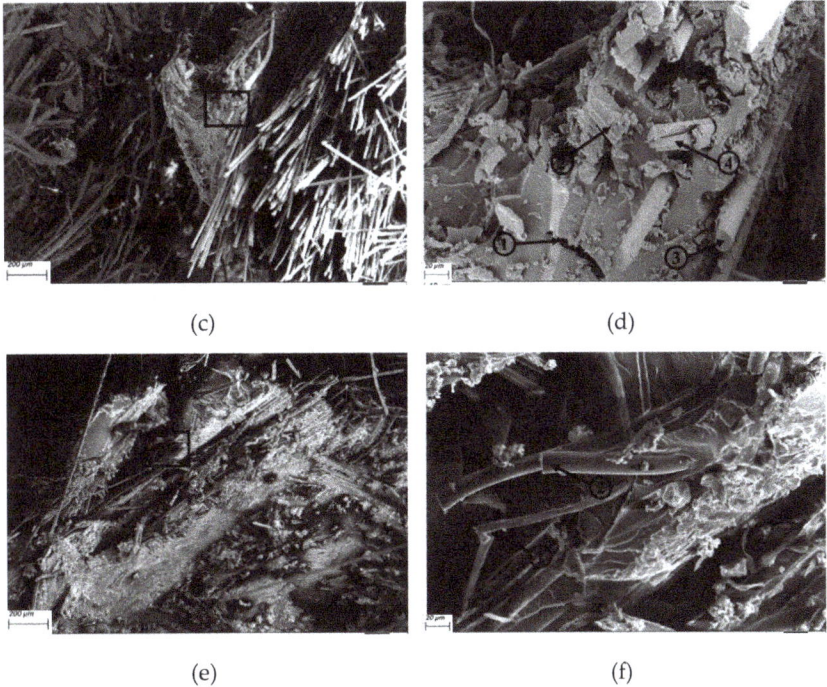

Figure 14. SEM images of flax/VE/basalt hybrid composites: (**a**) 150× magnification with 50 J, (**b**) 1000× magnification with 50 J, (**c**) 150× magnification with 60 J, (**d**) 1000× magnification with 60 J, (**e**) 150× magnification with 70 J, and (**f**) 1000× magnification with 70 J.

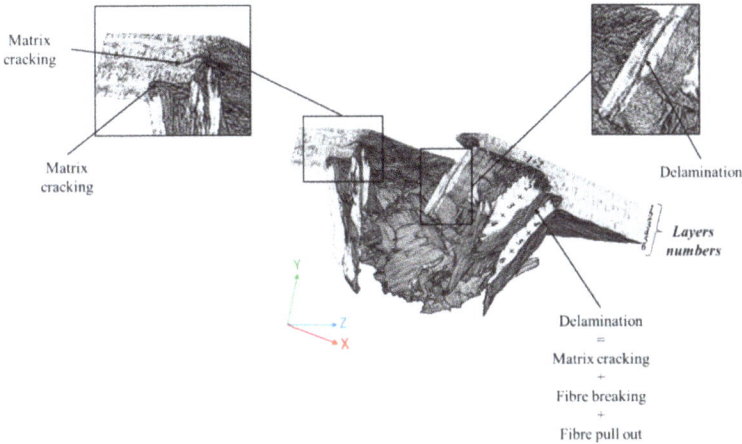

Figure 15. Micro CT scan picture of a flax composite specimen (3/4 of the impact hole) after an impact test, with a 50 J impact energy, showing different failure modes in the specimen.

Figure 16 shows a 3D view of the impacted flax/basalt hybrid sample. Unlike the image illustrated in Figure 15 for flax/VE without hybridisation, the hybrid sample shows less damage and different failure modes, such as delamination.

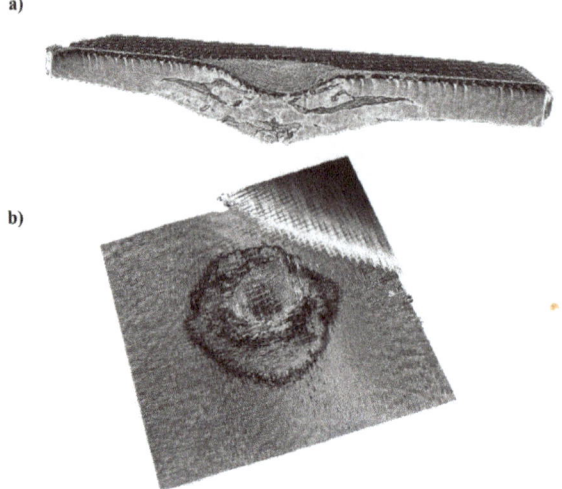

Figure 16. Micro CT scan pictures of a flax/basalt hybrid composite after an impact test, with a 50 J impact energy: (**a**) half-view of the specimen with impact failure, in the form of delamination, and (**b**) top view of the damaged area's shape.

Figure 17 illustrates a 3D half-view of a hybrid composite. It can be clearly observed that the basalt hybridised sample exhibited larger delamination within the top and adjacent layer. In comparison to the flax/VE composite, the flax/VE/basalt hybrid composite does not display penetration. Indeed, the basalt layer has absorbed a higher impact energy.

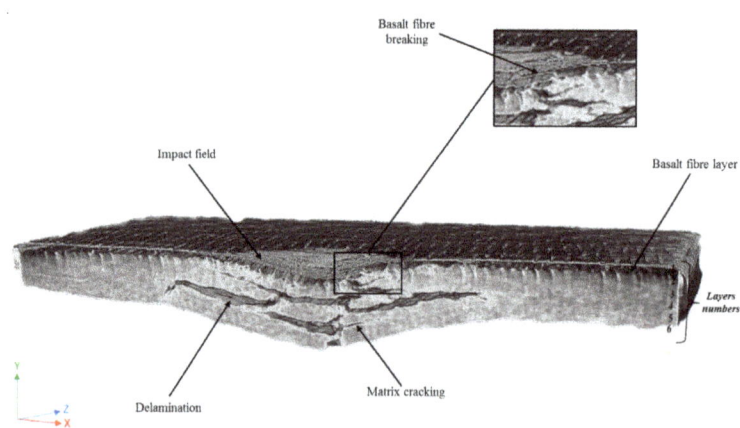

Figure 17. Micro CT scan picture of a flax/basalt hybrid composite specimen (half-view) after an impact test, with a 50 J impact energy, with different failures displayed by the specimen.

4. Conclusions

The effects of basalt fibre hybridisation on the low-velocity falling weight impact behaviour of flax/VE bio-based composites have been investigated following water immersion at room temperature at three incident impact energies: 50, 60, and 70 Joules. The results show that the incident energy has a significant influence on the load bearing capability and total energy absorption characteristics. For a short period of water immersion, it was found that water immersion did not result in a reduction in the impact load and energy absorption. Additionally, the basalt fibre hybridisation at the front side of the laminates significantly enhanced the impact load and total energy of flax/VE composites, showing the potential of flax/VE bio-based composites for semi-structural or structural applications. The damage mechanisms following the X-ray micro CT examination and SEM characterisations performed reveal the greater resistance to penetration and perforation by basalt hybrid composites, which is an indication of their higher impact performance, offering balanced properties of environmental benefits and enhanced impact behaviour. The damage modes for flax/VE composites were matrix cracking, fibre breakage, and fibre pull out. Comparatively, for the flax/basalt hybrid composites, the predominant failure mode was matrix cracking and delamination.

Author Contributions: H.N.D. contributed in design, conceptualization, overall leadership, project supervision and writing. E.L.M. and M.F. conducted the experimental work, analysis of results and writing of the preliminary report. C.J. contributed in conducting experimental works and data analysis. Z.Z. contributed in experimental design and analysis. All authors have read and agreed to the published version of the manuscript.

Funding: This research received no external funding.

Conflicts of Interest: The authors declare no conflict of interest.

References

1. Slaper, T.F.; Hall, T.J. The Triple Bottom Line: What Is It and How Does It Work? *Indiana Bus. Rev.* **2011**, *86*, 4–8.
2. Commission Directive (EU). *2017/2096 of 15 November 2017 Amending Annex II to Directive 2000/53/EC of the European Parliament and of the Council on End-of Life Vehicles (Text with EEA Relevance)*; EU: Brussels, Belgium, 2017.
3. Faruk, O.; Bledzki, A.K.; Fink, H.P.; Sain, M. Biocomposites reinforced with natural fibres:2000-2010. *Prog. Polym. Sci.* **2012**, *37*, 1552–1596. [CrossRef]
4. Bledzki, A.; Gassan, J. Composites reinforced with cellulose based fibres. *Prog. Polym. Sci.* **1999**, *24*, 221–274. [CrossRef]
5. Dhakal, H.; Zhang, Z.; Richardson, M. Effect of water absorption on the mechanical properties of hemp fibre reinforced unsaturated polyester composites. *Compos. Sci. Technol.* **2007**, *67*, 1674–1683. [CrossRef]
6. Pickering, K.; Efendy, M.A.; Le, T. A review of recent developments in natural fibre composites and their mechanical performance. *Compos. Part A Appl. Sci. Manuf.* **2016**, *83*, 98–112. [CrossRef]
7. Bourmaud, A.; Morvan, C.; Bouali, A.; Placet, V.; Perré, P.; Baley, C. Relationships between micro-fibrillar angle, mechanical properties and biochemical composition of flax fibers. *Ind. Crop. Prod.* **2013**, *44*, 343–351. [CrossRef]
8. Dhakal, H.; Skrifvars, M.; Adekunle, K.; Zhang, Z. Falling weight impact response of jute/methacrylated soybean oil bio-composites under low velocity impact loading. *Compos. Sci. Technol.* **2014**, *92*, 134–141. [CrossRef]
9. De Rosa, I.; Dhakal, H.; Santulli, C.; Sarasini, F.; Zhang, Z. Post-impact static and cyclic flexural characterisation of hemp fibre reinforced laminates. *Compos. Part B Eng.* **2012**, *43*, 1382–1396. [CrossRef]
10. Go, S.H.; Lee, M.-S.; Hong, C.-G.; Campesi, I.; Kim, H.G. Correlation between Drop Impact Energy and Residual Compressive Strength According to the Lamination of CFRP with EVA Sheets. *Polymers* **2020**, *12*, 224. [CrossRef]
11. Pil, L.; Bensadoun, F.; Pariset, J.; Verpoest, I. Why are designers fascinated by flax and hemp fibre composites? *Compos. Part A Appl. Sci. Manuf.* **2016**, *83*, 193–205. [CrossRef]
12. Duc, F.; Bourban, P.; Plummer, C.; Månson, J.-A. Damping of thermoset and thermoplastic flax fibre composites. *Compos. Part A Appl. Sci. Manuf.* **2014**, *64*, 115–123. [CrossRef]
13. Sombatsompop, N.; Chaochanchaikul, K. Effect of moisture content on mechanical properties, thermal and structural stability and extrudate texture of poly(vinyl chloride)/wood sawdust composites. *Polym. Int.* **2004**, *53*, 1210–1218. [CrossRef]
14. Dhakal, H.; Zhang, Z. *The Use of Hemp Fibres as Reinforcements in Composites*; Elsevier BV: Amsterdam, The Netherlands, 2015; pp. 86–103.

15. Mitrevski, T.; Marshall, I.; Thomson, R. The influence of impactor shape on the damage to composite laminates. *Compos. Struct.* **2006**, *76*, 116–122. [CrossRef]
16. Cantwell, W.; Morton, J. The impact resistance of composite materials—A review. *Composites* **1991**, *22*, 347–362. [CrossRef]
17. Wisheart, M.; Richardson, M.O.W. Low velocity response of a complex geometry pultruded glass/polyester composite. *J. Mater. Sci.* **1999**, *34*, 1107–1116. [CrossRef]
18. Huda, M.; Drzal, L.; Ray, D.; Mohanty, A.K.; Mishra, M. Natural-fiber composites in the automotive sector. In *Properties and Performance of Natural-Fibre Composites*; Woodhead Publishing; Elsevier: Amsterdam, The Netherlands, 2008; pp. 221–268.
19. Davoodi, M.; Sapuan, S.; Ahmad, D.; Ali, A.; Khalina, A.; Jonoobi, M.; Sapuan, S.M. Mechanical properties of hybrid kenaf/glass reinforced epoxy composite for passenger car bumper beam. *Mater. Des.* **2010**, *31*, 4927–4932. [CrossRef]
20. Bachmann, J.; Wiedemann, M.; Wierach, P. Flexural Mechanical Properties of Hybrid Epoxy Composites Reinforced with Nonwoven Made of Flax Fibres and Recycled Carbon Fibres. *Aerospace* **2018**, *5*, 107. [CrossRef]
21. Flynn, J.; Amiri, A.; Ulven, C. Hybridized carbon and flax fiber composites for tailored performance. *Mater. Des.* **2016**, *102*, 21–29. [CrossRef]
22. Fiore, V.; Scalici, T.; Calabrese, L.; Valenza, A.; Proverbio, E. Effect of external basalt layers on durability behaviour of flax reinforced composites. *Compos. Part B Eng.* **2016**, *84*, 258–265. [CrossRef]
23. Fiore, V.; Scalici, T.; Badagliacco, D.; Enea, D.; Alaimo, G.; Valenza, A. Aging resistance of bio-epoxy jute-basalt hybrid composites as novel multilayer structures for cladding. *Compos. Struct.* **2017**, *160*, 1319–1328. [CrossRef]
24. Dhakal, H.; Sarasini, F.; Santulli, C.; Tirillò, J.; Zhang, Z.; Arumugam, V. Effect of basalt fibre hybridisation on post-impact mechanical behaviour of hemp fibre reinforced composites. *Compos. Part A Appl. Sci. Manuf.* **2015**, *75*, 54–67. [CrossRef]
25. Dhakal, H.; Sain, M. Enhancement of Mechanical Properties of Flax-Epoxy Composite with Carbon Fibre Hybridisation for Lightweight Applications. *Materials* **2019**, *13*, 109. [CrossRef]
26. Paturel, A.; Dhakal, H. Influence of Water Absorption on the Low Velocity Falling Weight Impact Damage Behaviour of Flax/Glass Reinforced Vinyl Ester Hybrid Composites. *Molecus* **2020**, *25*, 278. [CrossRef]
27. George, M.; Chae, M.; Bressler, D.C. Composite materials with bast fibres: Structural, technical, and environmental properties. *Prog. Mater. Sci.* **2016**, *83*, 1–23. [CrossRef]
28. Dhand, V.; Mittal, G.; Rhee, K.Y.; Park, S.J.; Hui, D. A short review on basalt fiber reinforced polymer composites. *Compos. Part B* **2015**, *73*, 166–180. [CrossRef]
29. Kozlowski, R.; Władyka-Przybylak, M. Flammability and fire resistance of composites reinforced by natural fibers. *Polym. Adv. Technol.* **2008**, *19*, 446–453. [CrossRef]
30. Yin, X.; Liu, Y.; Miao, Y.; Xian, G. Water Absorption, Hydrothermal Expansion, and Thermomechanical Properties of a Vinylester Resin for Fiber-Reinforced Polymer Composites Subjected to Water or Alkaline Solution Immersion. *Polymers* **2019**, *11*, 505. [CrossRef]
31. BS EN ISO. *Plastics–Determination of Water Absorption*; British Standard: London, UK, 1999; Volume 62, pp. 1–8.
32. BS EN ISO 6603-2. *Determination of Puncture Impact Behaviour of Rigid Plastics. British Standard, Part 2: Instrumented Puncture Testing*; British Standard: London, UK, 2000.
33. Almansour, F.; Dhakal, H.; Zhang, Z. Effect of water absorption on Mode I interlaminar fracture toughness of flax/basalt reinforced vinyl ester hybrid composites. *Compos. Struct.* **2017**, *168*, 813–825. [CrossRef]
34. Dhakal, H.; Zhang, Z.; Guthrie, R.; MacMullen, J.; Bennett, N. Development of flax/carbon fibre hybrid composites for enhanced properties. *Carbohydr. Polym.* **2013**, *96*, 1–8. [CrossRef]
35. Jefferson, A.J.; Srinivasan, S.M.; Arokiarajan, A.; Dhakal, H.N. Parameters influencing the impact response of fibre-reinforced polymer matrix materials: A critical review. *Compos. Struct.* **2019**, *224*, 111007.
36. Almansour, F.; Dhakal, H.; Zhang, Z.; Ghasemnejad, H. Effect of hybridization on the mode II fracture toughness properties of flax/vinyl ester composites. *Polym. Compos.* **2015**, *38*, 1732–1740. [CrossRef]

© 2020 by the authors. Licensee MDPI, Basel, Switzerland. This article is an open access article distributed under the terms and conditions of the Creative Commons Attribution (CC BY) license (http://creativecommons.org/licenses/by/4.0/).

Article

Effect of Carbon Nanostructures and Fatty Acid Treatment on the Mechanical and Thermal Performances of Flax/Polypropylene Composites

Pietro Russo [1], Libera Vitiello [1], Francesca Sbardella [2,*], Jose I. Santos [3], Jacopo Tirillò [2], Maria Paola Bracciale [2], Iván Rivilla [2,4] and Fabrizio Sarasini [2,*]

1. Institute for Polymers, Composites and Biomaterials, National Council of Research, 80078 Pozzuoli, Italy; pietro.russo@ipcb.cnr.it (P.R.); liberavitiello29@gmail.com (L.V.)
2. Department of Chemical Engineering Materials Environment & UdR INSTM, Sapienza-Università di Roma, 00184 Roma, Italy; jacopo.tirillo@uniroma1.it (J.T.); mariapaola.bracciale@uniroma1.it (M.P.B.); ivan.rivilla@ehu.es (I.R.)
3. SGIker-UPV/EHU, Centro "Joxe Mari Korta", Tolosa Hiribidea, 72, 20018 Donostia−San Sebastián, Spain; joseignacio.santosg@ehu.es
4. Donostia International Physics Center, 20018 Donostia-San Sebastián, Spain
* Correspondence: francesca.sbardella@uniroma1.it (F.S.); fabrizio.sarasini@uniroma1.it (F.S.); Tel.: +39-0644585314 (F.S.); +39-0644585408 (F.S.)

Received: 31 January 2020; Accepted: 10 February 2020; Published: 13 February 2020

Abstract: Four different strategies for mitigating the highly hydrophilic nature of flax fibers were investigated with a view to increase their compatibility with apolar polypropylene. The effects of two carbon nanostructures (graphene nanoplatelets (GNPs) and carbon nanotubes (CNTs)), of a chemical modification with a fatty acid (stearic acid), and of maleated polypropylene on interfacial adhesion, mechanical properties (tensile and flexural), and thermal stability (TGA) were compared. The best performance was achieved by a synergistic combination of GNPs and maleated polypropylene, which resulted in an increase in tensile strength and modulus of 42.46% and 54.96%, respectively, compared to baseline composites. Stearation proved to be an effective strategy for increasing the compatibility with apolar matrices when performed in an ethanol solution with a 0.4 M concentration. The results demonstrate that an adequate selection of surface modification strategies leads to considerable enhancements in targeted properties.

Keywords: polymer matrix composites; flax fibers; surface treatments; adhesion

1. Introduction

Natural-fiber-reinforced composites have received attention over the recent years because of their potential ability to replace their synthetic counterparts in an attempt to meet the new regulations that promote the use of more sustainable and recyclable materials [1,2]. The high specific mechanical properties and the carbon dioxide neutrality of natural fibers have already stimulated the replacement of glass fibers in several sectors, especially the automotive and construction ones, but usually as secondary load-bearing structures [3,4].

A step forward is their use in structural applications, but some challenges still need to be properly faced and solved [5]. The variability in physical and mechanical properties, due to their natural origin, is difficult to manage unless the fiber supply chain is carefully controlled and the manufacturing processes are optimized [6]. Goudenhooft et al. [7] recently showed that tensile properties of flax fibers are not significantly affected over time, regardless of the fiber yield and variety, and that the resulting dispersion in the specific mechanical properties is in the same range as that of glass fibers. Another significant issue is related to the processing conditions of the composites (temperature, dwell time, pressure), which have a major impact on the final mechanical properties, especially for thermoplastic-based composites [8–10].

The last topic of considerable interest is the extent of fiber/matrix interfacial adhesion. It is well known that the mechanical properties of composites are dictated not only by the inherent properties of the constituents, but also by the fiber/matrix interface. The poor compatibility with polymer matrices (especially thermoplastics) due to their hydrophilic behavior still represents a major limitation for a wider industrial exploitation of natural fibers [11]. Several efforts to enhance the interfacial adhesion of natural fibers have been proposed, including chemical [12–16] and physical treatments [17–20], but their industrial implementations are often complicated by the large amounts of chemicals involved or the multiple processing steps required. A more recent approach deals with the grafting of nanostructures onto fiber surfaces to increase the adhesion with the polymer matrix. This strategy has been widely exploited for synthetic fibers, such as glass [21–23] and carbon fibers [24–26], but has attracted less attention in the field of natural fibers. Wang et al. [27] modified the surfaces of flax fibers by grafting TiO_2 nanoparticles using a silane coupling agent. The authors reported an increase in tensile strength and interfacial strength with an epoxy matrix of 23.1% and 40.5%, respectively. Copper nanoparticles on flax fibers were found to produce significant improvements in fiber tensile modulus and strength, equal to 50% and 75%, respectively [28]. Ajith et al. [29] modified flax yarns with hydrous zirconia nanoparticles synthesized by hydrolysis of a zirconium oxychloride solution. The presence of these nanoparticles resulted in an increase in single fiber tensile strength and interfacial strength with an epoxy matrix of 85% and 65%, respectively. Lakshmanan and Chakraborty [30] synthesized and deposited silver nanoparticles on jute fibers without deteriorating the mechanical properties of the fibers. In addition, the modified fabrics exhibited good antibacterial properties. In [31], the authors reported a simple spray-coating process to deposit carbon nanotubes (CNTs) over the surfaces of ramie fibers. This coating enhanced the flexural strength and modulus of an epoxy-based composite by 38.4% and 36.8%, respectively, while a microdebonding test highlighted an increase in the interfacial shear strength of 25.7%. Sarker et al. [32] coated graphene materials, i.e., graphene oxide (GO) and graphene flakes (G), on alkali-treated jute fibers, and an interfacial shear strength enhancement of ~236% compared to untreated fibers was achieved. In [33], the authors coupled a jute fiber individualization procedure with the grafting of GO and subsequent hot pressing to get preforms that were then vacuum-infused with epoxy matrix. The graphene coating resulted in a dramatic increase in tensile modulus and strength of the jute–epoxy composites compared to untreated composites of 324% and 110%, respectively. Grafting of nanometer-sized materials can therefore be considered as an effective method for improving fiber/matrix interfacial adhesion, thus leading to the manufacturing of high-performance natural-fiber-reinforced composites. Another positive feature of this strategy is the possibility of adding functionalities to the resulting composites. Zhuang et al. [34] deposited multi-walled carbon nanotubes (MWCNTs) on the surfaces of jute fibers, and the epoxy-based composites exhibited multifunctional sensing abilities for temperature, moisture, and strain. In [35], graphene nanoplatelets (GNPs) and carbon black were used to make flax yarns electrically conductive; these were then used to fabricate stretchable strain sensors with gauge factors ranging from 1.46 up to 5.62, and a reliability for sensing strains of up to 60%.

The need to optimize the interfacial adhesion in natural-fiber composites is even more important with thermoplastic-based composites due to the non-polar nature of most of them. In particular, polypropylene (PP) is one of the most widely used polyolefins. Its low density, low price, good mechanical properties, good processability, and recyclability make it a popular material as a matrix for natural fiber composites [10,36,37]. Flax fibers currently account for about half of the natural fibers used in automotive applications, followed by kenaf and hemp [4], and the combination of PP/flax has been widely investigated in literature, highlighting the dramatic incompatibility between these two constituents. In an attempt to tailor the properties of natural fibers for their subsequent successful application in high-performance plant-fiber composites but with limited costs and environmental impact, in this work, we investigated the interfacial interactions in flax/PP composites through two different approaches: (i) The grafting of carbon nanostructures (CNTs and GNPs) and (ii) the chemical modification with a fatty acid (stearic acid). In both cases, the addition of a maleic-anhydride-modified polypropylene (MAPP) was also used to tune the interfacial adhesion. In particular, stearic acid, a long alkyl chain fatty acid, was used to lower the hydrophilic character of flax fibers. This surface modification treatment has already been used with limited success in other studies [38–41], even though a detailed investigation on the effects of its concentration on the surface properties of flax fibers has not been reported so far. In addition, grafting of nanostructures for improving interfacial adhesion has mostly been exploited for thermoset-based composites, and scarcely with thermoplastic polymers [42]. The morphology and the thermal stability of flax/PP composites were characterized, and the impacts of the surface treatments or compatibilization with MAPP on their mechanical performance were addressed.

2. Materials and Methods

2.1. Materials

The composites investigated in this study are based on a polypropylene (PP) matrix (Hyosung Topilene PP J640, MFI@230 °C, 2.16 kg: 10 g/10 min) supplied by Songhan Plastic Technology Co. Ltd. (Shanghai City, China) and a commercial 2 × 2 twill flax fabric (areal weight: 200 g/m^2) commercialized without any specific sizing agent and supplied by Composites Evolution (Chesterfield, UK). The matrix was used as received or pre-modified by inclusion of 2 wt.% of a coupling agent, Polybond 3000 (maleic-anhydride-modified PP, MFI@190 °C, 2.16 kg: 400 g/10 min) from Chemtura (Cologne, Germany). The stearic acid (SA), ethanol, and toluene were of analytical grade and used without further purifications. Carbon nanotubes (CNTs) with average length < 1 µm, average outer diameter < 9.5 nm, and bulk density of 100 g/L were provided by Nanocyl SA (Sambreville, Belgium) with the code NC3150. The graphene nanoplatelets (GNPs) supplied by Nanesa s.r.l. (Arezzo, Italy) in the form of black powder have an average flake thickness of 40 nm corresponding to 40 stacked layers, an average particle size of 30 µm, a bulk density of 20–42 g/L, and a specific surface area (BET) > 30 m^2/g.

2.2. Surface Treatment of Flax Fabrics with Carbon Nanostructures

The adopted procedure involved the preparation of an aqueous dispersion of the carbonaceous filler with a concentration equal to 0.5 wt.%. In the case of carbon nanotubes, the dispersion was performed in the presence of 1% by weight of a Triton X-100 surfactant supplied by Sigma Aldrich (Milano, Italy). Layers of flax fabric, already cut to such dimensions as to be used for the preparation of the laminated samples, were immersed for 30 min at room temperature in these dispersions and pre-sonicated for 180 min at room temperature. Finally, the wet fabric layers were subjected to drying for 30 min at 80 °C in a ventilated oven.

2.3. Surface Treatment of Flax Fabrics with Stearic Acid

Different concentrations (0.1, 0.2, 0.3, and 0.4 M) of stearic acid in toluene or ethanol were prepared. These solutions were heated at temperatures close to the boiling points of the solvents, 100 °C for toluene and 65 °C for ethanol, respectively. Once this temperature was reached, the reaction mixture, including the flax fabric, was maintained for 3 h and then washed three times with deionized water and dried at room temperature (Scheme 1). The carboxyl group (–COOH) is supposed to react with the hydroxyl groups of the fiber through an esterification reaction and, hence, the treatment should reduce the number of hydroxyl groups available for bonding with water molecules. Furthermore, the long hydrocarbon chain of stearic acid (18 carbon atoms) provides an extra protection from water due to its hydrophobic nature [43].

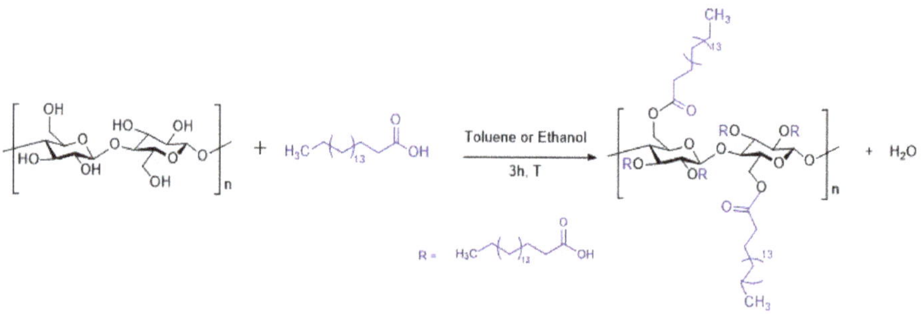

Scheme 1. Synthetic route for the modification of the flax fabric with stearic acid (SA).

2.4. Composite Manufacturing

Composite samples with a symmetrical stacking sequence [0/90] and consisting of 8 plies were obtained by alternately stacking polymer films and layers of as-received or pre-treated flax fabrics, pre-conditioned in a vacuum oven at 70 °C for 2 h, and subsequently underwent hot compression at 210 °C. This last step was carried out using a Collin GmbH (Edersberg, Germany) model P400E press according to a pre-optimized pressure cycle: 2 min—0 bar, 2 min—5 bar, 2 min—15 bar, 1 min—25 bar, 2 min—35 bar, 2 min—40 bar. Finally, the cooling of the composite plates to 30 °C was conducted at a constant pressure of 40 bar before releasing the pressure and extracting the sample. These process conditions provided the laminates with an average thickness of 2.5 mm and with a fiber content of approximately 45% by volume. Films of PP and PPC (PP modified with coupling agent) with a thickness approximately equal to 80 μm were obtained with a Collin Teach-Line E 20-T single-head extruder and Collin CR 72T calender (Ebersberg, Germany), setting a screw speed of 55 rpm and a temperature profile from the hopper to the die equal to: 180–190–200–190–185 °C.

2.5. Characterization Techniques

The mechanical properties of the flax fabrics before and after surface modification treatments were assessed according to ASTM D5035. Tensile tests were carried out at room temperature by means of a Zwick/Roell Z010 (Ulm, Germany) equipped with a 10 kN load cell. A gauge length of 75 mm was used for specimens with a width equal to 25 mm. Tests were performed in displacement control at a crosshead speed of 100 mm/min to ensure failure within 20 ± 3 s. At least five tests were performed for each fabric.

The mechanical properties of the composites were investigated in quasi-static tensile and flexural tests with a Zwick/Roell Z010. For tensile measurements, a gauge length of 60 mm and a cross-head speed of 5 mm/min were set in accordance with ASTM D3039, while flexural tests were conducted in a three-point bending configuration with a span of 76 mm and a speed of loading equal to 5 mm/min as per ASTM D790. Five tests were carried out for each composite formulation.

FT-IR spectra were carried out with a Bruker Vertex 70 spectrometer (Bruker Optik GmbH, Ettlingen, Germany) equipped with a single reflection Diamond ATR (Attenuated Total Reflectance) cell. The ATR-FTIR spectrum was recorded with a 3 cm^{-1} spectral resolution in the mid-infrared range (350–4000 cm^{-1}) using 256 scans.

CPMAS (Cross Polarization/Magic Angle Spinning) NMR spectra were recorded on a 9.4T (400 MHz) Bruker (Billerica, USA) system equipped with a 4 mm MASDVT Double Resonance HX MAS probe. Larmor frequencies were 400.17 MHz and 100.63 MHz for ^1H and ^{13}C nuclei, respectively. Chemical shifts were calibrated indirectly with glycin, with a carbonyl peak at 176 ppm. The sample rotation frequency was 10 kHz and the relaxation delay was 5 s. The number of scans was 4096. Polarization transfer was achieved with RAMP cross-polarization (ramp on the proton channel) with a contact time of 5 ms. High-power SPINAL 64 heteronuclear proton decoupling was applied during acquisition.

X-ray diffraction (XRD) analysis was performed with a diffractometer X'Pert PRO by Philips (Malvern, UK) (CuKα radiation = 1.54060 Å; 40 kV and 40 mA) at room temperature. XRD patterns were collected in the range of 2θ = 10°–80° with a step size of 0.02° scan and a time per step of 3 s.

Surface wettability tests were performed measuring the contact angles of water droplets on the twill fabric surface using an optical analyzer (OCA15Pro, DataPhysics Instruments, Filderstadt, Germany). The static sessile method with a droplet volume of 3 μL was selected and a Milli-Q ultrapure water was used as the testing liquid. A minimum of ten droplets localized on different areas of the flax fabric samples were analyzed. Contact angle values were determined by drop shape analysis using the DataPhysics SCA 20 software module.

A scanning electron microscope, FEG Mira3 (Tescan, Brno, Czech Republic), was used to analyze the morphologies of neat and pre-treated flax fabrics as well as those of fractured surfaces of composite laminates. Prior to observation, the fracture surfaces were sputter-coated with gold.

3. Results and Discussion

3.1. Characterization of Composites Reinforced with Flax Fabrics Decorated with Carbon Nanostructures

In an attempt to reduce the use of chemicals and to make the process easy and industrially scalable, a simple dip-coating method was used to decorate the flax fibers. SEM micrographs were taken to investigate the morphologies of coated and uncoated fabrics (Figures 1–3). Comparing the surfaces of the untreated (Figure 1) and treated flax fabrics with CNTs (Figure 2), it is possible to observe the formation of an interconnected MWCNT network (white arrows in Figure 2c) on the fiber surface, though the dispersion was not uniform with the presence of agglomerates (Figure 2d). Untreated flax fibers (Figure 1) showed an almost smooth and featureless surface, with the presence of some impurities because no pre-treatment was applied.

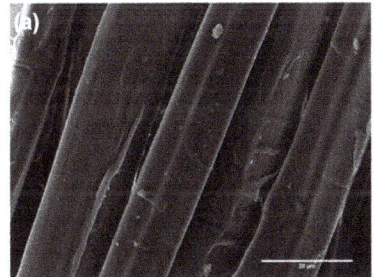 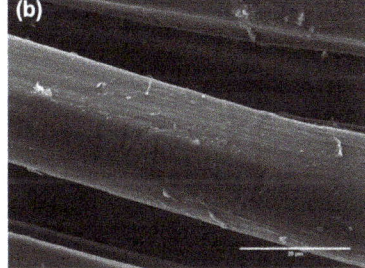

Figure 1. SEM micrographs showing untreated (**a**) flax yarn and (**b**) a close-up view of a single flax fibre.

The same conclusions hold for flax fabrics decorated with GNPs (Figure 3). The distribution is not completely uniform over the fiber surface, and agglomerates can be easily detected (white arrows in Figure 3c,d). These results indicate that the bonding between CNTs, GNPs, and flax fibers may only result from weak van der Waals forces with no covalent bonds. In principle, natural fibers exhibit the unique feature of having a set of hydroxyl groups in cellulose that are reactive and available for potential interactions with host nanostructures, coupled with mechanical interlocking with the rough grooves that characterize the surfaces of natural fibers. This feature was used by Sarker et al. [32] for decorating jute fibers with a uniform layer of graphene oxide (GO) by exploiting the oxygen functional groups of GO and the effects produced by an alkali pre-treatment that removed the cementing layer and exposed the hydroxyl groups of cellulose. When less-reactive graphene flakes were used, these were not fixed on the jute fiber surface. Wang et al. [31] observed a uniform deposition of CNTs on ramie fibers, but also, in this case, the fibers were subjected to an alkali pre-treatment and CNTs were prepared in a solution containing a silane coupling agent and a dispersant (polyvinylpyrrolidone). The results highlighted the positive role played by the silane coupling agent, which was essential in order to avoid agglomeration and to promote the formation of Si–O–C covalent bonds between the silane molecule and the hydroxyl groups of ramie fibers.

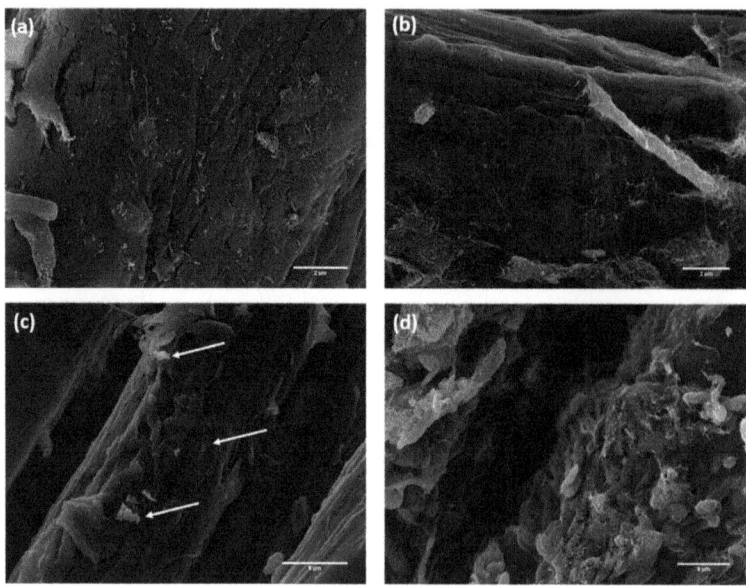

Figure 2. SEM micrographs showing the flax fabrics decorated with carbon nanotubes (CNTs) at different magnifications.

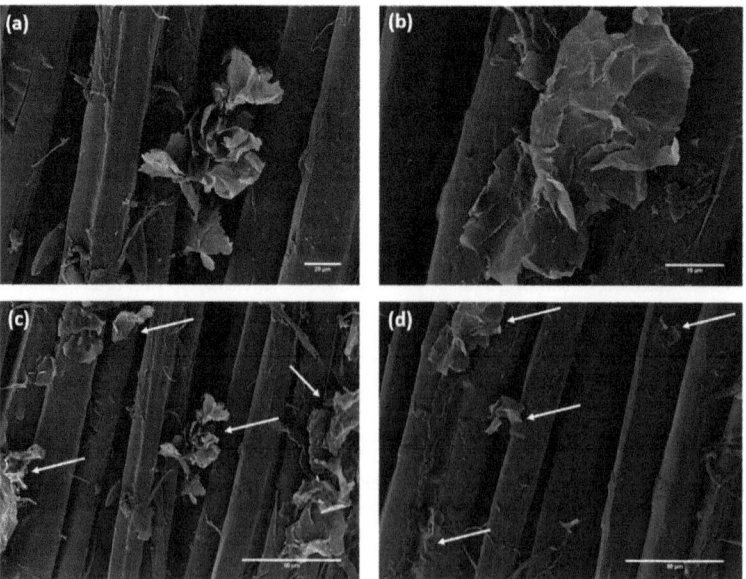

Figure 3. SEM micrographs showing the flax fabrics decorated with graphene nanoplatelets (GNPs) at different magnifications.

In the present work, to keep the number of processing steps and the amounts of chemicals at a minimum, no pre-treatments were used, and this resulted in a certain degree of agglomeration of both carbon nanostructures, which were not functionalized.

The convenient dip-coating method used did not lead to a significant reduction in the mechanical properties of the fabrics if one considers the natural variability in the mechanical responses of natural fibers, as can be inferred from the tensile tests on flax fabrics (Figure 4), thus excluding any degradation effects of the cellulose and of the cell wall materials.

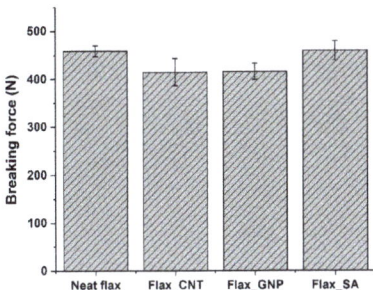

Figure 4. Breaking forces of untreated and surface-modified flax fabrics.

GO and graphene flakes significantly increased both the tensile strength and the Young's modulus of single jute fibers [32]. This better mechanical performance was ascribed by the authors to a kind of healing effect played by these nanostructures, in that they were able to remove stress concentrations on the fiber surfaces due to their homogeneous and uniform coating. A similar increase in tensile strength was reported in [29] on single flax fibers, again attributed to the removal of surface defects, even though, in this case, the distribution of zirconia particles was not sufficiently homogeneous.

Despite the non-optimal distribution of carbon nanostructures on the flax fabrics, the surface-modified fabrics were used for manufacturing composites based on the PP matrix. To tailor the interfacial adhesion, a standard coupling agent (MAPP) was used to explore any synergistic effects. The mechanical properties of tension and bending are summarized in Table 1, along with the percentage of variation in comparison with untreated PP/Flax composites. The incorporation of a polypropylene-grafted maleic anhydride (PPC/Flax) improved both the flexural and tensile properties due to an increased interfacial adhesion between the flax fibers and the PP matrix.

Table 1. Summary of the tensile and flexural properties of the composite materials based on polypropylene (PP) and modified flax fabrics (YM = Young's modulus; TS = Tensile strength; FM = Flexural modulus; FS = Flexural strength; percentage variation is in reference to the corresponding property of PP/Flax).

Specimen ID	YM (GPa)	TS (MPa)	Percentage Variation, YM (%)	Percentage Variation, TS (%)	FM (GPa)	FS (MPa)	Percentage Variation, FM (%)	Percentage Variation, FS (%)
PP/Flax	6.75 ± 0.50	66.09 ± 4.83	-	-	5.42 ± 0.96	49.47 ± 6.77	-	-
PPC/Flax	9.98 ± 0.78	89.46 ± 2.33	+47.85	+35.36	7.80 ± 0.43	94.34 ± 3.68	+43.91	+90.70
PP/Flax_GNP	6.18 ± 0.17	59.54 ± 1.01	−8.44	−9.91	4.29 ± 0.25	51.25 ± 1.32	−20.85	+3.60
PPC/Flax_GNP	10.46 ± 0.29	94.15 ± 1.70	+54.96	+42.46	7.98 ± 0.32	106.36 ± 0.63	+47.23	+114.99
PP/Flax_CNT	5.17 ± 0.17	59.72 ± 3.01	−23.41	−9.64	3.69 ± 0.23	43.23 ± 2.38	−31.92	−12.61
PP/Flax_SA	13.70 ± 0.04	68.33 ± 3.73	+102.96	+3.39	8.77 ± 0.72	56.38 ± 1.11	+61.81	+13.97
PPC/Flax_SA	13.41 ± 1.69	73.28 ± 1.71	+98.67	+10.88	10.88 ± 0.88	87.44 ± 3.31	+100.74	+76.87

The maleic anhydride polar groups create covalent and hydrogen bonds with the flax fiber surfaces, while the polypropylene chains of MAPP form compatible blends with the bulk PP matrix through co-crystallization [37]. The improved interfacial adhesion can be readily confirmed by comparing the fracture surfaces of untreated (Figure 5) and compatibilized (Figure 6) composites.

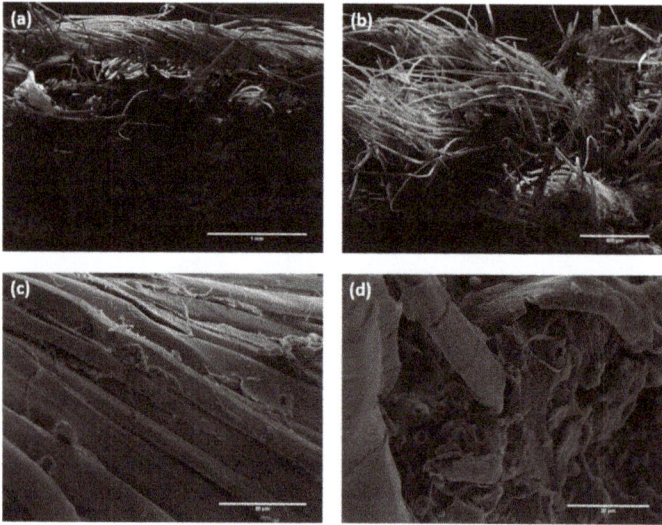

Figure 5. SEM micrographs of the fracture surfaces of PP/Flax composites at different magnifications.

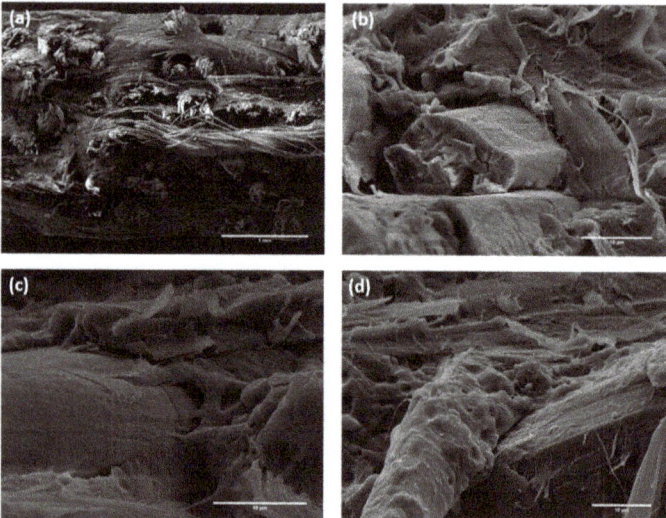

Figure 6. SEM micrographs of the fracture surface of PPC (PP modified with coupling agent)/Flax composites at different magnifications.

In non-compatibilized composites, flax yarns are completely pulled out from the matrix and interfacial debonding turns out to be the dominant failure mechanism, thus suggesting a rather low fiber/matrix adhesion. Clear gaps at the interfaces between the fibers and the PP matrix can be easily observed (Figure 5d), along with the clean surfaces of the flax fibers (Figure 5c) with no matrix residues.

On the contrary, a lower degree of fiber pull-out and no gaps were found between the fibers and the matrix (Figure 6b–d), with a large amount of matrix adhering to the flax surface (Figure 6d) and fiber fractures (Figure 6b).

The presence of CNTs was not beneficial, with resulting composites that exhibited reduced flexural and tensile properties to a great extent compared to untreated composites. This behavior can be ascribed to the agglomerations of CNTs which acted as stress concentrations and to their poor compatibility with the PP matrix [44,45]. Figure 7 shows the fracture surfaces of such composites, where carbon nanotubes are entangled and defects along the flax fibers can be observed (Figure 7d). The fibers are still scarcely covered by the matrix after pull-out (Figure 7b,d), and in the grooves created by the pulled-out fibers (Figure 7c), CNTs clusters can easily be seen, thus confirming the occurrence of weak van der Waals bonds between the CNTs and the flax fibers. Due to the poor mechanical results and fiber/matrix adhesion, composites with CNTs were not investigated further. In fact, it is often suggested that the existence of hydrogen bonds or other dipole–dipole interactions between maleic anhydride and modified nanotubes with carbonyl and carboxyl groups increases CNT dispersion and the properties of PP-based nanocomposites [46,47].

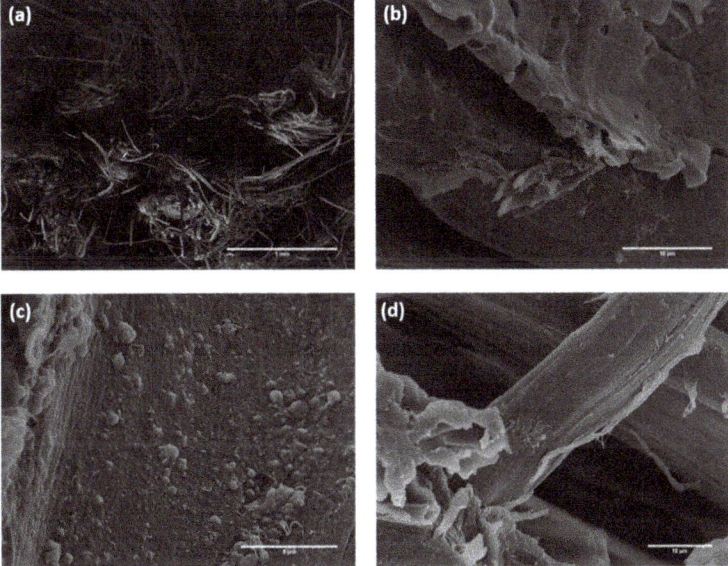

Figure 7. SEM micrographs of the fracture surfaces of PP/Flax_CNT composites at different magnifications.

On the other hand, the decrease in mechanical properties caused by the presence of GNPs was lower compared to CNTs. The fracture surfaces of the resulting composites (Figure 8) showed the same features as those of CNT-reinforced composites, but with a slightly better interfacial compatibility, confirmed by layers of matrix material pulled out together with the flax fibers (Figure 8b,c), where GNPs are well embedded in the polymer matrix (white arrows in Figure 8c). The relative chemical inertness of GNPs [48] did not allow the exploitation of their full potential; therefore, these composites were modified with the MAPP coupling agent.

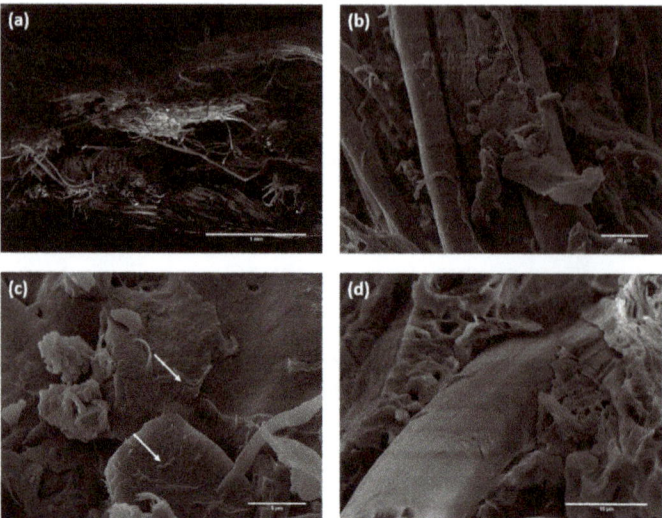

Figure 8. SEM micrographs of the fracture surfaces of PP/Flax_GNP composites at different magnifications.

These composites (PPC/Flax_GNP) outperformed the tensile and flexural behaviors of untreated PP/Flax composites, showing values of tensile and flexural strength even higher than those of PPC/Flax composites. As already found for GNPs in polypropylene matrices [48,49], MAPP is able to increase the chemical compatibility between PP and the polar groups of GNPs, providing a better anchoring of the GNPs into the PP matrix, which results in the enhanced adhesion between them and in a higher constraint of the polymer chains. These effects are visible in the fracture surfaces of the corresponding composites (Figure 9). In this case, it is much more difficult to differentiate the flax fibers from the PP matrix, and the extent of fiber pull-out is significantly reduced. In addition, the fibers are, in many cases, completely covered by the matrix reinforced with GNPs (Figure 9d). It is interesting to note that the tensile and flexural moduli of these composites are higher than those reported in literature for flax/PP composites [10,50,51] when considering similar fiber volume fractions, even higher than low-twisted and unidirectional MAPP-treated flax yarns in a polypropylene matrix, for which a tensile modulus of 9.26 ± 0.4 GPa was reported [52]. In addition, the tensile and flexural strength values are comparable with those of similar unidirectional composites [52]. Figure 10 shows the thermal stability (TGA) curves of the composites, and the results point out that the presence of carbon nanostructures did not markedly affect the degradation profile in comparison with untreated flax/PP composites.

While the first weight loss around 340–360 °C is due to the degradation of flax fibers and, in particular, of the cellulose [53]; the second significant weight loss (380–470 °C) is related to the degradation of PP [44]. The increased decomposition temperatures (inset of Figure 10) of the composites can be ascribed to the physical–chemical absorption of the decomposed products [54]. The physical absorption of PP molecules on the carbon nanostructures induces a delay in their volatilization, but the decomposition temperature of PPC/Flax_GNP composites is the highest among the other formulations. This significant increase cannot be due only to physical absorption, but also to a chemical absorption, thus confirming the higher level of interfacial adhesion.

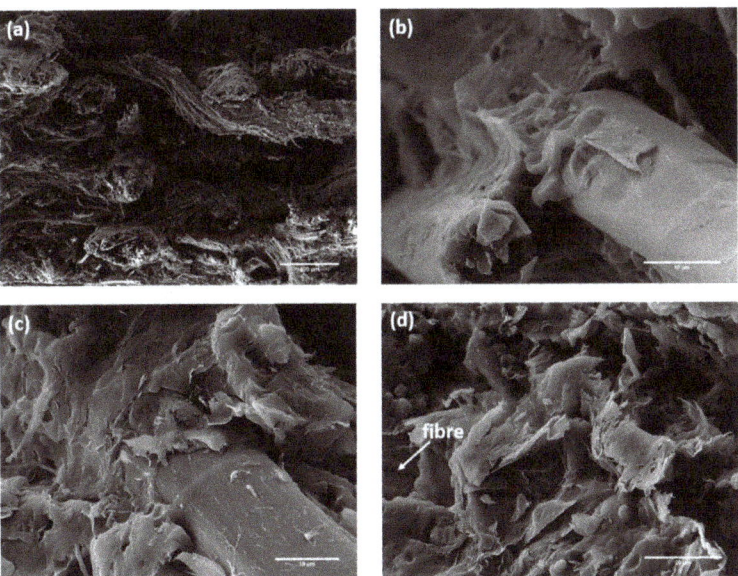

Figure 9. SEM micrographs of the fracture surfaces of PPC/Flax_GNP composites at different magnifications.

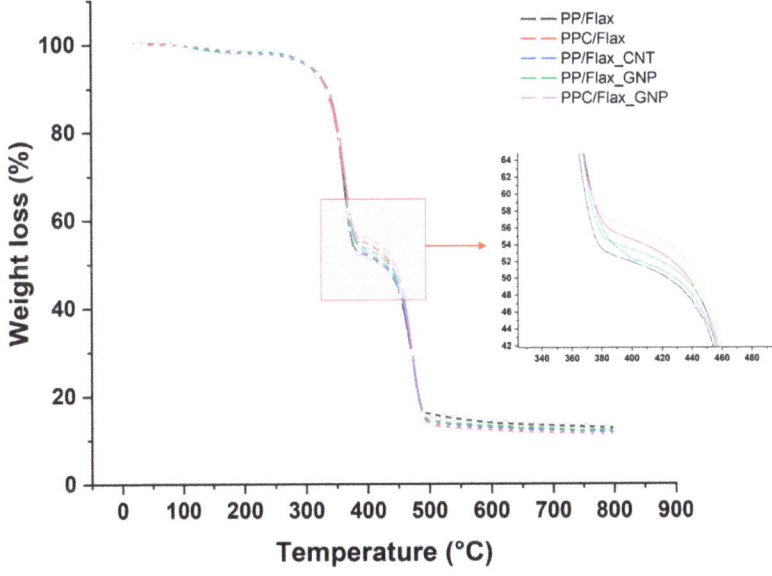

Figure 10. Thermograms for the different PP/flax composites.

3.2. Characterization of Composites Reinforced with Flax Fabrics Treated with Stearic Acid

Organic acids, and in particular fatty acids, are extensively used in surface treatments for particulate mineral fillers. The resulting modification causes the filler surface to become hydrophobic, thus reducing the moisture adsorption during storage and improving the incorporation of polar mineral fillers in non-polar polymer matrix melts, with reduced melt viscosity and associated enhanced dispersion [55]. These commercially available fatty acids are generally sourced from plant or animal sources and contain mixtures of mainly even-carbon-number acids. The use of stearic acid as a surface modification treatment for natural fibers has been investigated in other studies dealing with PP. This treatment is usually performed via a solution process, in which the stearic acid is dissolved in a suitable solvent, via a vapor phase [56], or by dry-blending [39]. In the first case, different solvents have been suggested in literature, including from acetone [38], toluene [57], and ethanol [58]. During the modification, it is expected that the carboxyl group of the stearic acid reacts with the hydroxyl groups of the natural fibers, but the OH groups of the different solvents, characterized by different reactivity, could also be involved. This explains why it was decided to investigate the effects of two different solvents, toluene and ethanol.

At first, all of the treated flax samples were analyzed with the FT-IR and CPMAS NMR spectroscopies to understand the chemical structure. In the infrared spectra (Figure 11), the adsorption bands at about 2916 and 2848 cm^{-1} are attributed to the asymmetric ($\nu_{as}(CH_2)$) and symmetric ($\nu_s(CH_2)$) methylene vibration, while the carbonyl absorption of the carboxylic acid dimer ($\nu C=O$) for stearic acid appeared clearly at 1703 cm^{-1}. This last band at 1703 cm^{-1} is a strong stretching vibrational mode of modified cellulose, which can be attributed to the ester –C=O moieties present. These are formed by esterification between –CO$_2$H in stearic acid and –OH in modified cellulose, indicating that stearic acid undergoes a chemical reaction with cellulose. The main bands between 815 and 1469 cm^{-1} were attributed to the δOH, νC–O, deformation bands of (–CH$_2$–)$_n$, and the out-of-plane vibration bands of O–H of stearic acid dimer [59,60]. The bands at 3024–3650 cm^{-1} correspond to the stretching vibrations of the OH of cellulose, which is the main element of flax created via β1-4 linked D-glucose. The corresponding vibrational bands of C=O and OH are gradually affected as the concentration of SA increases from 0.1 to 0.4 M. This fact indicates the strong intermolecular hydrogen bond interactions between cellulose and SA. The interactions between SA and cellulose were further studied by solid-state ^{13}C CP/MAS NMR (Figure 11c (ethanol) and d (toluene)) at 0.4 M. The spectrum exhibited some characteristic peaks at 181.3 and 181.0 ppm, corresponding to C from the –CO– group to free stearic acid (Figure 11c.1,d.1) and the ester group (Figure 11d.3,c.3), respectively, 104.2, 88.3, 74.4, and 71.5 ppm, corresponding to C1, C4, C5, and (C3, C2), respectively, and the peak for C6 at 64.3 ppm. These peaks could be assigned to the cellulose. In addition, 32.3 (c.3 and d.3) and 32.14 (c.1 and d.1), 24.7 and 14.5 ppm could be assigned to the aliphatic chain to SA [61,62]. The slight shift of the resonance corresponding to the group –CO– with respect to the same group –CO– of the stearic acid, together with the broadening of the signals corresponding to the –CH$_2$– groups of the aliphatic chain of the stearic acid, as well as a slight chemical shift, would indicate the binding or formation of ester groups in the flax fabric after treatment. These data could indicate that, in both cases, using toluene or ethanol as solvents and at two different temperatures, the cellulose that makes up the flax fabric is functionalized with stearic acid through its OH groups.

An assessment of the wettability of the flax fabric surfaces was performed on the flax fabrics treated with stearic acid + ethanol (0.1–0.4 M) and stearic acid + toluene (0.1–0.4 M). In Figure 12, the contact angles for the samples treated with 0.4 M of stearic acid are reported, both in toluene (b) and in ethanol (c). The images were taken from the videos at a fixed time of 10 s after water contact with the fabrics, so that all of the treated fabrics could be compared with each other. The values of the contact angles at different concentrations of stearic acid for the fabrics treated in ethanol and in toluene are reported in Table 2.

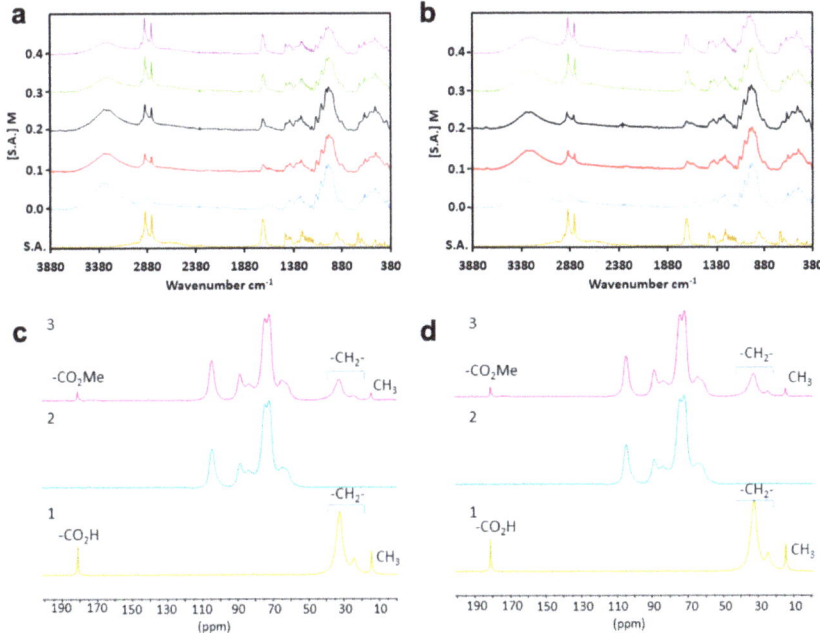

Figure 11. FTIR and NMR spectra for flax modified by stearic acid. (**a**) (Ethanol) and (**b**) (toluene) FTIR spectra ([SA] = 0.1, 0.2, 0.3, and 0.4 M). ^{13}C CP/MAS NMR spectrum of stearic acid **1** (SA), flax fabric (blue) **2**, and, in purple, the flax fabric treated with 0.4 M of SA in ethanol (**c**) and in toluene (**d**).

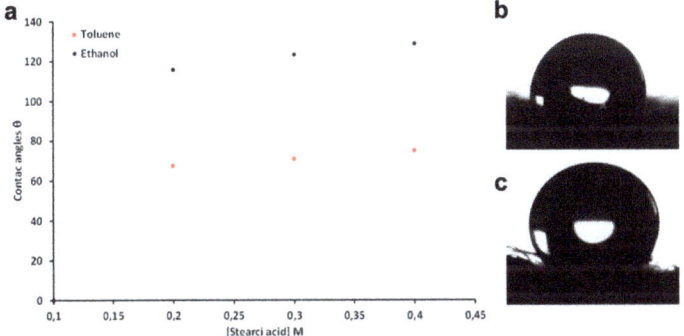

Figure 12. (**a**) Plot of contact angle vs. stearic acid concentration. Picture of a drop of water on flax fabric treated with 0.4 M stearic acid in toluene (**b**) or ethanol (**c**).

Table 2. Contact angles for flax fabrics treated with stearic acid in different solvents.

Stearic Acid [M]	Contact angle, θ [°]	
	Solvent	
	Toluene	Ethanol
0.1	-	-
0.2	67.6 ± 0.05	115.8 ± 0.27
0.3	71 ± 0.32	123.3 ± 0.31
0.4	75.2 ± 0.02	128.8 ± 0.01

The results shown in Table 2 demonstrate that the surface modification carried out in ethanol reaches higher values in terms of contact angle, up to 128.8° relative to a 0.4 M concentration of stearic acid, thus obtaining highly hydrophobic surfaces; likewise, it is evident that the synthesis carried out in toluene as a solvent led to lower contact angle values, leaving the surface of the flax mostly hydrophilic [63]. The differences in hydrophobicity shown in Figure 12 and Table 2 by the flax fabric samples treated in different solvents may be due to different phenomena. The reaction between an acid and an alcohol, which is known as Fischer–Speier esterification [64], produces an ester by refluxing a carboxylic acid and an alcohol in the presence of an acid catalyst. In our case, in none of the reactions was an acid used as a catalyst, but EtOH was able to yield H^+ to the reaction media and then to act as catalyst and reaction solvent at the same time. In fact, the pKa of EtOH is 15.9, while that of toluene is 43. This fact makes EtOH a stronger acid with respect to toluene. In addition, this reaction, which takes place through cationic type intermediates, is most favored in polar media such as EtOH [65]. On the other hand, EtOH itself could give esterification processes, whereby not only SA, but also esters would be formed with the OH groups of the cellulose. This fact could significantly reduce the number of OH groups present in the sample, increasing hydrophobicity due to the loss of OH groups that can interact with water via H-bonds. Finally, the relatively high temperature at which the reaction is carried out with toluene, 100 °C, could cause some degradation of the molecular structure of the cellulose of flax, making it less reactive. The high level of hydrophobicity reached with a treatment performed in ethanol at 0.4 M allowed us to select this treatment condition for the modification of flax fabrics to be used as reinforcement in the PP matrix.

To determine the structure and dispersion of stearic acid on flax fabrics before composite manufacturing, the morphologies of the resulting treated fabrics were investigated by SEM (Figure 13).

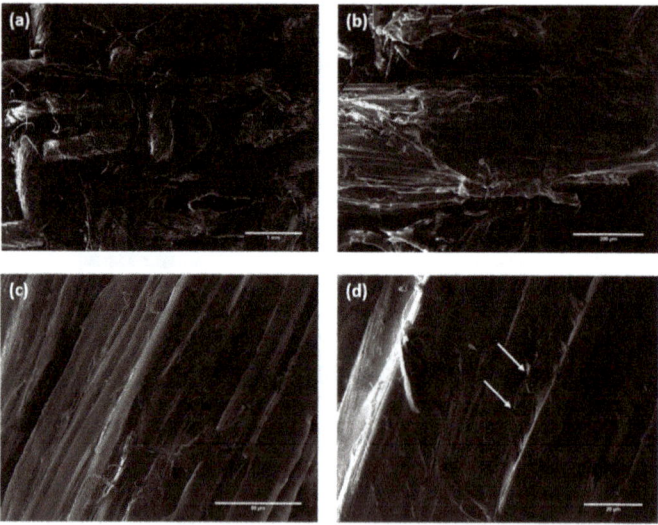

Figure 13. SEM micrographs showing the flax fabrics modified with 0.4 M SA at different magnifications.

The fibers appear to be covered by a thin layer of stearic acid with some micro-sized waxy protrusions (white arrows in Figure 13d), indicating a quite uniform distribution of stearic acid on the flax fiber surface. Figure 14 shows the XRD spectra of pure stearic acid and modified flax fabrics. The main characteristic peaks of the untreated flax fabric were located at 2θ = 14.7°, 16.4°, 22.6°, and 34.5°, which can be assigned to cellulose I [66], for planes $(1\bar{1}0)$, (110), (200), and (004), respectively. In the surface-modified flax fabric, additional peaks located 21.6° and 24.0° can be clearly seen, which correspond to the interplanar spacings of stearic acid, thus suggesting that the stearic acid exists in its crystal form in the modified fabric. These values can be assigned to the stearic acid monoclinic C-form, which is in line with the crystallized form obtained from solution [67].

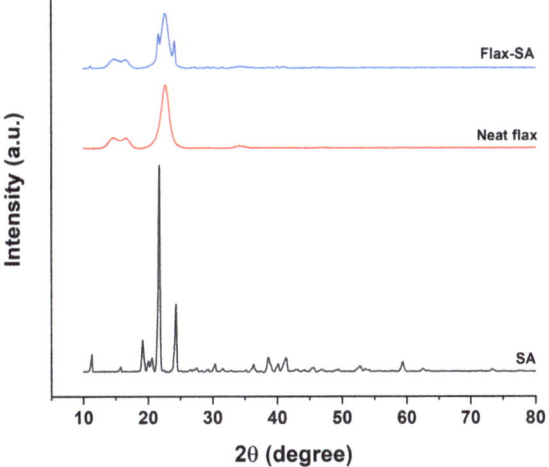

Figure 14. XRD patterns of pure stearic acid (SA), untreated flax fabric (Neat flax), and flax fabric modified with 0.4 M SA (Flax–SA).

The effect of stearic acid treatment on the thermal stability of flax fabrics was investigated by thermogravimetric analysis, and the corresponding thermograms are reported in Figure 15. The thermogram of the untreated flax fabric showed the typical three peaks in the derivative curve. The first mass loss, at about 60–120 °C, is due to the release of water, a shoulder at about 240–280 °C is ascribed to the decomposition of the non-cellulosic components such as pectin and hemicellulose, and the third mass-loss peak, at about 340–360 °C, is due to the cellulose degradation [68]. A one-step mass loss of pure stearic acid was observed, with an onset weight loss temperature higher than 230 °C, which is thus compatible with the manufacturing process of PP-based composites. The thermal stability of the modified flax fabric was not significantly affected, with the exception of an additional decomposition step at a lower temperature (>230 °C) due to the vaporization of stearic acid [69], which again confirms the successful deposition of stearic acid on the flax fabric. In addition, the stearic acid treatment did not degrade the mechanical properties of the modified flax fabric, as can be inferred from the results included in Figure 4.

The reduced polar character of the modified flax fabrics resulted in composites characterized not only by higher moduli compared to composites reinforced with GNPs, but also by lower strength values, which, in any case, are higher than those exhibited by PP/Flax composites (Table 1). It is worth noting the synergistic effect on the modulus and, to a lesser extent, on the strength played by the further use of MAPP, an effect already observed by Spoljaric et al. in microcrystalline cellulose composites [57]. An improved interfacial bond strength can be the reason for this behavior. Zafeiropoulos et al. [70] reported a slight increase in stress transfer efficiency in flax fibers treated with stearic acid in the vapor phase after 36 h treatment with a PP matrix. They ascribed this improvement to the inter-entanglement of the stearic acid chains with the PP chains. The same authors also reported the development of a transcrystalline layer in stearic-acid-treated cellulose fibers [71], which was suggested to increase the interfacial adhesion, as assessed by fiber fragmentation tests.

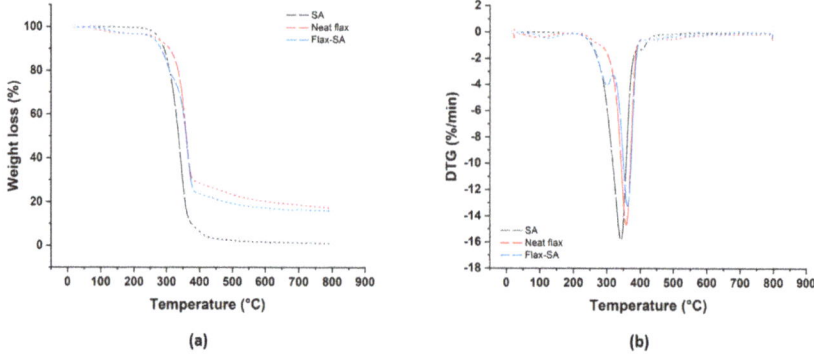

Figure 15. (a) Thermal stability (TGA) thermograms and (b) derivative thermogravimetry (DTG) curves of flax fabrics before and after the treatment with stearic acid (SA).

In the present work, a better fiber/matrix adhesion was induced by the surface treatment between the hydrophobic chains of stearic acid and the polypropylene matrix and between the hydrophilic carboxyl group of the stearic acid and the flax fibers. This was confirmed by SEM analysis of the fracture surfaces of the resulting composites, reported in Figures 16 and 17 for non-compatibilized (PP/Flax_SA) and compatibilized (PPC/Flax_SA) composites with stearic-acid-treated flax fibers, respectively. A strong interfacial adhesion was found in PP/Flax_SA composites, which increased after the addition of MAPP. It is possible to note the presence of a significant number of stearic acid plate-like crystals on the flax fiber surface (for instance, the white arrows in Figure 16c,d). These are supposed to create a rough surface on the flax surface, thus promoting mechanical interlocking and hindrance to polymer chain mobility, which supports the significant increase in stiffness. Flax fibers can be hardly seen in Figure 17a,b, as they are densely covered with matrix residues. The dramatic increases in the moduli were not accompanied by similar increases in strength. It is speculated that the large amount of stearic acid deposited on the fiber surface might have acted, at stresses higher than those needed to evaluate the tensile and flexural moduli, more as a lubricant than as a compatibilizer. Stearic acid is, in fact, commonly used as a processing aid to increase homogeneity and processability [39].

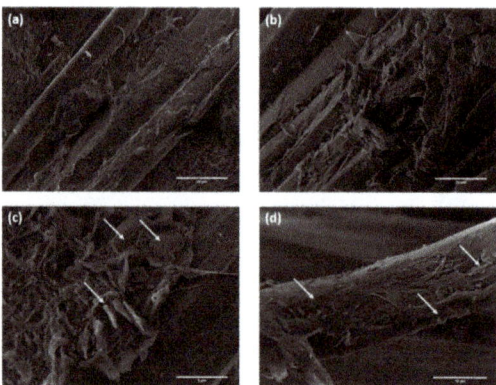

Figure 16. SEM micrographs of the fracture surfaces of PP/Flax_SA composites at different magnifications.

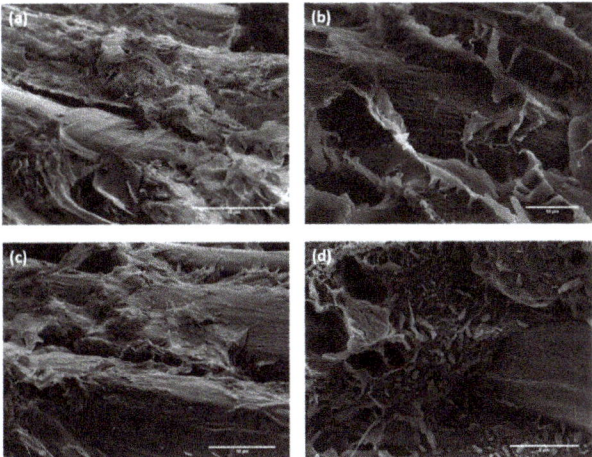

Figure 17. SEM micrographs of the fracture surfaces of PPC/Flax_SA composites at different magnifications.

This consideration suggests not only the potential of stearic acid, but also the need to optimize its content in order to balance these opposing effects.

Compared to untreated PP/Flax composites, the treatment with stearic acid did not modify the overall degradation profile (Figure 18), even though a lower onset temperature of thermal instability occurred in composites with stearic acid. A slight shift toward higher temperatures for the two peak mass-loss temperatures was detected after the incorporation of stearic acid. The mass-loss rate for compatibilized systems was slightly reduced, thus confirming the higher level of interfacial adhesion.

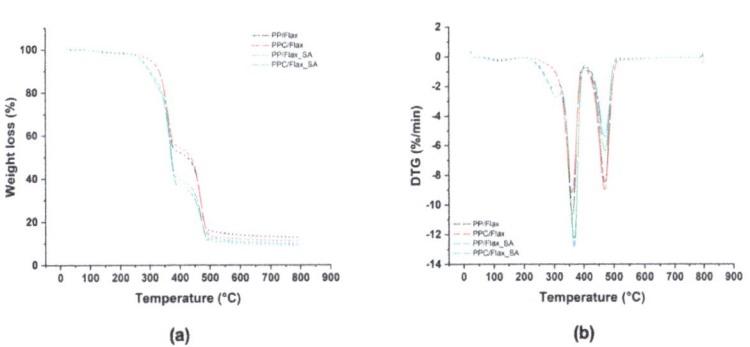

Figure 18. (a) TGA thermograms and (b) derivative thermogravimetry (DTG) curves of stearic-acid-treated flax fibers in a non-compatibilized (PP) or compatibilized (PPC) matrix.

4. Conclusions

Four different surface modification treatments, including grafting of GNPs and CNTs, stearation, and incorporation of maleated polypropylene, were developed and applied on flax fibers to produce high-performance polypropylene-based composites. The grafting of carbon nanostructures by a simple and cost-effective dip-coating process was implemented in order to try to limit the amounts of chemicals and the number of processing steps. This resulted in a non-optimal distribution of carbon nanostructures on the fiber surface. GNPs were found to be much more effective than CNTs, leading to composites with an increased Young's modulus and tensile strength of 54.96% and 42.46% compared to the reference ones when combined with maleated PP. These results are comparable to those obtained for unidirectional PP/Flax composites developed in other studies. The stearation treatment was optimized in terms of the solvent type and the amount of stearic acid. A 0.4 M concentration of stearic acid in ethanol provided the highest reduction in the polarity of flax fibers without altering their degradation profile and mechanical properties. The higher compatibility with apolar PP resulted in enhanced mechanical properties in tension by 102.96% and 3.39% for modulus and strength, respectively, and in bending by 61.81% and 13.97% for modulus and strength, respectively, compared to baseline composites; these were further improved by the addition of maleated polypropylene. These simple treatments, potentially prone to further optimization, can represent a step toward producing natural fiber composites with mechanical profiles compatible with semi-structural applications.

Author Contributions: All authors have read and agreed to the published version of the manuscript. Conceptualization, F.S. (Francesca Sbardella), F.S. (Fabrizio Sarasini), and I.R.; methodology, F.S. (Francesca Sbardella), P.R., and J.T.; validation, F.S. (Francesca Sbardella), F.S. (Fabrizio Sarasini), and I.R.; formal analysis, F.S. (Francesca Sbardella), F.S. (Fabrizio Sarasini), and J.T.; investigation, J.T., I.R., F.S. (Francesca Sbardella), L.V., M.P.B., and J.I.S.; data curation, F.S. (Francesca Sbardella), I.R., and F.S. (Fabrizio Sarasini); writing—original draft preparation, F.S. (Francesca Sbardella), F.S. (Fabrizio Sarasini), and I.R.; writing—review and editing, P.R., J.T., and M.P.B.; visualization, F.S. (Francesca Sbardella), I.R., F.S. (Fabrizio Sarasini); supervision, F.S. (Fabrizio Sarasini), P.R., and J.T. All authors have read and agreed to the published version of the manuscript.

Funding: This research received no external funding.

Conflicts of Interest: The authors declare no conflict of interest.

References

1. Koronis, G.; Silva, A.; Fontul, M. Green composites: A review of adequate materials for automotive applications. *Compos. Part B Eng.* **2013**, *44*, 120–127. [CrossRef]
2. Pickering, K.L.; Aruan Efendy, M.G.; Le, T.M. A review of recent developments in natural fibre composites and their mechanical performance. *Compos. Part A Appl. Sci. Manuf.* **2016**, *83*, 98–112. [CrossRef]
3. Ahmad, F.; Choi, H.S.; Park, M.K. A Review: Natural Fiber Composites Selection in View of Mechanical, Light Weight, and Economic Properties. *Macromol. Mater. Eng.* **2015**, *300*, 10–24. [CrossRef]
4. Akampumuza, O.; Wambua, P.M.; Ahmed, A.; Li, W.; Qin, X. Review of the applications of biocomposites in the automotive industry. *Polym. Compos.* **2017**, *38*, 2553–2569. [CrossRef]
5. Shah, D.U. Developing plant fibre composites for structural applications by optimising composite parameters: A critical review. *J. Mater. Sci.* **2013**, *48*, 6083–6107. [CrossRef]
6. Baley, C.; Gomina, M.; Breard, J.; Bourmaud, A.; Davies, P. Variability of mechanical properties of flax fibres for composite reinforcement. A review. *Ind. Crops Prod.* **2019**, 111984. [CrossRef]
7. Goudenhooft, C.; Bourmaud, A.; Baley, C. Varietal selection of flax over time: Evolution of plant architecture related to influence on the mechanical properties of fibers. *Ind. Crops Prod.* **2017**, *97*, 56–64. [CrossRef]
8. Ausias, G.; Bourmaud, A.; Coroller, G.; Baley, C. Study of the fibre morphology stability in polypropylene-flax composites. *Polym. Degrad. Stab.* **2013**, *98*, 1216–1224. [CrossRef]
9. Lebaupin, Y.; Chauvin, M.; Hoang, T.-Q.T.; Touchard, F.; Beigbeder, A. Influence of constituents and process parameters on mechanical properties of flax fibre-reinforced polyamide 11 composite. *J. Thermoplast. Compos. Mater.* **2017**, *30*, 1503–1521. [CrossRef]
10. Dobah, Y.; Zampetakis, I.; Ward, C.; Scarpa, F. Thermoformability characterisation of Flax reinforced polypropylene composite materials. *Compos. Part B Eng.* **2020**, *184*, 107727. [CrossRef]
11. Bourmaud, A.; Beaugrand, J.; Shaf, D.U.; Placet, V.; Baley, C. Towards the design of high-performance plant fibre composites. *Prog. Mater. Sci.* **2018**, *97*, 347–408. [CrossRef]
12. Bulut, Y.; Aksit, A. A comparative study on chemical treatment of jute fiber: Potassium dichromate, potassium permanganate and sodium perborate trihydrate. *Cellulose* **2013**, *20*, 3155–3164. [CrossRef]
13. Amiri, A.; Ulven, C.; Huo, S. Effect of Chemical Treatment of Flax Fiber and Resin Manipulation on Service Life of Their Composites Using Time-Temperature Superposition. *Polymers* **2015**, *7*, 1965–1978. [CrossRef]
14. Van de Weyenberg, I.; Chi Truong, T.; Vangrimde, B.; Verpoest, I. Improving the properties of UD flax fibre reinforced composites by applying an alkaline fibre treatment. *Compos. Part A Appl. Sci. Manuf.* **2006**, *37*, 1368–1376. [CrossRef]

15. Doan, T.T.L.; Brodowsky, H.; Mäder, E. Jute fibre/epoxy composites: Surface properties and interfacial adhesion. *Compos. Sci. Technol.* **2012**, *72*, 1160–1166. [CrossRef]
16. Hill, C.A.S.; Khalil, H.P.S.A.; Hale, M.D. A study of the potential of acetylation to improve the properties of plant fibres. *Ind. Crops Prod.* **1998**, *8*, 53–63. [CrossRef]
17. Bozaci, E.; Sever, K.; Sarikanat, M.; Seki, Y.; Demir, A.; Ozdogan, E.; Tavman, I. Effects of the atmospheric plasma treatments on surface and mechanical properties of flax fiber and adhesion between fiber-matrix for composite materials. *Compos. Part B Eng.* **2013**, *45*, 565–572. [CrossRef]
18. Yuan, X.; Jayaraman, K.; Bhattacharyya, D. Plasma treatment of sisal fibres and its effects on tensile strength and interfacial bonding. *J. Adhes. Sci. Technol.* **2002**, *16*, 703–727. [CrossRef]
19. Seghini, M.C.; Touchard, F.; Sarasini, F.; Chocinski-Arnault, L.; Tirillò, J.; Bracciale, M.P.; Zvonek, M.; Cech, V. Effects of oxygen and tetravinylsilane plasma treatments on mechanical and interfacial properties of flax yarns in thermoset matrix composites. *Cellulose* **2020**, *27*, 511–530. [CrossRef]
20. Seki, Y.; Sarikanat, M.; Sever, K.; Erden, S.; Gulec, H.A. Effect of the low and radio frequency oxygen plasma treatment of jute fiber on mechanical properties of jute fiber/polyester composite. *Fibers Polym.* **2010**, *11*, 1159–1164. [CrossRef]
21. Chen, J.; Zhao, D.; Jin, X.; Wang, C.; Wang, D.; Ge, H. Modifying glass fibers with graphene oxide: Towards high-performance polymer composites. *Compos. Sci. Technol.* **2014**, *97*, 41–45. [CrossRef]
22. Park, B.; Lee, W.; Lee, E.; Min, S.H.; Kim, B.-S. Highly Tunable Interfacial Adhesion of Glass Fiber by Hybrid Multilayers of Graphene Oxide and Aramid Nanofiber. *ACS Appl. Mater. Interfaces* **2015**, *7*, 3329–3334. [CrossRef]
23. An, Q.; Rider, A.N.; Thostenson, E.T. Hierarchical Composite Structures Prepared by Electrophoretic Deposition of Carbon Nanotubes onto Glass Fibers. *ACS Appl. Mater. Interfaces* **2013**, *5*, 2022–2032. [CrossRef]
24. Yu, B.; Jiang, Z.; Tang, X.Z.; Yue, C.Y.; Yang, J. Enhanced interphase between epoxy matrix and carbon fiber with carbon nanotube-modified silane coating. *Compos. Sci. Technol.* **2014**, *99*, 131–140. [CrossRef]
25. Hung, P.Y.; Lau, K.T.; Fox, B.; Hameed, N.; Lee, J.H.; Hui, D. Surface modification of carbon fibre using graphene–related materials for multifunctional composites. *Compos. Part B Eng.* **2018**, *133*, 240–257. [CrossRef]
26. Zhang, S.; Liu, W.B.; Hao, L.F.; Jiao, W.C.; Yang, F.; Wang, R.G. Preparation of carbon nanotube/carbon fiber hybrid fiber by combining electrophoretic deposition and sizing process for enhancing interfacial strength in carbon fiber composites. *Compos. Sci. Technol.* **2013**, *88*, 120–125. [CrossRef]
27. Wang, H.; Xian, G.; Li, H. Grafting of nano-TiO2 onto flax fibers and the enhancement of the mechanical properties of the flax fiber and flax fiber/epoxy composite. *Compos. Part A Appl. Sci. Manuf.* **2015**, *76*, 172–180. [CrossRef]
28. Sherief, Z.; Xian, G.; Thomas, S.; Ajith, A. Effects of surface grafting of copper nanoparticles on the tensile and bonding properties of flax fibers. *Sci. Eng. Compos. Mater.* **2017**, *24*, 651–660. [CrossRef]
29. Ajith, A.; Xian, G.; Li, H.; Sherief, Z.; Thomas, S. Surface grafting of flax fibres with hydrous zirconia nanoparticles and the effects on the tensile and bonding properties. *J. Compos. Mater.* **2016**, *50*, 627–635. [CrossRef]
30. Lakshmanan, A.; Chakraborty, S. Coating of silver nanoparticles on jute fibre by in situ synthesis. *Cellulose* **2017**, *24*, 1563–1577. [CrossRef]
31. Wang, W.; Xian, G.; Li, H. Surface modification of ramie fibers with silanized CNTs through a simple spray-coating method. *Cellulose* **2019**, *26*, 8165–8178. [CrossRef]
32. Sarker, F.; Karim, N.; Afroj, S.; Koncherry, V.; Novoselov, K.S.; Potluri, P. High-Performance Graphene-Based Natural Fiber Composites. *ACS Appl. Mater. Interfaces* **2018**, *10*, 34502–34512. [CrossRef]
33. Sarker, F.; Potluri, P.; Afroj, S.; Koncherry, V.; Novoselov, K.S.; Karim, N. Ultrahigh Performance of Nanoengineered Graphene-Based Natural Jute Fiber Composites. *ACS Appl. Mater. Interfaces* **2019**, *11*, 21166–21176. [CrossRef]
34. Zhuang, R.C.; Doan, T.T.L.; Liu, J.W.; Zhang, J.; Gao, S.L.; Mäder, E. Multi-functional multi-walled carbon nanotube-jute fibres and composites. *Carbon* **2011**, *49*, 2683–2692. [CrossRef]
35. Souri, H.; Bhattacharyya, D. Wearable strain sensors based on electrically conductive natural fiber yarns. *Mater. Des.* **2018**, *154*, 217–227. [CrossRef]
36. Joly, C.; Gauthier, R.; Chabert, B. Physical chemistry of the interface in polypropylene/cellulosic-fibre composites. *Compos. Sci. Technol.* **1996**, *56*, 761–765. [CrossRef]

37. Sojoudiasli, H.; Heuzey, M.C.; Carreau, P.J. Rheological, morphological and mechanical properties of flax fiber polypropylene composites: Influence of compatibilizers. *Cellulose* **2014**, *21*, 3797–3812. [CrossRef]
38. George, G.; Tomlal Jose, E.; Jayanarayanan, K.; Nagarajan, E.R.; Skrifvars, M.; Joseph, K. Novel bio-commingled composites based on jute/polypropylene yarns: Effect of chemical treatments on the mechanical properties. *Compos. Part A Appl. Sci. Manuf.* **2012**, *43*, 219–230. [CrossRef]
39. Dányádi, L.; Móczó, J.; Pukánszky, B. Effect of various surface modifications of wood flour on the properties of PP/wood composites. *Compos. Part A Appl. Sci. Manuf.* **2010**, *41*, 199–206. [CrossRef]
40. Gonzalez, L.; Lafleur, P.; Lozano, T.; Morales, A.B.; Garcia, R.; Angeles, M.; Rodriguez, F.; Sanchez, S. Mechanical and thermal properties of polypropylene/montmorillonite nanocomposites using stearic acid as both an interface and a clay surface modifier. *Polym. Compos.* **2014**, *35*, 1–9. [CrossRef]
41. Kumar, V.; Dev, A.; Gupta, A.P. Studies of poly(lactic acid) based calcium carbonate nanocomposites. *Compos. Part B Eng.* **2014**, *56*, 184–188. [CrossRef]
42. Xie, L.; Shan, B.; Sun, X.; Tian, Y.; Xie, H.; He, M.; Xiong, Y.; Zheng, Q. Natural Fiber-Anchored Few-Layer Graphene Oxide Nanosheets for Ultrastrong Interfaces in Poly(lactic acid). *ACS Sustain. Chem. Eng.* **2017**, *5*, 3279–3289. [CrossRef]
43. Salem, I.A.S.; Rozyanty, A.R.; Betar, B.O.; Adam, T.; Mohammed, M.; Mohammed, A.M. Study of the effect of surface treatment of kenaf fiber on chemical structure and water absorption of kenaf filled unsaturated polyester composite. *J. Phys. Conf. Ser.* **2017**, *908*, 012001. [CrossRef]
44. Jin, S.H.; Kang, C.H.; Yoon, K.H.; Bang, D.S.; Park, Y.-B. Effect of compatibilizer on morphology, thermal, and rheological properties of polypropylene/functionalized multi-walled carbon nanotubes composite. *J. Appl. Polym. Sci.* **2008**. [CrossRef]
45. Ezat, G.; Kelly, A.; Mitchell, S.; Youseffi, M.; Coates, P.D. Influence of maleic anhydride compatibiliser on properties of polpropylene/multiwalled carbon nanotube composites. *Plast. Rubber Compos.* **2011**, *40*, 438–448. [CrossRef]
46. Ambrogi, V.; Gentile, G.; Ducati, C.; Oliva, M.C.; Carfagna, C. Multiwalled carbon nanotubes functionalized with maleated poly(propylene) by a dry mechano-chemical process. *Polymer* **2012**, *53*, 291–299. [CrossRef]
47. Lee, G.W.; Jagannathan, S.; Chae, H.G.; Minus, M.L.; Kumar, S. Carbon nanotube dispersion and exfoliation in polypropylene and structure and properties of the resulting composites. *Polymer* **2008**, *49*, 1831–1840. [CrossRef]
48. Roh, J.U.; Ma, S.W.; Lee, W., II; Thomas Hahn, H.; Lee, D.W. Electrical and mechanical properties of graphite/maleic anhydride grafted polypropylene nanocomposites. *Compos. Part B Eng.* **2013**, *45*, 1548–1553. [CrossRef]
49. Pedrazzoli, D.; Pegoretti, A.; Kalaitzidou, K. Synergistic effect of graphite nanoplatelets and glass fibers in polypropylene composites. *J. Appl. Polym. Sci.* **2014**. [CrossRef]
50. Kannan, T.G.; Wu, C.M.; Cheng, K.B.; Wang, C.Y. Effect of reinforcement on the mechanical and thermal properties of flax/polypropylene interwoven fabric composites. *J. Ind. Text.* **2013**, *42*, 417–433. [CrossRef]
51. Huang, G.; Liu, L. Research on properties of thermoplastic composites reinforced by flax fabrics. *Mater. Des.* **2008**, *29*, 1075–1079. [CrossRef]
52. Bar, M.; Alagirusamy, R.; Das, A. Properties of flax-polypropylene composites made through hybrid yarn and film stacking methods. *Compos. Struct.* **2018**, *197*, 63–71. [CrossRef]
53. Van de Velde, K.; Baetens, E. Thermal and Mechanical Properties of Flax Fibres as Potential Composite Reinforcement. *Macromol. Mater. Eng.* **2001**, *286*, 342–349. [CrossRef]
54. Yang, J.; Lin, Y.; Wang, J.; Lai, M.; Li, J.; Liu, J.; Tong, X.; Cheng, H. Morphology, thermal stability, and dynamic mechanical properties of atactic polypropylene/carbon nanotube composites. *J. Appl. Polym. Sci.* **2005**, *98*, 1087–1091. [CrossRef]
55. Liauw, C.M. Filler surface modification with organic acids. *Plast. Addit. Compd.* **2000**, *2*, 26–29.
56. Zafeiropoulos, N.E.; Williams, D.R.; Baillie, C.A.; Matthews, F.L. Engineering and characterisation of the interface in flax fibre/polypropylene composite materials. Part, I. Development and investigation of surface treatments. *Compos. Part A Appl. Sci. Manuf.* **2002**, *33*, 1083–1093. [CrossRef]
57. Spoljaric, S.; Genovese, A.; Shanks, R.A. Polypropylene-microcrystalline cellulose composites with enhanced compatibility and properties. *Compos. Part A Appl. Sci. Manuf.* **2009**, *40*, 791–799. [CrossRef]
58. Kiattipanich, N.; Kreua-ongarjnukool, N.; Pongpayoon, T.; Phalakornkule, C. Properties of polypropylene composites reinforced with stearic acid treated sugarcane fiber. *J. Polym. Eng.* **2007**, *27*, 411–428. [CrossRef]

59. Fan, H.; Wang, X.; Wang, Y. Rapid determination on cellulose content of wood by using FTIR spectrometry. *Wood Proc. Mach.* **2014**, *4*, 33–37.
60. Webber, J.; Zorzi, J.E.; Perottoni, C.A.; Moura e Silva, S.; Cruz, R.C.D. Identification of α-Al2O3 surface sites and their role in the adsorption of stearic acid. *J. Mater. Sci.* **2016**, *51*, 5170–5184. [CrossRef]
61. Arata, S.; Kurosu, H.; Kuroki, S.; Ando, I. Structural characterization of stearic acid in the crystalline state by the cross polarization in solid state 13C NMR. *J. Mol. Struct.* **1999**, *513*, 133–138. [CrossRef]
62. Ribeiro, G.A.C.; Silva, D.S.A.; Dos Santos, C.C.; Vieira, A.P.; Bezerra, C.W.B.; Tanaka, A.A.; Santana, S.A.A. Removal of Remazol brilliant violet textile dye by adsorption using rice hulls. *Polimeros* **2017**, *27*, 16–26. [CrossRef]
63. He, M.; Xu, M.; Zhang, L. Controllable Stearic Acid Crystal Induced High Hydrophobicity on Cellulose Film Surface. *ACS Appl. Mater. Interfaces* **2013**, *5*, 585–591. [CrossRef] [PubMed]
64. Fischer, E.; Speier, A. Darstellung der Ester. *Ber. Dtsch. Chem. Ges.* **1895**, *28*, 3252–3258. [CrossRef]
65. Bouguerra Neji, S.; Trabelsi, M.; Frikha, M. Esterification of Fatty Acids with Short-Chain Alcohols over Commercial Acid Clays in a Semi-Continuous Reactor. *Energies* **2009**, *2*, 1107–1117. [CrossRef]
66. Nam, S.; French, A.D.; Condon, B.D.; Concha, M. Segal crystallinity index revisited by the simulation of X-ray diffraction patterns of cotton cellulose Iβ and cellulose II. *Carbohydr. Polym.* **2016**, *135*, 1–9. [CrossRef]
67. Malta, V.; Celotti, G.; Zannetti, R.; Martelli, A.F. Crystal structure of the C form of stearic acid. *J. Chem. Soc. B Phys. Org.* **1971**, 548–553. [CrossRef]
68. Mazian, B.; Bergeret, A.; Benezet, J.C.; Malhautier, L. Influence of field retting duration on the biochemical, microstructural, thermal and mechanical properties of hemp fibres harvested at the beginning of flowering. *Ind. Crops Prod.* **2018**, *116*, 170–181. [CrossRef]
69. Fu, Z.; Dai, L.; Yi, Y.; Luo, J.; Li, B. Structure and thermal properties of stearic acid/silica composites as form-stable phase change materials. *J. Sol-Gel Sci. Technol.* **2018**, *87*, 419–426. [CrossRef]
70. Zafeiropoulos, N.E.; Baillie, C.A.; Hodgkinson, J.M. Engineering and characterisation of the interface in flax fibre/polypropylene composite materials. Part II. The effect of surface treatments on the interface. *Compos. Part A Appl. Sci. Manuf.* **2002**, *33*, 1185–1190. [CrossRef]
71. Zafeiropoulos, N.E.; Baillie, C.A.; Matthews, F.L. Study of transcrystallinity and its effect on the interface in flax fibre reinforced composite materials. *Compos. Part A Appl. Sci. Manuf.* **2001**, *32*, 525–543. [CrossRef]

© 2020 by the authors. Licensee MDPI, Basel, Switzerland. This article is an open access article distributed under the terms and conditions of the Creative Commons Attribution (CC BY) license (http://creativecommons.org/licenses/by/4.0/).

Article

Modeling the Stiffness of Coupled and Uncoupled Recycled Cotton Fibers Reinforced Polypropylene Composites

Albert Serra [1], Quim Tarrés [1,2], Miquel-Àngel Chamorro [3], Jordi Soler [3], Pere Mutjé [1,2], Francesc X. Espinach [4,*] and Fabiola Vilaseca [5]

- [1] LEPAMAP Group, Department of Chemical Engineering, University of Girona, 17003 Girona, Spain; albert.serra@udg.edu (A.S.); joaquimagusti.tarres@udg.edu (Q.T.); pere.mutje@udg.edu (P.M.)
- [2] Càtedra de Processos Industrials Sostenibles, University of Girona, 17003 Girona, Spain
- [3] Department of Architecture and Construction, 17003 Girona, Spain; mangel.chamorro@udg.edu (M.-À.C.); jordi.soler@udg.es (J.S.)
- [4] Design, Development and Product Innovation, Dept. of Organization, Business, University of Girona, 17003 Girona, Spain
- [5] Advanced Biomaterials and Nanotechnology, Dpt. of Chemical Engineering, University of Girona, 17003 Girona, Spain; fabiola.vilaseca@udg.edu
- * Correspondence: francisco.espinach@udg.edu

Received: 3 October 2019; Accepted: 17 October 2019; Published: 21 October 2019

Abstract: The stiffness of a composite material is mainly affected by the nature of its phases and its contents, the dispersion of the reinforcement, as well as the morphology and mean orientation of such reinforcement. In this paper, recovered dyed cotton fibers from textile industry were used as reinforcement for a polypropylene matrix. The specific dye seems to decrease the hydrophilicity of the fibers and to increase its chemical compatibility with the matrix. The results showed a linear evolution of the Young's moduli of the composites against the reinforcement contents, although the slope of the regression line was found to be lower than that for other natural strand reinforced polypropylene composites. This was blamed on a growing difficulty to disperse the reinforcements when its content increased. The micromechanics analysis returned a value for the intrinsic Young's modulus of the cotton fibers that doubled previously published values. The use of two different micromechanics models allowed evaluating the impact of the morphology of the fibers on the Young's modulus of a composite.

Keywords: recycled cotton fibers; stiffness; micromechanics; Young's modulus

1. Introduction

The use of fibrous industrial byproducts as reinforcement for polymer-based composites has increasingly been attracting the attention of researchers. The use of byproducts is in line with the principles of green chemistry and the actual demands of the society for greener materials and more environmentally friendly products [1–3]. The literature shows the opportunity to use agroforestry wastes such as prunings, used paper fibers, or textile byproducts [4–8]. These studies reveal how the nature of the reinforcements has a high impact on the mechanical properties of its composites. In this sense, artificial fibers like glass fibers, aramids, or carbon fibers show the highest strengthening and stiffening abilities [9,10]. Natural fiber strands and wood fibers also show notable capabilities as polyolefin reinforcements. Nonetheless, strands like jute or hemp showed higher stiffening potential than wood fibers [11–15]. In this sense, cotton strands have been used as polyolefin reinforcement successfully [16–18]. While some of the studies have used raw cotton as reinforcing fibers, a vast majority prefer to use recycled fibers from the textile industry [7,13,19,20], however, the number of published studies are still limited.

Cotton fibers recovered from the textile industry have some advantages, such as low cost and availability, but also some apparent drawbacks, since usually these fibers contain textile dyes [7,20]. Additionally, there is a large quantity of discarded textiles that are directly landfilled [21]. Moreover, landfilled textiles contribute to the formation of 'leachate' that can contaminate ground and underground waters [22]. Thus, the use of such textiles as composite reinforcement can contribute to widen the value chain of the textile sector on the one hand, and to decrease landfilling and contamination on the other hand.

Cotton fibers are almost 100% cellulosic fibers, and thus have a high presence of hydroxyl groups in their surface, and a high potential to create hydrogen bonds under favorable conditions. Therefore, they tend to aggregate, making their individualization and dispersion on a polymeric phase difficult [8]. In addition, cotton fibers are hydrophilic, while the vast majority of polymeric matrices are hydrophobic.

A previous study revealed that the presence of dyes diminished the hydrophilicity of cotton fibers, allowing the obtaining of better interphases without using any coupling agent [7,20]. The same dyes eased the dispersion of the fibers without any treatment. Nonetheless, the tensile strength of the composites reinforced with dyed cotton fibers were lower than those obtained with other natural fiber strands. Some authors claim that the same dyes hindered the action of coupling agents [7], although these researchers did not publish any results concerning the stiffness of the composites. According to the literature, the intrinsic Young's modulus of cotton fibers is found between 5 and 13 GPa, however, this value seems too low compared with the values obtained for other strands [11,12,23–28].

This paper examines the Young's modulus of cotton fiber (CF) reinforced polypropylene (PP) composites. CFs were recovered from a yarning process where all the fibers with lengths below 10mm were discarded. The byproduct has the shape of cotton dyed flocks that must be individualized prior to its use as reinforcement [20]. Composite materials adding CF percentages ranging from 20 to 50 wt% were formulated. Two batches of every formulation were prepared, one with 6 wt% of coupling agent added, and the other without. The composites were mold injected to obtain the standard specimens, and later on tested under tensile conditions. The Young's moduli of the materials were evaluated and discussed. The Young's moduli of the composites were not coherent with the intrinsic Young's modulus for CFs found in the literature. Therefore, a micromechanical analysis was proposed to analyze the properties of CF. First, the Hirsch model provided a value for the intrinsic Young's modulus of CF that doubled those on the literature [29]. Then, the efficiency factors allowed discussing a possible poor dispersion of the fibers at high reinforcement contents. Finally, the Tsai and Pagano model in combination with Halpin and Tsai equations [30,31] allowed incorporating the morphology of the reinforcements to back-calculate a theoretical Young's modulus for the composites. The paper actualizes the value of the intrinsic Young's modulus of cotton fibers, and proposes a series of composites that reuse textile byproducts, and thus avoids their landfilling or incineration.

2. Materials and methods

2.1. Materials

The cotton fibers (CF) used as reinforcement were recovered from cotton flock residues. These cotton flocks are textile industry byproducts and are composed of entangled cotton fibers with lengths too short for spinning. The flocks were previously treated with a reactive dye and were kindly supplied by Fontfilva S. L. (Olot, Girona, Spain).

A polypropylene (PP) Isplen PP090 62M by Repsol-YPF (Tarragona, Spain) was kindly supplied by its producer and used as the polymeric matrix. The use of a coupling agent was proposed in order to prevent chemical incompatibilities between the hydrophilic reinforcements and the hydrophobic matrix. Epolene G3015 polypropylene functionalized with maleic anhydride (MAPP) by Eastman

Chemical Products (San Roque, Spain) was purchased for this purpose. This reactive has an acid number of 15 mg KOH/g and a Mn of 24800.

Decalin (decahydronaphthalene) was acquired from Fischer Scientific (Madrid, Spain) and had a 190 °C boiling point and 97% purity. This reagent was used to dissolve the PP matrix in the fiber extraction from composites.

Figure 1 shows the workflow for the research, from the production of cotton flocks by the textile industry to the measurement and evaluation of the mechanical properties.

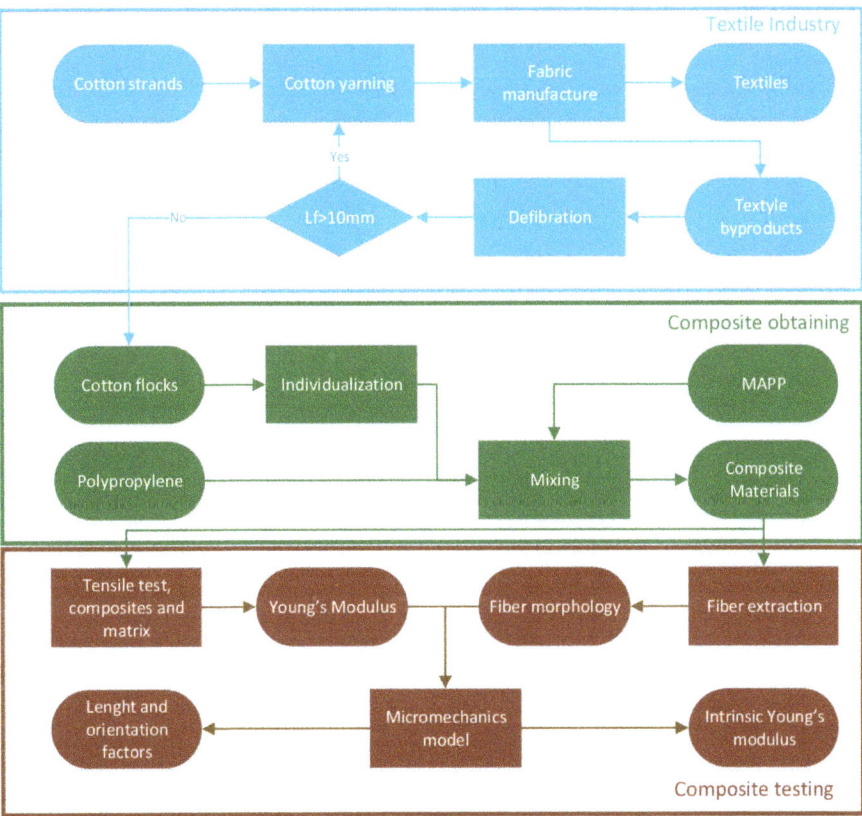

Figure 1. Workflow of the research, including the production of cotton flock byproducts, composite mixing and material testing.

2.2. Cotton Flocks Treatment and Composites Preparation

The cotton residues were passed through a blade mill in order to individualize the entangled fibers. The mill was provided with a 1mm sieve to obtain cotton fibers able to attain a good dispersion. These CFs were mixed with the PP in a Brabender Plastograph kinetic mixer by Brabender® (Duisburg, Germany). The coupled composites added a 6 wt% of MAPP at the same time than the other phases. The process took 10 min, at 185 °C and at a speed of 80 rpm. Coupled and uncoupled composites with CF contents ranging from 20 to 50 wt% were prepared. The obtained blends were cut down to 8 mm pellets able to be mold injected. These pellets were stored for 24 h in an oven at 80 °C to eliminate the humidity.

2.3. Composite and Standard Specimen Preparation

The composite pellets were injection molded in the shape of standard dog bone specimens, in agreement with ASTM D638. The injection molding machine was a Meteor-40 by Mateu & Solé (Barcelona, Spain). The machine has three heating areas that were operated at 175 °C, 175 °C, and 190 °C, corresponding to the highest to the nozzle. The injection pressure was 120 kg/cm^{-2} and the maintaining pressure was 37.5 kg/cm^{-2}. A steel mold with a cavity in the shape of the standard specimen was used, and at least ten specimens for every one of the composite formulations were obtained.

2.4. Mechanical Test

Prior to any mechanical test, the specimens were stored in a conditioning chamber by Dycometal. The stabilization of the specimens took 48 h, and the conditions were at 23 °C with 50% relative humidity.

The specimens were placed in the gauges of an Instron TM 1122 universal testing machine. The machine was fitted with a 5 Kn load cell. The test was performed in agreement with ASTM D790 standard. An extensometer MFA2 was used to measure the strains with addequate precision. The measurements were the result of testing at least 5 samples for every composite formulation.

2.5. Morphologic Analysis of the Reinforcements

Some micromechanics models use the morphology of the reinforcements as input. As soon as the literature accepts that the morphology of such reinforcements changes noticeably during composite preparation, the study was performed to reinforcements extracted from the polymeric matrix. The extraction was obtained by matrix solubilization using a Soxhlet apparatus and using Decalin as a solvent. Composite material pieces approximately 10 × 10mm were placed inside a cellulose filter into the Soxhlet equipment. The process lasted 24 h until the matrix was totally dissolved. Then, the fibers were rinsed with acetone and distilled water.

The morphology of the fibers was measured in a FS-300 Kajanni analyzer. The equipment measured from 2500 to 3000 fibers and returned a fiber length distribution, mean length, and diameter and the percentage of fines (fibers shorter than 70 µm).

3. Results and Discussion

3.1. Young's Modulus of the Composites

Table 1 shows the Young's moduli of the coupled and uncoupled composites (E_t^C) reinforced with CF contents ranging from 20 to 50 wt%. The table also shows the tensile strength of the composites (σ_t^C), the percentage of reinforcement in weight (W^F), and its volume fraction (V^F).

Table 1. Young's modulus and tensile strength of the cotton fiber (CF)/polypropylene (PP) composites.

		0%MAPP		6%MAPP	
W^F	V^F	E_t^C (GPa)	σ_t^C (MPa)	E_t^C (GPa)	σ_t^C (MPa)
0	0	1.5 ± 0.1	27.6 ± 0.5	1.5 ± 0.1	27.6 ± 0.5
20%	0.131	3.2 ± 0.1	35.0 ± 0.5	3.3 ± 0.1	41.7 ± 0.7
30%	0.205	3.9 ± 0.2	38.2 ± 0.8	3.9 ± 0.1	47.1 ±0.7
40%	0.287	4.7 ± 0.2	41.7 ± 0.8	4.8 ± 0.2	53.6 ± 1.0
50%	0.376	5.6 ± 0.2	45.4 ± 1.1	5.4 ± 0.2	58.3 ± 1.2

It was found that the use of a coupling agent had a low effect on the Young's modulus of the composites. In fact, an ANOVA analysis (at 95% confidence rate) reveals that the differences between the Young's moduli of the composites with the same percentage of reinforcement, despite adding or not adding a coupling agent, were not statistically relevant. This result was expected as it has been reported previously in the literature [5,32]. The same materials revealed totally different behaviors in the case of the tensile strength, where the presence of MAPP considerably increased the strength of the materials [7,20]. Thus, while the coupling agent has a noticeable effect on the tensile strength of the composites, its impact is not statistically relevant in the case of the Young's modulus. Some authors prefer to state that the strength of the interphase between the matrix and the reinforcements has a limited impact on the stiffness of the composites [6,11]. Other authors prefer to justify the differences on the methods used to evaluate the strength and the modulus. While the strength is measured at the maximum strain, where the interphase has been fully put to test, the Young's modulus is measured at low strains [32].

The Young's modulus of a semi-aligned short fiber reinforced composite is mainly affected by the properties of the phases and its contents, the morphology of the reinforcement, its mean orientation, and its grade of dispersion. In the case of a correctly dispersed reinforcement, the increase of the Young's modulus against reinforcement content was expected to be linear (Figure 2) [32].

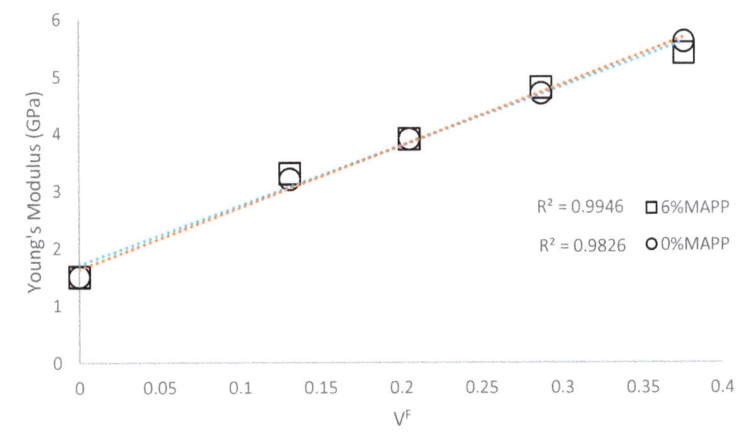

Figure 2. Young's modulus of the coupled and uncoupled CF-PP composites against reinforcement content.

Both, coupled and uncoupled composites showed a linear evolution of its Young's moduli against CF content. Thus, a proper dispersion of the reinforcement was assumed. Nonetheless, the higher the percentage of reinforcement the harder it becomes to obtain a good or proper dispersion. In the case of the coupled composite at 50 wt% of CF, the Young's modulus seems to start to decrease under the regression line. Besides, CF incorporates a textile dye that seems to increase the strength of the interphase and ease the dispersion at low reinforcement rates. Nonetheless, it was impossible to corroborate this evolution due to the impossibility of preparing materials with higher reinforcement contents that are still able to be mold injected. From now on, and due to the equivalence between coupled and uncoupled CF-based composites, the analysis will be referring to the coupled materials.

The literature shows multiple studies on the evolution of the Young's modulus against the fiber contents. We have chosen stone groundwood fibers (SGW), commonly used for papermaking, hemp strands (HS), as a byproduct of agroforestry, old newspaper fibers (ONPF), as recycled fibers, and glass fibers (GF) as an industrial commodity and the most commonly used reinforcement [11,14,33]. Table 2 shows the Young's moduli of SGW, HS, ONPF and GF reinforced PP composites.

Table 2. Young's moduli of stone groundwood, hemp strands, and glass fiber reinforced polypropylene coupled composites.

	SGW	HS	ONPF	GF
20%	2.7 ± 0.1	2.8 ± 0.1	2.8 ± 0.1	4.1 ± 0.1
30%	3.5 ± 0.1	3.8 ± 0.1	3.8 ± 0.1	5.7 ± 0.1
40%	4.3 ± 0.1	5.2 ± 0.1	4.2 ± 0.1	7.7 ± 0.1
50%	5.2 ± 0.1	6.3 ± 0.1	5.3 ± 0.1	-

The Young's moduli of natural fiber reinforced polypropylene composites are similar, with slight advantages for those reinforced with strands, especially at high reinforcement contents. Cotton fibers showed Young's moduli as superior to SGW and ONPF, and in line with the other strands. Nonetheless, cotton fibers are recycled and a byproduct of the textile industry, while hemp strands can be considered virgin materials. ONPF are recycled fibers that come from the disintegration of used newspaper. The Young's moduli of ONPF and SGW based composites are very similar, showing that recovering the fibers from the paper had little effect on the stiffening potential of the reinforcements. Moreover, CF showed higher Young's moduli than ONPF based composites. All the natural-based composites showed a linear evolution of their Young's moduli against fiber contents, but different slopes on their regression lines.

On the other hand, GF-based materials showed noticeably higher Young's moduli than natural fiber-based composites. At the same reinforcement contents, Young's modulus of CF-based composites is noticeably lower than GF-based ones. It was necessary to increase 20 wt% the amount of CF to obtain a Young's moduli similar to GF.

3.2. Neat Contribution of the Reinforcements

Attending to the above-mentioned parameters that affect the Young's modulus of a composite, the differences must be related with the morphology of the reinforcements, its mean orientation, or the intrinsic properties of the phases. The modified rule of mixtures (RoM) for the Young's modulus summarizes all these parameters (Equation (1)):

$$E_t^C = \eta_e \cdot E_t^F \cdot V^F + (1 - V^F) \cdot E_t^M \tag{1}$$

where E_t^C, E_t^F, and E_t^M are the Young's moduli of the composite, reinforcement, and matrix, respectively. V^F represents the reinforcement volume fraction, and η_e is a modulus efficiency factor that equalizes the contribution of the reinforcements to the Young's modulus of the composite. This efficiency factor is seldom presented as a length efficiency factor times an orientation efficiency factor ($\eta_e = \eta_l \cdot \eta_o$). At the exception of the intrinsic Young's modulus of the reinforcements and the modulus efficiency factor, the rest of the values can be easily obtained during the tensile test of the composites. Clearly, the RoM can only be used if the Young's modulus of the composite evolves linearly against reinforcement content.

In any case, the neat contribution of the reinforcements to the Young's modulus of the composite is represented by $\eta_e \cdot E_t^F$ in the RoM. Thus, the RoM can be rearranged to account for such neat contribution as:

$$\eta_e \cdot E_t^F = \frac{E_t^C - (1 - V^F) \cdot E_t^M}{V^F} \tag{2}$$

Then, if the neat contribution is represented against the reinforcement volume fraction, a regression line is obtained, and the slope of such a line has been referred to in the literature as a fiber tensile modulus factor (FTMF) [6,32]. This factor can be used as a measure of the stiffening capabilities of a reinforcement. Figure 3 shows the FTMF for different fibers as polypropylene reinforcement.

The FTMF of CF was between HS and SGW. This value ensures good stiffening abilities for CF as PP reinforcement because the literature shows possible applications for materials with similar FTMF for building or product design purposes [34,35]. Moreover, some researchers used an ONPF-based composite to substitute a GF-based one [36]. On the other hand, GF showed higher stiffening capabilities than the rest of the reinforcements. This is not a surprise, having in account that GF is a man-made material with more stable intrinsic properties and a regular morphology. The FTMF of the reinforcements shows a similar behavior than the Young's moduli of its composites. Thus, the differences between such moduli seem to be focused on the neat contribution of the fibers, specifically, the intrinsic Young's modulus of the reinforcement and the modulus efficiency factor. In order to analyze such differences, the researchers propose a micromechanics analysis.

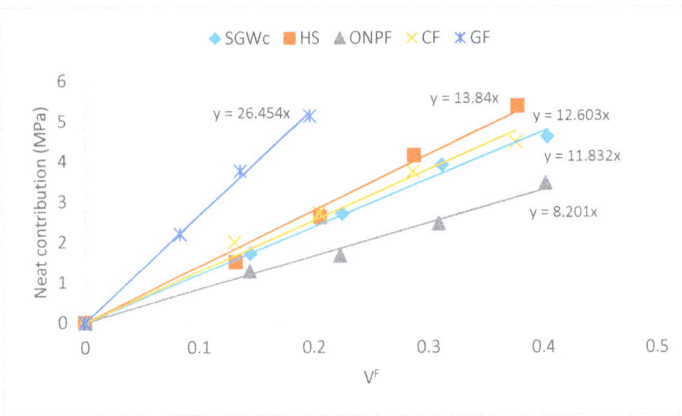

Figure 3. Neat contribution of the reinforcements to the Young's modulus of the polymers.

3.3. Micromechanics Analysis of the Young's Modulus

The RoM (Equation (1)) shows two unknowns that coincide with the neat contribution of the fibers: $\eta_e \cdot E_t^F$. While it is possible to measure the intrinsic Young's modulus of the fibers, and more so in the case of the strands, some authors defend the use of micromechanics methods as an alternative [11,37,38]. In addition, a high number of experiments are necessary due to the foreseeable standard deviations of the mechanical properties of natural fiber reinforcements. Thus, the Hirsh model was proposed as a means to evaluate the intrinsic Young's modulus of CF.

$$E_t^C = \beta \cdot (E_t^F \cdot V^F + E_t^M(1 - V^F)) + (1 - \beta) \frac{E_t^F \cdot E_t^M}{E_t^F \cdot V^F + E_t^m(1 - V^F)} \quad (3)$$

where β is a parameter that modules the stress transference between both phases of the composite material. In the case of semi-aligned short fibers reinforced composites β has a value of 0.4 [14]. Table 3 shows the micromechanical parameters obtained after the analysis.

Table 3. Micromechanics of the Young's moduli of CF reinforced polypropylene coupled composites.

V^F	E_t^F (GPa)	η_e	η_l	η_o	α_o
0.131	31.48	0.52	0.89	0.58	48.8
0.205	28.06	0.47	0.89	0.53	53.3
0.287	26.48	0.45	0.89	0.51	55.1
0.376	25.46	0.45	0.90	0.49	56.2
Mean	27.87 ± 2.63	0.47 ± 0.03	0.89 ± 0.01	0.53 ± 0.04	53.3 ± 3.3

The mean intrinsic Young's modulus of CF was found to be 27.87 ± 2.63 GPa, similar to HS, with a value of 26.8 GPa [11]. This coincidence agrees with the already similarities found in the neat contributions of such fibers (Figure 3). Nonetheless, the computed intrinsic Young's modulus of CF contrasts heavily with the neatly inferior values found in the literature. Some authors place this intrinsic Young's modulus in the range from 5 to 13 GPa [26–28]. Using such values with the RoM is not possible to reach the obtained experimental values without using modulus efficiency factors outside the usual range.

On the other hand, the value for CF is noticeably higher than the value obtained for SGW and ONPF, 21.2 ± 1.9 and 22.8 ± 1.8 GPa, respectively [14,39]. This also agrees with the neat contributions of such fibers. Similarly, GF showed an intrinsic Young's modulus of 76 GPa, justifying the differences obtained in the Young's modulus of its composites [11,40].

Consequently, the intrinsic Young's moduli of the different reinforcements affected heavily the Young's moduli of its composites. Nonetheless, CF showed a higher intrinsic Young's modulus than HS, but HS-based composites showed higher Young's moduli, at the same reinforcement contents than CF-based composites. Thus, the differences are expected to be found in the modulus efficiency factor (Table 3).

The values for the modulus efficiency factor were obtained by using all the experimental data (Table 1) and the mean intrinsic Young's modulus of CF (Table 3). The mean value was found to be 0.47 ± 0.03. The value is inside the usual range of values, between 0.45 and 0.56 for such factor [11,12,14,40]. Nonetheless, it is worth noting that the obtained value is in the lower half bound of the expected values. Thus, presumably, CF based composites have not taken advantage of the stiffening capabilities of CF. Particular values decrease when the CF contents increase (Table 3). The composite with a 20 wt% of CF exhibits a modulus efficiency factor higher than the other CF-based composites, and also a higher intrinsic Young's modulus, indicating a higher yield on the stiffening capabilities of CF. The reasons must be found on the mean orientation of the fibers, the morphology of the reinforcements or its dispersion.

In the case of HS the modulus efficiency factor was evaluated at 0.50 ± 0.02. This value is higher than CF and can compensate the difference between the intrinsic Young's moduli of the reinforcements and justify the higher moduli of the HS-based composites. In the case of ONPF, the value of η_e was evaluated at 0.49 ± 0.04, a value similar to HS. Finally, SGW showed the highest values for η_e, with a mean of 0.56 ± 0.02.

In order to find the impact of the morphology and the mean orientation of CF, the morphology and orientation efficiency factors were computed. The length efficiency factor was calculated according to Cox-Krenchel's model (Equations (4) and (5)) [41]:

$$\eta_l = 1 - \frac{\tanh\left(\frac{\mu \cdot L^F}{2}\right)}{\left(\frac{\mu \cdot L^F}{2}\right)} \tag{4}$$

with

$$\mu = \frac{1}{r^F} \sqrt{\frac{E_t^M}{E_t^F \cdot (1-v) \cdot Ln\left(\sqrt{\pi/4 \cdot V^F}\right)}} \tag{5}$$

where L^F and r^F are the reinforcement mean weighed length and radius, respectively. The Poisson's ratio of the matrix is represented by v and μ is a coefficient of the stress concentration rate at the end of the fibers. The Poisson ratio was 0.36, as found in the literature [22]. The orientation factor η_o was obtained from $\eta_o = \eta_l/\eta_e$. Table 3 shows the obtained values.

The length efficiency factor remained almost the same for all the composite formulations, with a mean value of 0.89 ± 0.01. Usually this factor decreases when the percentage of reinforcement increases [5]. This is due to the changes in the mean length of the reinforcements during compounding, when reinforcements are exposed to attrition phenomena and tend to break. There is a decrease in the mean length of such reinforcements as the reinforcement content increases [32]. Thus, these changes are expected to affect the Young's moduli of the composites. In the case of CF based composites, the impact of the morphology of the fibers seems to little impact the Young's modulus, although the reinforcements decreased their mean length from 293 to 185 µm [7]. This hypothesis will be put to test later on by applying a different micromechanics model.

On the other hand, the orientation efficiency factor clearly changed with the amount of reinforcement (Table 3). This factor showed a mean value of 0.53 ± 0.04 and ranged from 0.58 to 0.49. Usually, the orientation efficiency factor is more stable than the length efficiency factor, because the mean orientation of the fibers is heavily impacted by the geometry of the injection mold and the parameters used during the mold injection [42,43].

Fukuda and Kawata [44] studied the tensile modulus of short fiber reinforced thermoplastics, and the orientation of the fibers inside the composites. The authors proposed different fiber distributions, but based on the literature, a rectangular distribution (square packing) renders adequate results for short fiber semi-aligned reinforced composites [32,45]. The authors present an equation that computes the orientation efficiency factor from a mean orientation angle (α):

$$\eta_o = \frac{sin(\alpha)}{\alpha}\left(\frac{3-v}{4}\frac{sin(\alpha)}{\alpha} + \frac{1-v}{4}\frac{sin(3\alpha)}{3\alpha}\right) \tag{6}$$

Equation (6) was used to compute the mean orientation angles of the reinforcements (Table 3). The mean orientation angle was found to be 53.3 ± 3.3°. This value is in line with other natural fiber reinforced composites, thought in the upper bounds, meaning that the reinforcements are les oriented than the expected. It must be stressed that the orientation decreased with the amount of reinforcement, from 48.8° for the composite containing 20 wt% of CF to 56.2° for the composite containing 50 wt%. In other cases, it was found that the composites with higher fiber contents showed higher orientations [11]. In these studies, it was assumed that the shorter fibers were easily aligned than the longer ones. Nonetheless, the RoM does not incorporate a factor taking in account the dispersion of the fibers, and as has been previously commented, the authors suspect that the dispersion of the composites at high-reinforcement contents was improvable.

3.4. Effect of the Morphology of the Reinforcements

While the rule of mixtures (Equation (1)) and Hirsch's equation (Equation (3)) are elegant models that allow predicting the Young's modulus of a composite from a variety of parameters, they do not incorporate any morphological property of the reinforcements. The morphology of the reinforcements is known to greatly impact the mechanical properties of a composite, especially the ratio between its mean length and diameter, known as aspect ratio [12,46,47]. Thus, the authors propose using the Tsai and Pagano model (Equation (7)) in combination with Halpin and Tsai equations (Equations (8)–(10)) to evaluate a theoretical Young's modulus of the composites.

In agreement with Tsai and Pagano model, the stiffness in the fiber direction is given by:

$$E_t^C = \frac{3}{8}E^{11} + \frac{5}{8}E^{22} \tag{7}$$

where, E^{11} and E^{22} are the longitudinal and transversal elastic modulus, calculated by the Halpin–Tsai equations [11]:

$$E^{11} = \frac{1 + 2(L^F/2r^F)\cdot\lambda_l V^F}{1 - \lambda_l V^F} E_t^M \tag{8}$$

with

$$\lambda_l = \frac{(E_t^F/E_t^M) - 1}{(E_t^F/E_t^M) + 2(L^F/2r^F)} \tag{9}$$

and:

$$E^{22} = \frac{1 + 2\lambda_t V^F}{1 - \lambda_t V^F} E_t^M \tag{10}$$

with

$$\lambda_t = \frac{\left(E_t^F/E_t^M\right) - 1}{\left(E_t^F/E_t^M\right) + 2} \tag{11}$$

Table 4 presents the obtained values. The values obtained by using the micromechanics model show a good alignment with the experimental values, especially at reinforcement contents higher than 20 wt%. In these cases, the error goes to a maximum of 5.6%, which is assumable when modeling a mechanical property of a natural fiber reinforcement composite. In fact, the standard deviation of the experimental values is higher than this maximum of 5.6% (Table 1). The higher errors found for the composite with a 20 wt% of CF can be explained by the intrinsic Young's modulus derived from these composites (Table 2). The value of such a parameter is 11.5% higher than the mean value. If the 31.48 GPa value is used as input for the Tsai and Pagano model, the theoretical Young's modulus increases to 3.02 GPa. This value is more similar to the value obtained experimentally. Moreover, the authors blame the differences between the experimental and theoretical values to deviations from a totally linear evolution of the Young's moduli against reinforcement contents—due to possible auto entanglements between the CF—and thus showing a slightly worse dispersion of the CF.

Table 4. Theoretical Young's moduli of the composites computed by using the Tsai and Pagano model in combination with Halpin and Tsai equations.

V^F	Experimental		Tsai-Pagano		Error (GPa)		Error (%)	
	0% MAPP	6% MAPP	0% MAPP	6% MAPP	0% MAPP	6% MAPP	0% MAPP	6% MAPP
0.131	3.2	3.3	2.9	2.9	0.3	0.4	9.4	12.1
0.205	3.9	3.9	3.7	3.7	0.2	0.2	5.1	5.1
0.287	4.7	4.8	4.7	4.6	0	0.2	0	4.2
0.376	5.6	5.4	5.8	5.7	−0.2	−0.3	−3.6	−5.6

The obtained values were plotted against the experimental ones to find the grade of correlation between both values (Figure 4).

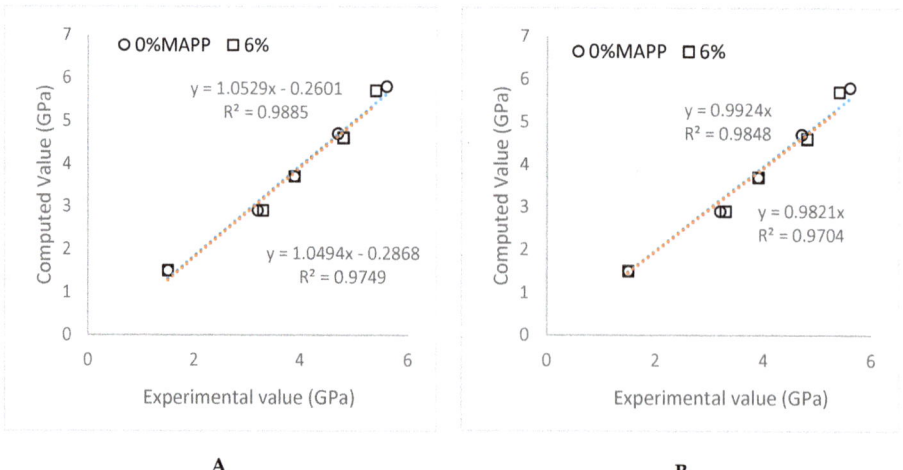

Figure 4. Correlation between the experimental Young's moduli of the composites and the computed ones by using the Tsai and Pagano model in combination with Halpin and Tsai equations: (**A**) Unweighted correlation; (**B**) correlation line adding the condition of such line going through the origin.

The regression lines for the uncoupled and coupled composites using the experimental and theoretic values (Figure 4A) show 1.05 slopes, with a high correlation. This slope shows a high coincidence between the computes and experimental values. A second regression line was proposed, passing thorough the origin (Figure 4B). In this case the slopes were found to be 0.99 and 0.98 for the uncoupled and coupled composite, respectively. As mentioned above, these results prove the accuracy of the values predicted from the model. The values also prove the impact of the morphology of the reinforcements in the Young's moduli of the polymers, as Tsai and Pagano model uses these values as the input.

4. Conclusions

Composite materials with CF as reinforcement and PP as the matrix were formulated with 20 to 50 wt% CF contents. Two batches were prepared, one including a 6 wt% of coupling agent and another without.

The Young's moduli of the composites were little impacted by the presence of coupling agents and thus by the strength of the interphase. Thus, for applications where stiffness is paramount, uncoupled composites can be used with the same response under load as the coupled composites. On the other hand, in the case of semi-structural uses, the coupled composites ensure a higher tensile strength and similar deformations under load than the uncoupled ones.

The Young's moduli of the composites were similar to those obtained for natural strands reinforced composites and higher than wood fibers reinforced composites. The presence of a textile dye in the CF decreased its hydrophilicity but also seemed to increase the difficulty in obtaining a good dispersion of the fibers inside the composite, and the decrease in the quality of the dispersion decreased the stiffening yield of CF.

A value of the intrinsic Young's modulus of CF was obtained by using Hirsh's model. The mean value of the modulus was found to be 27.87 GPa, lower than other strands. Nonetheless, this value doubles the values previously published. The value obtained for the CF in a composite with 20 wt% of reinforcement reached 31.48 GPa, a value more proper to other natural strands like hemp.

The micromechanics properties of the Young's modulus of the composites showed the effect of the orientation and the morphology of the fibers. Nonetheless, the authors found some deviations from a linear behavior of the Young's modulus against CF contents. The authors assume that the dispersion of the fibers can be improved to increase the Young's moduli of the composites with high rates of reinforcement. Nonetheless, further research is needed to prove this point.

This paper shows the opportunity of recovering textile cotton fibers, which are useless for the textile industry, to obtain composite materials able to replace glass fiber reinforced materials. In doing so, the dumping and incinerating of such fibers is avoided and the value chain of the textile industry is widened.

Author Contributions: A.S. performed the experiments and wrote the first version of the manuscript; M.-À.C. and J.S. conceived and designed the experiments; F.X.E. and Q.T. performed the calculus and represented the data; F.V. and P.M. guided the project. All the authors contributed to writing and correcting the document.

Funding: The authors wish to acknowledge the financial support of the Càtedra de Processos Industrials Sostenibles of the University of Girona.

Conflicts of Interest: The authors declare no conflict of interest.

References

1. Anastas, P.T.; Warner, J. *Green Chemistry: Theory and Practice*; Oxford University Press: Oxford, UK, 1998.
2. Schumann, A. Plastics-the Environmentally Friendly Design Material. *ATZ Heavy Duty Worldw.* **2019**, *12*, 74. [CrossRef]
3. Ferreira, F.V.; Pinheiro, I.F.; Mariano, M.; Cividanes, L.S.; Costa, J.C.; Nascimento, N.R.; Kimura, S.P.; Neto, J.C.; Lona, L.M. Environmentally friendly polymer composites based on PBAT reinforced with natural fibers from the amazon forest. *Polym. Compos.* **2019**. [CrossRef]

4. Reixach, R.; Espinach, F.X.; Franco-Marquès, E.; Ramirez de Cartagena, F.; Pellicer, N.; Tresserras, J.; Mutjé, P. Modeling of the tensile moduli of mechanical, thermomechanical, and chemi-thermomechanical pulps from orange tree pruning. *Polym. Compos.* **2013**, *34*, 1840–1846. [CrossRef]
5. Serrano, A.; Espinach, F.X.; Tresserras, J.; del Rey, R.; Pellicer, N.; Mutje, P. Macro and micromechanics analysis of short fiber composites stiffness: The case of old newspaper fibers-polypropylene composites. *Mater.Des.* **2014**, *55*, 319–324. [CrossRef]
6. Espinach, F.X.; Chamorro-Trenado, M.A.; Llorens, J.; Tresserras, J.; Pellicer, N.; Vilaseca, F.; Pèlach, A. Study of the Flexural Modulus and the Micromechanics of Old Newspaper Reinforced Polypropylene Composites. *BioResources* **2019**, *14*, 3578–3593.
7. Serra, A.; Tarrés, Q.; Llop, M.; Reixach, R.; Mutjé, P.; Espinach, F.X. Recycling dyed cotton textile byproduct fibers as polypropylene reinforcement. *Text. Res. J.* **2019**, *89*, 2113–2125. [CrossRef]
8. Petrucci, R.; Nisini, E.; Puglia, D.; Sarasini, F.; Rallini, M.; Santulli, C.; Minak, G.; Kenny, J. Tensile and fatigue characterisation of textile cotton waste/polypropylene laminates. *Compos. Pt B-Eng.* **2015**, *81*, 84–90. [CrossRef]
9. Kallel, T.K.; Taktak, R.; Guermazi, N.; Mnif, N. Mechanical and structural properties of glass fiber-reinforced polypropylene (PPGF) composites. *Polym. Compos.* **2018**, *39*, 3497–3508. [CrossRef]
10. Thomason, J.; Jenkins, P.; Yang, L. Glass Fibre Strength—A Review with Relation to Composite Recycling. *Fibers* **2016**, *4*, 18. [CrossRef]
11. Espinach, F.X.; Julian, F.; Verdaguer, N.; Torres, L.; Pelach, M.A.; Vilaseca, F.; Mutje, P. Analysis of tensile and flexural modulus in hemp strands/polypropylene composites. *Compos. Pt. B-Eng.* **2013**, *47*, 339–343. [CrossRef]
12. Vilaseca, F.; Valadez-Gonzalez, A.; Herrera-Franco, P.J.; Pelach, M.A.; Lopez, J.P.; Mutje, P. Biocomposites from abaca strands and polypropylene. Part I: Evaluation of the tensile properties. *Bioresource Technol.* **2010**, *101*, 387–395. [CrossRef] [PubMed]
13. Mishra, R.; Behera, B.; Militky, J. Recycling of Textile Waste Into Green Composites: Performance Characterization. *Polym. Compos.* **2014**, *35*, 1960–1967. [CrossRef]
14. Lopez, J.P.; Mutje, P.; Pelach, M.A.; El Mansouri, N.E.; Boufi, S.; Vilaseca, F. Analysis of the tensile modulus of PP composites reinforced with Stone grounwood fibers from softwood. *BioResources* **2012**, *7*, 1310–1323. [CrossRef]
15. Espinach, F.X.; Mendez, J.A.; Granda, L.A.; Pelach, M.A.; Delgado-Aguilar, M.; Mutje, P. Bleached kraft softwood fibers reinforced polylactic acid composites, tensile and flexural strengths. In *Natural Fibre-Reinforced Biodegradable and Bioresorbable Polymer Composites*; Lau, A., Ed.; Woodhead Publishing Limited: Cambridge, UK, 2017.
16. Araújo, R.S.; Rezende, C.C.; Marques, M.F.V.; Ferreira, L.C.; Russo, P.; Emanuela Errico, M.; Avolio, R.; Avella, M.; Gentile, G. Polypropylene-based composites reinforced with textile wastes. *J. Appl. Polym. Sci.* **2017**, *134*, 45060. [CrossRef]
17. Monteiro, S.N.; Lopes, F.P.D.; Barbosa, A.P.; Bevitori, A.B.; Da Silva, I.L.A.; Da Costa, L.L. Natural Lignocellulosic Fibers as Engineering Materials-An Overview. *Metall. Mater. Trans. A-Phys. Metall. Mater. Sci.* **2011**, *42*, 2963–2974. [CrossRef]
18. Reddy, N.; Yang, Y.Q. Properties and potential applications of natural cellulose fibers from the bark of cotton stalks. *Bioresource Technol.* **2009**, *100*, 3563–3569. [CrossRef]
19. De Silva, R.; Byrne, N. Utilization of cotton waste for regenerated cellulose fibres: Influence of degree of polymerization on mechanical properties. *Carbohyd. Polym.* **2017**, *174*, 89–94. [CrossRef]
20. Serra, A.; Tarrés, Q.; Claramunt, J.; Mutjé, P.; Ardanuy, M.; Espinach, F. Behavior of the interphase of dyed cotton residue flocks reinforced polypropylene composites. *Compos. Pt. B-Eng.* **2017**, *128*, 200–207. [CrossRef]
21. Ryu, C.; Phan, A.N.; Sharifi, V.N.; Swithenbank, J. Combustion of textile residues in a packed bed. *Exp. Therm. Fluid Sci.* **2007**, *31*, 887–895. [CrossRef]
22. Jha, M.K.; Kumar, V.; Maharaj, L.; Singh, R.J. Studies on leaching and recycling of zinc from rayon waste sludge. *Ind. Eng. Chem. Res.* **2004**, *43*, 1284–1295. [CrossRef]
23. Bledzki, A.K.; Franciszczak, P.; Osman, Z.; Elbadawi, M. Polypropylene biocomposites reinforced with softwood, abaca, jute, and kenaf fibers. *Ind. Crop. Prod.* **2015**, *70*, 91–99. [CrossRef]

24. Lopez, J.P.; Boufi, S.; El Mansouri, N.E.; Mutje, P.; Vilaseca, F. PP composites based on mechanical pulp, deinked newspaper and jute strands: A comparative study. *Compos. Pt. B-Eng.* **2012**, *43*, 3453–3461. [CrossRef]
25. Bledzki, A.K.; Mamun, A.A.; Faruk, O. Abaca fibre reinforced PP composites and comparison with jute and flax fibre PP composites. *Express Polym. Lett.* **2007**, *1*, 755–762. [CrossRef]
26. Bledzki, A.K.; Gassan, J. Composites reinforced with cellulose based fibres. *Prog. Polym. Sci.* **1999**, *24*, 221–274. [CrossRef]
27. Rogovina, S.; Prut, E.; Berlin, A. Composite Materials Based on Synthetic Polymers Reinforced with Natural Fibers. *Polym. Sci. Ser. A* **2019**, *61*, 417–438. [CrossRef]
28. Faruk, O.; Bledzki, A.K.; Fink, H.P.; Sain, M. Progress report on natural fiber reinforced composites. *Macromol. Mater. Eng.* **2014**, *299*, 9–26. [CrossRef]
29. Hirsch, T. Modulus of elasticity of concrete affected by elastic moduli of cement paste matrix and aggregate. *J. Am. Concr. Inst.* **1962**, *59*, 427–451. [CrossRef]
30. Halpin, J.C.; Tsai, S.W. *Effects of Environmental Factors on Composite Materials*; Technical Report AFML-TR-67-423; National Institute of Occupational Safety and Health: Cincinnati, OH, USA, 1969.
31. Halpin, J.C.; Pagano, N.J. The Laminate Approximation for Randomly Oriented Fibrous Composites. *J. Compos. Mater.* **1969**, *3*, 720–724. [CrossRef]
32. Granda, L.A.; Espinach, F.X.; Mendez, J.A.; Tresserras, J.; Delgado-Aguilar, M.; Mutje, P. Semichemical fibres of Leucaena collinsii reinforced polypropylene composites: Young's modulus analysis and fibre diameter effect on the stiffness. *Compos. Pt. B-Eng.* **2016**, *92*, 332–337. [CrossRef]
33. Tarrés, Q.; Vilaseca, F.; Herrera-Franco, P.J.; Espinach, F.X.; Delgado-Aguilar, M.; Mutjé, P. Interface and micromechanical characterization of tensile strength of bio-based composites from polypropylene and henequen strands. *Ind. Crop. Prod.* **2019**, *132*, 319–326. [CrossRef]
34. Julian, F.; Mendez, J.A.; Espinach, F.X.; Verdaguer, N.; Mutje, P.; Vilaseca, F. Bio-based composites from stone groundwood applied to new product development. *BioResources* **2012**, *7*, 5829–5842. [CrossRef]
35. Oliver-Ortega, H.; Chamorro-Trenado, M.À.; Soler, J.; Mutjé, P.; Vilaseca, F.; Espinach, F.X. Macro and micromechanical preliminary assessment of the tensile strength of particulate rapeseed sawdust reinforced polypropylene copolymer biocomposites for its use as building material. *Constr. Build. Mater.* **2018**, *168*, 422–430. [CrossRef]
36. Serrano, A.; Espinach, F.X.; Tresserras, J.; Pellicer, N.; Alcala, M.; Mutje, P. Study on the technical feasibility of replacing glass fibers by old newspaper recycled fibers as polypropylene reinforcement. *J. Clean. Prod.* **2014**, *65*, 489–496. [CrossRef]
37. Shah, D.U.; Nag, R.K.; Clifford, M.J. Why do we observe significant differences between measured and 'back-calculated' properties of natural fibres? *Cellulose* **2016**, *23*, 1481–1490. [CrossRef]
38. Facca, A.G.; Kortschot, M.T.; Yan, N. Predicting the elastic modulus of natural fibre reinforced thermoplastics. *Compos. Pt. A-Appl. Sci.* **2006**, *37*, 1660–1671. [CrossRef]
39. Serrano, A.; Espinach, F.X.; Julian, F.; del Rey, R.; Mendez, J.A.; Mutje, P. Estimation of the interfacial shears strength, orientation factor and mean equivalent intrinsic tensile strength in old newspaper fiber/polypropylene composites. *Compos. Pt. B-Eng.* **2013**, *50*, 232–238. [CrossRef]
40. Vallejos, M.E.; Espinach, F.X.; Julian, F.; Torres, L.; Vilaseca, F.; Mutje, P. Micromechanics of hemp strands in polypropylene composites. *Compos. Sci. Technol.* **2012**, *72*, 1209–1213. [CrossRef]
41. Krenchel, H. *Fibre Reinforcement*; Akademisk Forlag: Copenhagen, Denmark, 1964.
42. Del Rey, R.; Serrat, R.; Alba, J.; Perez, I.; Mutje, P.; Espinach, F.X. Effect of Sodium Hydroxide Treatments on the Tensile Strength and the Interphase Quality of Hemp Core Fiber-Reinforced Polypropylene Composites. *Polymers* **2017**, *9*, 377. [CrossRef]
43. Reixach, R.; Franco-Marquès, E.; El Mansouri, N.-E.; de Cartagena, F.R.; Arbat, G.; Espinach, F.X.; Mutjé, P. Micromechanics of Mechanical, Thermomechanical, and Chemi-Thermomechanical Pulp from Orange Tree Pruning as Polypropylene Reinforcement: A Comparative Study. *BioResources* **2013**, *8*, 3231–3246. [CrossRef]
44. Fukuda, H.; Kawata, K. On Young's modulus of short fibre composites. *Fibre Sci. Technol.* **1974**, *7*, 207–222. [CrossRef]
45. Espinach, F.X.; Julián, F.; Alcalà, M.; Tresserras, J.; Mutjé, P. High stiffness performance alpha-grass pulp fiber reinforced thermoplastic starch-based fully biodegradable composites. *BioResources* **2013**, *9*, 738–755. [CrossRef]

46. Thomason, J.L. The influence of fibre length and concentration on the properties of glass fibre reinforced polypropylene: 5. Injection moulded long and short fibre PP. *Compos. Pt. A-Appl. Sci.* **2002**, *33*, 1641–1652. [CrossRef]
47. Lopez, J.P.; Mendez, J.A.; Espinach, F.X.; Julian, F.; Mutje, P.; Vilaseca, F. Tensile Strength characteristics of Polypropylene composites reinforced with Stone Groundwood fibers from Softwood. *BioResources* **2012**, *7*, 3188–3200. [CrossRef]

© 2019 by the authors. Licensee MDPI, Basel, Switzerland. This article is an open access article distributed under the terms and conditions of the Creative Commons Attribution (CC BY) license (http://creativecommons.org/licenses/by/4.0/).

Article

Effect of Graphene Oxide Coating on Natural Fiber Composite for Multilayered Ballistic Armor

Ulisses Oliveira Costa, Lucio Fabio Cassiano Nascimento *, Julianna Magalhães Garcia, Sergio Neves Monteiro, Fernanda Santos da Luz, Wagner Anacleto Pinheiro and Fabio da Costa Garcia Filho

Department of Materials Science, Military Institute of Engineering-IME, Rio de Janeiro 22290270, Brazil
* Correspondence: lucio_coppe@yahoo.com.br; Tel.: +55-(21)-98500-7084

Received: 3 July 2019; Accepted: 12 August 2019; Published: 16 August 2019

Abstract: Composites with sustainable natural fibers are currently experiencing remarkably diversified applications, including in engineering industries, owing to their lower cost and density as well as ease in processing. Among the natural fibers, the fiber extracted from the leaves of the Amazonian curaua plant (*Ananas erectifolius*) is a promising strong candidate to replace synthetic fibers, such as aramid (Kevlar™), in multilayered armor system (MAS) intended for ballistic protection against level III high velocity ammunition. Another remarkable material, the graphene oxide is attracting considerable attention for its properties, especially as coating to improve the interfacial adhesion in polymer composites. Thus, the present work investigates the performance of graphene oxide coated curaua fiber (GOCF) reinforced epoxy composite, as a front ceramic MAS second layer in ballistic test against level III 7.62 mm ammunition. Not only GOCF composite with 30 vol% fibers attended the standard ballistic requirement with 27.4 ± 0.3 mm of indentation comparable performance to Kevlar™ 24 ± 7 mm with same thickness, but also remained intact, which was not the case of non-coated curaua fiber similar composite. Mechanisms of ceramic fragments capture, curaua fibrils separation, curaua fiber pullout, composite delamination, curaua fiber braking, and epoxy matrix rupture were for the first time discussed as a favorable combination in a MAS second layer to effectively dissipate the projectile impact energy.

Keywords: curaua fibers; graphene oxide coating; epoxy composites; ballistic performance

1. Introduction

In recent decades, the increasing efficiency of ballistic armors has emerged as a relevant factor in personal and vehicular security, for both civilian and military protection. The search for lighter and stronger armor materials has been increasing in proportion to the escalating power and sophistication in firearms development [1]. Research works are showing that polymer composites reinforced with natural lignocellulosic fibers (NLFs) present ballistic efficiency in multilayered armor systems (MAS), with front ceramic, comparable to synthetic aramid fabric, such as Kevlar™ [1–17]. In general, NLF composites have the advantage of environmental sustainability in association with cost-effectiveness, lower density, and easy fabrication as compared to synthetic fibers composites [18–21].

Together with ballistic protection, recent works on nano and micro cellulose [22–27], are also disclosing special applications for NLFs. Among the several papers on ballistic application of NFL composites for MAS second layer stands those using curaua fibers (CF) [1,6,7,10,11,17]. This fiber, native of the Amazonian region, is extracted from the leaves of a plant, *Ananas erectifolius*, sharing the pineapple family. It has attracted considerable interest as polymer composite reinforcement [28–34] owing to relatively lower density (0.96 g/cm^3) in comparison to glass (2.58 g/cm^3) and aramid (1.44 g/cm^3) synthetic fibers [35]. In consequence, the CF specific tensile strength (~2.2·GPa·cm^3/g) is higher than that of glass (~1.4 Gpa·cm^3/g) and close to that of aramid (~2.8·Gpa·cm^3/g) fibers.

As most NLFs applied in polymer composite [36–41], the curaua fiber also displays low interfacial shear strength, associated with poor fiber adhesion, while reinforcing a polymer matrix. This is due to their amorphous hemicellulose and lignin that act as natural hydrophilic wax adsorbing water on the fiber surface. Consequently, a weak bonding is expected to exist between the surface of the curaua fiber and the hydrophobic polymer such as polyester [29] and epoxy [37]. This affects the composite performance as MAS second layer for ballistic protection. Indeed, the impact of a high velocity projectile against a MAS with curaua composite results in different fracture mechanisms including delamination and matrix cracking pattern as well as fiber rupture and pullout [7,10,11]. Some of these mechanisms are essential for impact energy. However, others like delamination can impair the integrity of the composite target after a first ballistic shooting. This causes loss of its ability to protect against serial shootings as required by the standard [42].

In spite of the comparable ballistic performance to a same thickness Kevlar™ laminate as MAS second layer, the integrity of a NLF composite is always questionable. Lower amounts, usually less than 30 vol%, of fiber were found to result in composite shattering [4,5,8,9,11–17]. Even a 30 vol% NLF composite may be split by delamination, i.e., decohesion between fiber and matrix, which allows easy perforation of the projectile in case of a second shooting. Surface modification of NLFs has extensively been applied to improve the fiber matrix adherence [43,44]. This will be an effective way to prevent delamination.

Since the rise of graphene [45], it has increasingly been studied and investigated for possible technological applications. In particular, graphene has attracted a considerable attention for its superior performance as composite reinforcement owing to outstanding mechanical properties [46]. The direct oxidation of graphite is considered as an alternative route for producing substantial quantities of another remarkable material, the graphene oxide (GO). Studies conducted on the properties of GO revealed good chemical reactivity and easy handling owing to its intrinsic functional groups in association with amphiphilic behavior [47,48]. Among the several methods reported, to improve NLF composite adhesion and prevent lamination, only few have today been dedicated to graphene or graphene oxide coating [45,49,50].

To the knowledge of the authors of the present work, GO has not yet been applied as a coating onto NLFs to improve interfacial shear strength with respect to a polymer composite for armor application. More specifically, as a novel method to provide efficient fiber/matrix interface for impact energy dissipation. Therefore, the objective of this work is, for the first time, to investigate the ballistic performance of 30 vol% graphene oxide coated curaua fiber (GOCF) reinforced epoxy composite, as a MAS second layer against the treat of level III [42] high velocity projectile. In addition to the comparison of GOCF with both non-coated 30 vol% CF epoxy composite and same thickness Kevlar™, this work also investigates the integrity condition of these composites.

2. Materials and Methods

Curaua fibers, shown in Figure 1a, were supplied by the University of Pará (UFPA), Belém, Brazil. The polymer used as matrix was a commercially available epoxy resin, bisphenol A diglycidyl ether type (DGEBA), hardened with triethylene tetramine (TETA), using the stoichiometric ratio of 13 parts of hardener per 100 parts of resin, fabricated by Dow Chemical, São Paulo Brazil, and distributed by Resinpoxy Ltda (Rio de Janeiro, Rio de Janeiro).

Curaua fibers were used in two main conditions, namely: as-received, non-coated fibers (CF), and graphene oxide coated fibers (GOCF). Initially the as-received fibers were subjected to a mechanical treatment using a hard bristle brush for cleaning, separation, and fiber alignment. Then fibers were cut into 150 mm in length and placed in an oven at 80 °C for 24 h until the fiber weight remained stable. This corresponds to the as-received CF used to produce plain composite plates.

Figure 1. General macroscopic aspect of curaua fibers: (**a**) curaua fibers (CF); (**b**) graphene oxide coated fibers (GOCF); (**c**) their 30 vol% epoxy composites.

The GO used in this work was produced by the Hummers Offeman method, modified by Rourke et al. [47]. The CFs, that have already passed the brush and drying stages, were then immersed in a 0.56 mg/mL GO solution corresponding to 0.1% of weight of the fiber and kept under agitation for 30 min in a universal mechanical shaker, in order to guarantee and optimize the contact of the GO with the fiber. Thereafter the CF soaked with GO dispersion were placed in an oven at 80 °C for 24 h, obtaining at the end the GOCF. Raman spectroscopy analysis was conducted in a model NTEGRA Spectra equipment to certify the existence of GO layers on the fiber surfaces.

To fabricate the composite plates, a metal mold with dimensions of 150 × 120 × 12 mm was used. The plates were processed in a SKAY hydraulic press by applying a load of 5 tons for 24 h. For the CF, the density of 0.92 g/cm^3 [31] was used as the initial reference and 1.11 g/cm^3 for the epoxy resin [35]. The percentages of both CF and GOCF studied in this work was 30 vol%. Figure 1 shows the general macroscopic aspect of (a) CF, (b) GOCF, (c) and their corresponding epoxy composites.

Interfacial shear strength tests were performed to investigate the influence of GO coating onto curaua fiber in curaua-epoxy composites. For this, the method described by Kelly and Tyson [51] was used. The measured parameters were the critical length and the interfacial shear strength. Tensile tests of the individual fiber were carried out according to ASTM D 3822-01 standard [52]. The test used a support (frame) made of paper and plaster, in order to keep the fiber stretched and firmly attached to facilitate the positioning in the grips of the model 3365 Instron equipment. A 25 KN load cell and a strain rate of 5 mm/min were used to perform each individual fiber specimen tensile test. Ten specimens for each test condition were used for both CF and GOCF, with a gage length of 40 mm. The fiber diameter was measured by an optical microscope Olympus BX53M. Before starting the test, the paper is cut to avoid interference in the tensile results.

Ballistic tests were carried out to investigate both CF and GOFC composites capacity of dissipating kinetic energy of a high velocity projectile in a MAS. The MAS used in this work consist, of a front layer of ceramic, an intermediate layer made from both the CF and GOCF epoxy composites, Figure 2. The MAS is placed over a 50 mm thick clay witness (CORFIX™), which has a similar consistency as a human body. The ballistic test system is illustrated in Figure 3. The objective is to obtain the measurement of the trauma, also known as backface signature (indentation) caused by the impact of the 7.62 mm caliber ammunition on the MAS target. According to the NIJ 0101.04 standard [42] a ballistic armor will be effective if the indentation caused in the clay witness is equal to or less than 44 mm. Measurements were performed with a Q4X Banner digital laser sensor. The tests were carried out at the Brazilian Army Assessment Center (CAEx), Rio de Janeiro.

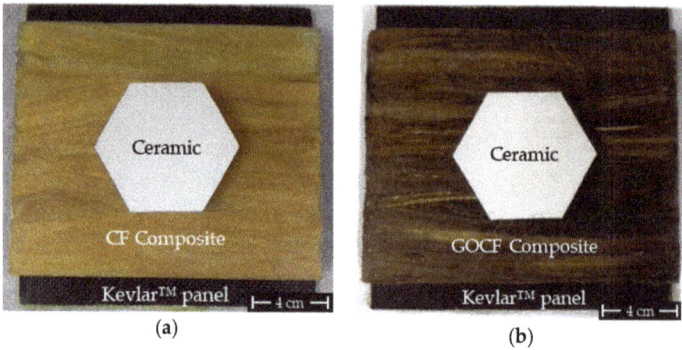

Figure 2. Multilayer armor system (MAS) mounted: (**a**) MAS with CF composite and (**b**) MAS with GOCF composite.

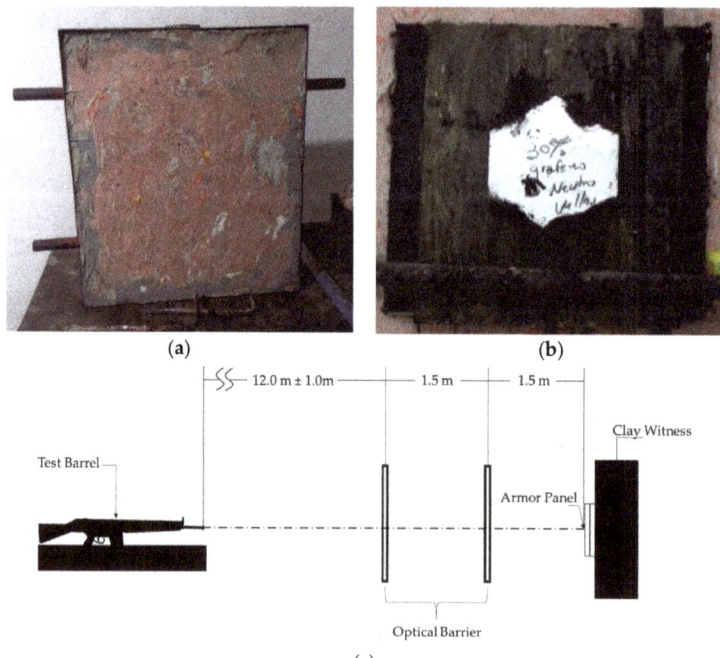

Figure 3. System used for ballistic tests: (**a**) Shooting support frame filled with clay witness; (**b**) MAS target ahead of the clay witness; (**c**) scheme of the system used for ballistic tests [42].

Microscopic analyses of the curaua fibers and fractured surface of the investigated composites were performed by scanning electron microscopy (SEM) in a model Quanta FEG 250 Fei microscope operating with secondary electrons between 5 and 10 KV. The energy dispersive spectroscopy (EDS) analyses were performed using a Bruker Nano GmbH XFlash 630M detector.

The FTIR technique was used to investigate the possible influences of GO on the functional groups of the curaua fibers, in an IR-Prestige-21 model spectrometer from Shimadzu, using the transmittance method with the KBr insert technique. For all samples, the same mass quantities of 2 mg of fiber and 110 mg of KBr were used.

For the analysis by thermogravimetry (TGA), the curauá fibers in CF, GOCF, and its composites were comminuted and placed in aluminum crucible of the TA Instruments, model Q 500 analyzer. Samples were subjected to a heating rate of 10°/min, starting at 30 °C up to 700 °C.

The thickness estimation of GO coating was obtained by atomic force microscope in a model Park systems XE7 atomic Force Microscope.

3. Results and Discussions

The Raman spectra of GO is shown in Figure 4. The intensity ratio of the D and G bands (ID/IG) revealed structural defects and the indication of disorder. The (ID/IG) ratio was calculated as 1.032:1, in accordance with previous authors [47,48]. Besides, a broad and shifted to higher wavenumber of 2D band was seen at 2720 cm^{-1} for GO in Figure 4. 2D band can be used to determine the layers of graphene (monolayer, double layer or multilayer) as it is highly sensitive to stacking of graphene layers. Thus, the location of 2D band confirms that the produced GO was multilayer. A monolayer graphene is normally observed at 2679 cm^{-1} from the spectra. In addition, the shifted location of 2D band, because of the presence of oxygen-containing functional groups, prevents the graphene layer to stack [49].

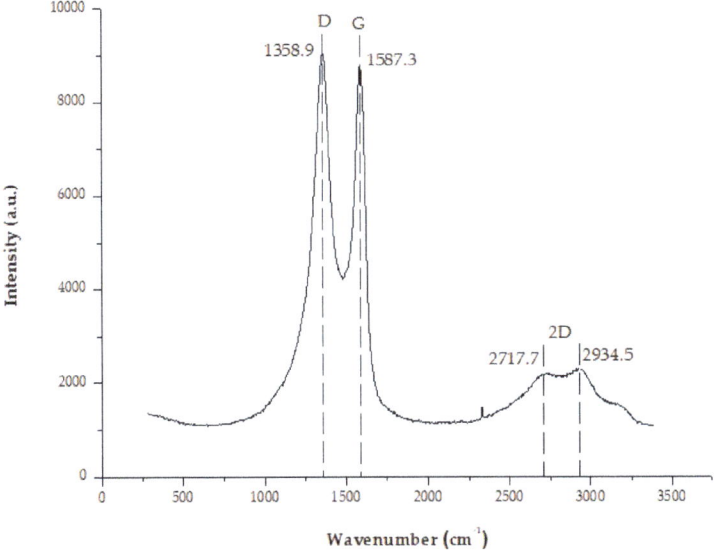

Figure 4. Raman spectra of GO colloid solution.

The main absorption bands of the CF fiber spectrum can be seen as: 3379 cm^{-1}, which is related to the elongation of OH groups present in cellulose and water. The 2916 cm^{-1} band can be attributed to the symmetrical and asymmetrical stretching (C–H) of the aliphatic chain, 1736 cm^{-1} corresponding to the acid elongation vibration (C=O); 1430 cm^{-1} (aliphatic C–H vibration) and 1110 cm^{-1} from the elongation vibration of the ether groups. Other bands refer to the existence of high content of oxygen functional groups on GOCF surface, such as (–C–O–C) and (–C–OOH) [53]. Chemical treatments or modifications of major fiber surface groups (–OH) can be very valuable in detecting and confirming the type of new bond established on the fiber surface and the interaction with the polymer in the case of fiber reinforced polymers [31].

With GO coating, even at low concentrations, several changes in the spectra can be seen in Figure 5. The relative intensities between some bands have changed, suggesting that the GO molecule may have linked to the functional groups such as those mentioned above, reducing almost all intensities of the spectrum. In addition, the absorption band at 1649 cm^{-1} may refer to vibrations of the present GO skeletal ring [54].

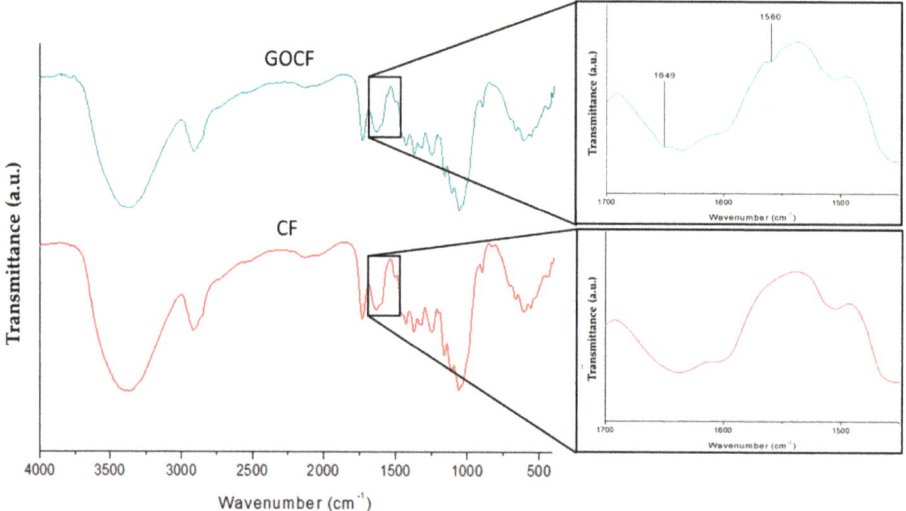

Figure 5. FTIR spectrum of CF and GOCF fibers.

The light band that can be seen at 1560 cm^{-1} can be attributed to the vibrations of benzene rings present in GO [55]. In addition, with the cure of GO curaua fibers, the absorption bands at 833 cm^{-1} (C–H out of plane for p-hydroxyphenyl units) reduced the intensity [56], suggesting that the GO caused changes in the CF fiber functional groups, such as the hydroxyl and carboxyl groups of GO sheets react with the hydroxyl groups of CF, resulting in better wettability between CF and epoxy matrix [49].

The onset of the degradation step was observed at approximately 64 to 150 °C in both CF and GOCF. This effect may indicate the evaporation of moisture absorbed by the fibers. The main mass degradation step was observed starting at 293 °C for CF and 300 °C for GOCF fibers as can be seen in Figure 6. According to some studies, this indicates the stages of hemicellulose, cellulose, and lignin degradation, respectively [33,57–59]. The residue generated by CF fibers was 15% and by GOCF it was 14%.

In the differential thermal analysis (DTA) curves as shown in Figure 7, for the CF fibers, three stages were observed: the first one was between 250 and 300 °C, referring to the decomposition of the hemicellulose, with maximum degradation rate at 272 °C. The second process occurred between 293 and 350 °C, with a maximum degradation rate around 327 °C, which may be related to decomposition of cellulose. Lignin decomposition occurred in the third stage, between 400 and 450 °C, with a maximum degradation rate of around 422 °C. However, a distinct behavior was presented by the GOCF fibers. Their degradation was shifted to higher temperatures and this effect may indicate an increase in the thermal stability of the fibers [33,57–59].

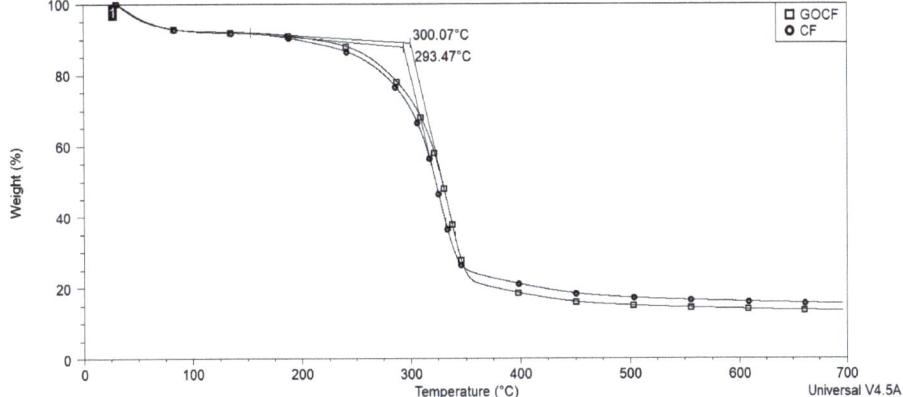

Figure 6. Thermogravimetry analysis (TGA) curves of CF and GOCF fibers.

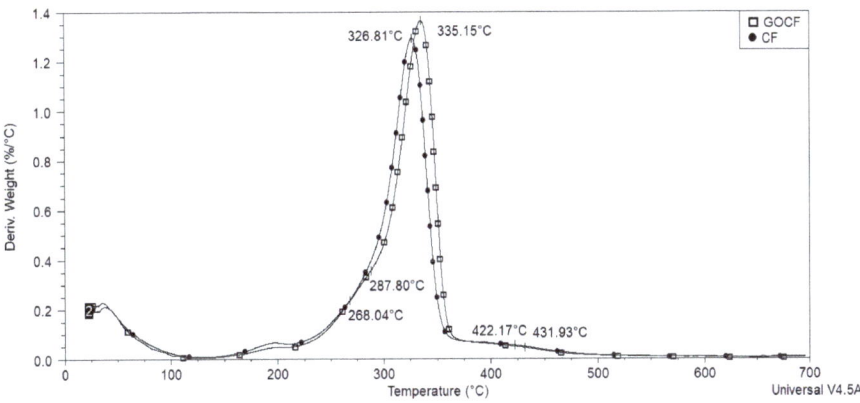

Figure 7. DTA curves of CF and GOCF fibers.

The degradation ratio of the different fibers, CF and GOCF, in Figures 6 and 7 indicate that by the presence of graphene oxide increases the thermal stability of the temperature at 7 °C, with the onset around 300 °C. The effect may be due to the formation of an insulation to heat propagation by the GO coating which retarded degradation and improved thermal stability [50].

Figure 8 shows SEM images of both curaua fibers investigated: (a) as-received non-coated (CF) and (b) graphene oxide coated (GOCF). Average diameter measurements conducted in 10 fibers for each case revealed values of 54.2 ± 14.3 µm for the CF and 51.1 ± 12.0 µm for the GOCF. As expected, these values are practically the same within the standard deviation. This indicates that the graphene oxide coating did not affect the fiber diameter. In fact, the GO coating was estimated to be approximately 10 nm. This would correspond to a negligible increase of less than 10^{-3} vol% in the composite volume fraction of curaua fiber. Since the thinner GO coating cannot affect the curaua fiber strength.

With higher magnification, Figure 9 illustrates the different surface aspects of CF and GOCF. One should notice the uniform smoother surface of the GOFC, Figure 9b, because of the graphene oxide coating as compared with the rougher CF surface in Figure 9a. The GO sheets, prepared by the modified Hummers method, form a stable and homogeneous suspension and exhibit a typical transparent wavy aspect, when coated on the fibers as shown in Figure 9b.

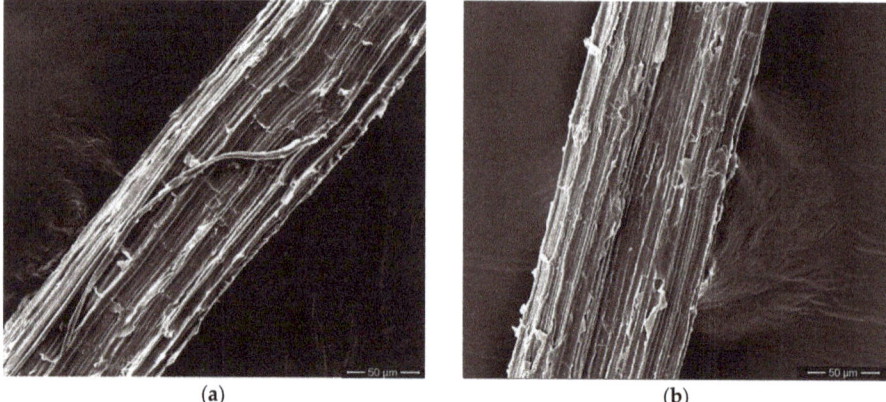

Figure 8. Scanning electron microscopy (SEM) micrographs of both investigated curaua fibers: (a) non-coated CF; (b) GOCF.

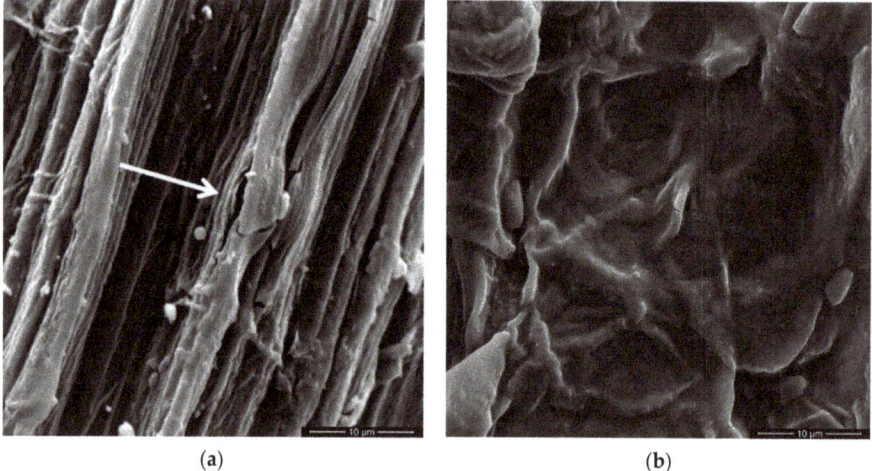

Figure 9. SEM surface images of both fibers: (a) CF; (b) GOCF.

With higher magnification, one notes that the CF fiber is not very stable under the electron beam, with cracks opening on its surface as can be seen in the Figure 9a and indicated by a white arrow. On the other hand, GOCF fiber is more thermally stable, not reacting with the heat generated by the electron beam during the image acquisition, which corroborates the TGA results.

Through the EDS analysis in Figure 10 it was possible to identify the elements present on the surface of both CF and GOCF fibers. For the CF, only carbon and oxygen were identified, the other peaks of the spectrum refer to the copper used in the covering of the fibers. For GOCF fibers, besides carbon and oxygen, it was identified phosphorus and sulfur, which are residues of the reagents used for the production of GO. It can be noted that for CF fibers the C/O ratio is 0.91 decreasing to 0.14 for GOCF fibers, due to the presence of oxygen-containing functional groups in GO [50].

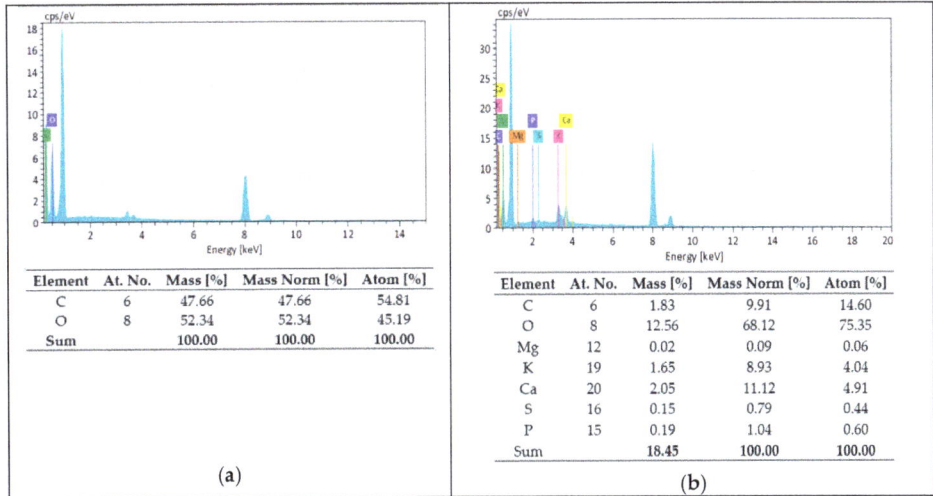

Figure 10. EDS pattern of CF and GOCF fibers: (a) CF; (b) GOCF.

Table 1 presents tensile test results for both curaua fibers, CF and GOCF, with regard to the ultimate stress (σ_f), total strain (ε), and Young's modulus (E). Regarding this table, it is important to mention that the values obtained for these properties agree with those reported in the literature [29,31].

Table 1. Mechanical properties of CF and GOCF.

Condition	σ_f (MPa)	ε (%)	E (GPa)
CF	3153 ± 970	13.48 ± 5.45	25.7 ± 11.3
GOCF	1834 ± 673	8.82 ± 3.10	38.0 ± 10.0

One may infer from the results in Table 1 that the GO coating caused an increase of 47.8% in the Young's modulus of the fibers, corroborating to other authors [49,50]. On the other hand, in maximum tension there was a 71.9% reduction showing a more brittle but more rigid behavior of GOCF compared to CF, possibly because of the relatively low amount of GO used, forming a very thin film on the surface. In addition, it will be shown that, as reinforcement of epoxy matrix composite, this coating is associated with relevant differences in terms of fiber/matrix adherence.

Figure 11 shows the pullout curves, based on the Kelly and Tyson method [51], the curve has three levels, corresponding to the failure mechanisms that occurs in the composite, the first one, for short embedded lengths, refers to the level where only the fiber pullout occurs. In the second stage, there are pullout and fiber rupture; at this stage the length of the fiber in the composite has already reached but not exceeded the critical length. However, when the critical length is exceeded, as the case of the third stage, the failure mechanism of the composite is only by rupture of the fibers, i.e., there are no longer the pullout mechanism. Thus, the critical length for the system fiber/matrix is defined by the maximum value associated with the first stage of the curve [30]. The value of the fiber critical length was calculated as l_c = 2 mm for the CF/epoxy l_c = 1 mm for the GOCF/epoxy. These values are much lower than that of l_c = 10.2 mm, reported for curaua fiber/polyester [30]. The GOCF/epoxy critical length is sensibly lower than non-coated CF/epoxy. Consequently, the interfacial shear strength of the GOCF/epoxy, τ_c = 27.5 MPa is more than 50% higher than that of the CF/epoxy, τ_c = 18.2 MPa. One may infer that for the same embedded length the CF fiber pullout voltage is greater than GOCF fiber, however, this behavior is due to the fact that the GOCF composite is now a new system with a new fiber/matrix interface [30]. Therefore, for each system, there is a strength and a certain critical length.

As the fibers had their tensile strength affected by the GO coating, it is expected that the fiber/matrix system strength also presents similar behavior.

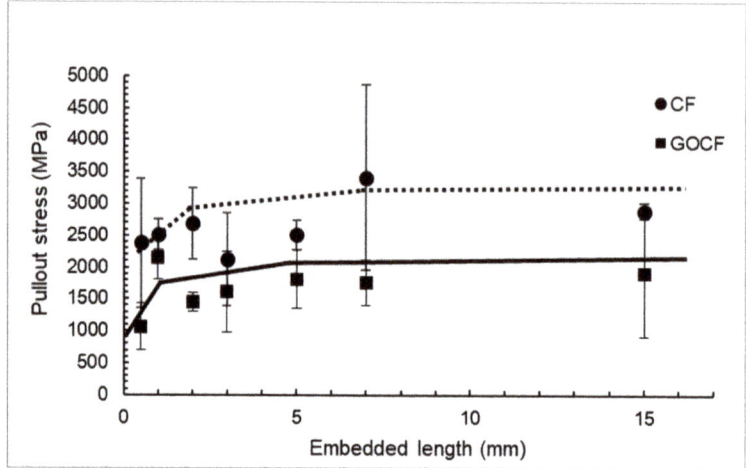

Figure 11. Pullout stress of both curaua fibers, CF and GOCF, versus epoxy embedded length curves.

In the present work, for the first time, ballistic tests were carried out to measure the trauma on the witness clay in MAS target with a second layer epoxy matrix composite reinforced with 30 vol% of curaua fiber both CF and GOCF. In none of the MAS tested, there was complete perforation of the 7.62 mm projectile and the indentation of the clay witness was less than 44 mm, a value considered to be non-lethal to humans by the standard [42]. The results obtained are presented in the Table 2 and visualized in Figure 8. They were also compared with other MAS using distinct fibers, as well as with a same thickness laminate of Kevlar™, as a second layer. The limit value established by the standard is shown as an upper dashed horizontal line, in Figure 12. These results were found to be in good agreement with other authors [4,15] and relatively better than those by Braga et al. [7].

In Figure 12, one should note a slight increase in the value of the indentation in the clay witness caused by the 7.62 mm projectile impact against a MAS target with GOCF epoxy composite as a second layer. Figure 13 illustrates the aspect of both MASs, with CF and GOCF composites, before and after the ballistic test. The integrity, an essential factor for practical applications, is shown to be better than the GOCF in comparison to MAS with CF epoxy composite as a second layer. Indeed, in this latter, the plate fractured into two large pieces as can be seen in Figure 13b. By contrast, MAS target with GOCF composite remained relatively intact in Figure 13d.

Table 2. Depth of indentation of MAS with natural fibers composites and same thickness Kevlar™ for comparison.

MAS Second Layer	Depth of Indentation (mm)	Reference
30 vol% curaua fiber/epoxy composite	25.6± 0.2	PW
30 vol% curaua fiber coated with GO/epoxy composite	27.4 ± 0.3	PW
Kevlar™	23 ± 3	[13]
30 vol% jute fabric/epoxy composite	21 ± 3	[15]
30% curaua non-woven mat/epoxy composite	28 ± 3	[7]
30 vol% jute non-woven mat/polyester composite	24 ± 7	[4]

PW—Present work.

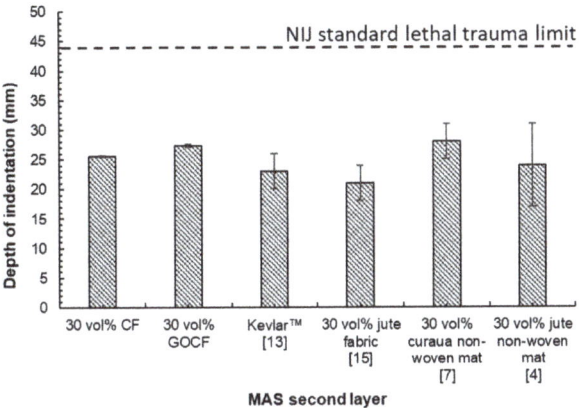

Figure 12. Depth indentation in clay witness of the reinforced composites with 30 vol%.

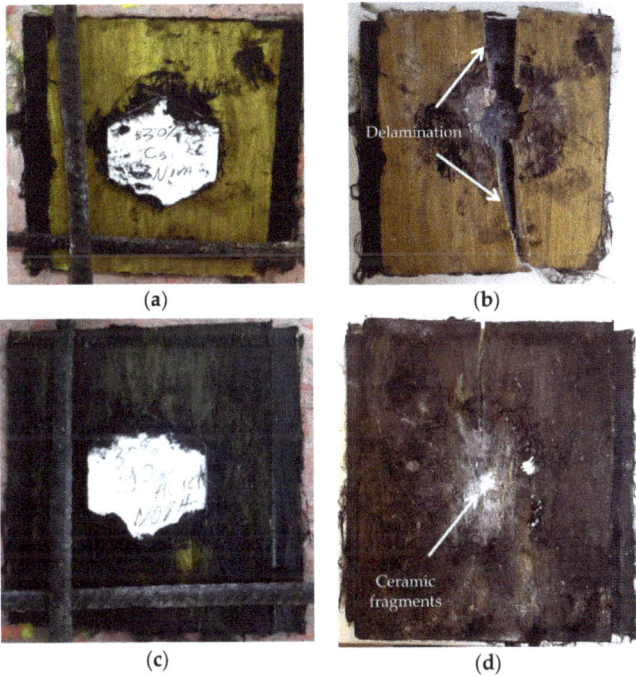

Figure 13. View of MAS target before (**a,c**) and after (**b,d**) the ballistic test: with second layer of (**a,b**) 30 vol% CF; (**c,d**) 30 vol% GOCF.

The smaller hexagonal ceramic tiles, front MAS layer in Figure 13a,c are completely destroyed, Figure 13b,d upon the projectile impact. In an actual armor vest, these tiles compose a mosaic to allow multiple shootings in which a single tile is hit at a time without compromising the armor protection. Figure 14 shows by SEM the ruptured surface of a tile ceramic totally destroyed. This rupture occurs by intergranular fracture absorbing most of the kinetic energy of the projectile. The magnified image in Figure 14b, displays in detail an intergranular microcrack associated with this mechanism of fracture, similar to what was verified by other authors [15,34].

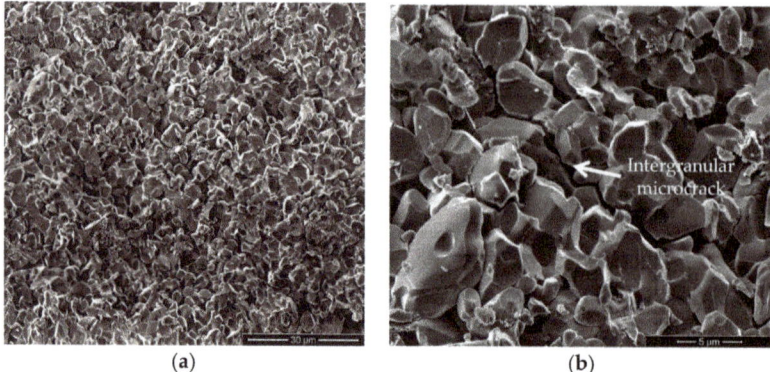

Figure 14. Surface of fracture of the ceramic tablets: (**a**) 3000×; (**b**) 10,000×.

Another important participation of the composite plate as MAS second layer is the capture of ceramic fragments resulting from the shattered front ceramic, Figure 14, which corresponds to a significant amount of the absorbed impact energy [60]. Figure 15 illustrates the capture of ceramic fragments by curaua fibrils that compose each curaua fiber in the epoxy composite. In this figure it is important to note not only the extensive incrustation of microfragments covering the fibrils but also effective fibrils separation. Indeed, as shown in Figures 15 and 16 like most LNFs a curaua fiber is composed of well-adhered fibrils that split apart when subjected to an applied stress [29]. The shock wave resulting from the projectile impact in the present ballistic tests, Figure 3, in addition to complete shatter the front ceramic, Figure 14, also caused separation of fibrils clearly shown in Figure 15. Therefore, for the first time, it is reported a whole view of the mechanisms responsible for dissipating the remaining energy, after the projectile impact against the front ceramic, by the curaua fiber composite as MAS second layer. The indentation results in Table 2, indicate that these mechanisms are responsible for a ballistic performance comparable to Kevlar™ laminate, which is a much stronger material. While the Kevlar™ mechanisms of energy absorption, as MAS second layer, is basically the capture of fragments [60], the curaua fiber composite is associated with several mechanisms with distinct participation of the GO coating. The combination of the following mechanisms makes both CF and GOCF epoxy composites in Table 2 as effective as Kevlar™.

Figure 15. Curaua fiber covered with ceramic fragments.

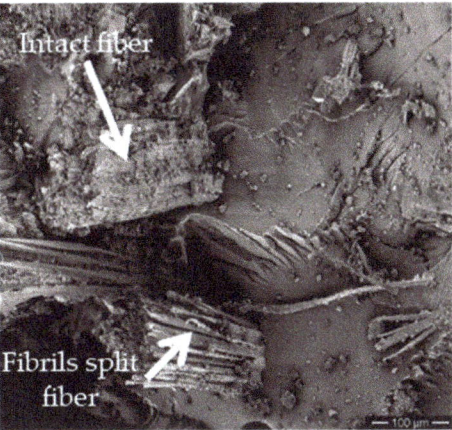

Figure 16. Fiber breaking of the GOCF composite fracture surfaces.

Capture of fragments, Figure 15, the same mechanism first shown in Kevlar™ [26] and later reported for curaua fiber [11,17] and non-woven curaua fabric [7] polymer composites. Apparently, this capture of fragments is not affected by the GO coating.

Fibrils separation, also illustrated in Figure 16, is a specific mechanism for stress-subjected curaua fibers [29], which contributes to dissipate energy by generating free surface area between fibrils. Observed evidences suggest that GO coating makes difficult the fibril separation and has, comparatively, a reduced dissipated energy. This separation in plain curaua fibers (CFs) might disclose individual nano and micro cellulose chains with special behavior [22–27].

Fiber pullout shown in Figure 17 in which a hole left in one site of the fracture surface was caused by a curaua fiber pullout. The insert with higher magnification revels a remaining attached fibril separated from the pulled fiber. In this case, energy is dissipated by the created hole/pulled-out fiber-free surface. No evidence of pullout was found in the GOCF composites, which also indicates a reduced impact energy absorption.

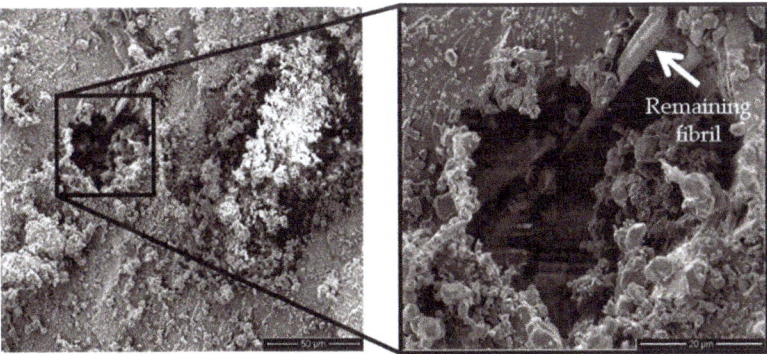

Figure 17. Fiber pullout of the CF composites.

Composite delamination, Figure 13b, which is a macro mechanism of energy dissipation involving the creation of relatively large free surface area associated with the extensive separation between curaua fiber/epoxy matrix. As aforementioned, delamination impairs the integrity of the 30 vol% CF composites despite the dissipated impact energy. In contrast, delamination is not effective in the

30 vol% GOCF. In this case, integrity is maintained as required by the standard for testing armor vests [42].

Fiber breaking, depicted in Figure 16, is a general mechanism common to natural and synthetic fibers, including the aramid fibers in Kevlar™ [61]. In principle, fiber breaking is an alternative to its pullout. In other words, a matrix well-adhered fiber will break instead of pulled-out. This is the case of GOCF composites in which the graphene oxide coating, Figure 9b, is expected to improve the curaua fiber adhesion to the epoxy matrix. Therefore, no pullout occurs in the GOCF fibers that comparatively dissipates more energy by breaking. It is interesting to observe in Figure 16 the rupture of an intact as well as a fibrils split curaua fibers, both indicated by corresponding arrows.

Matrix rupture exemplified in Figure 18 by a flat epoxy broken surface (right side) around a well-adhered GOCF fiber (left side). This is a specific mechanism for brittle polymer composites that undergo extensive matrix rupture upon a ballistic impact. A significant amount of energy is dissipated but enough well-adhered fibers, like in the present case of 30 vol% of GOCF, is important to avoid loss of integrity as shown in Figure 13.

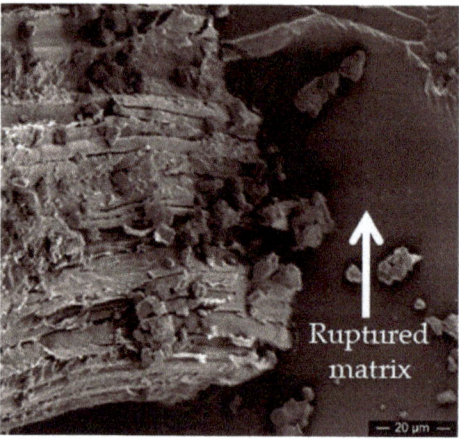

Figure 18. Matrix rupture of the GOCF Composite.

As a final remark, it is worth reminding that the combination of energy dissipation mechanisms guarantees to a 30 vol% curaua fiber (plain or graphene oxide coated) reinforced epoxy composite as MAS second layer, an acceptable ballistic performance, Table 2, similar to that of a Kevlar™ laminate with same thickness. This performance, given by the standard backface signature less than 44 mm [42], is slightly superior in the GOCF composites, Table 2 and Figure 12, owing to the better fiber/matrix adhesion provided by the GO coating, in some of the aforementioned mechanisms. On the other hand, this better adhesion supports the 30 vol% GOCF integrity, which is essential for MAS in armor vest.

4. Conclusions

- According to the FTIR analysis, the GO caused changes in the characteristic bands of the CF fibers, suggesting that bonds were formed as well as the appearance of new bands characteristic of the molecular structure of the GO.
- The thermal degradation of the GOCF fibers was retarded by the action of the GO coating, causing an insulation which contributes to higher temperature resistance, in relation to the CF fibers.
- Pullout test of untreated curaua fiber (CF) and graphene oxide coated curaua fiber (GOCF) embedded in epoxy matrix revealed a substantial reduction in the GOCF critical length in association with a more than 50 percent higher interfacial shear strength. This behavior is also superior to those of other material fibers.

- Epoxy composite plates reinforced with 30 vol% of either CF or GOCF, applied as 10 mm thick second layer in a front ceramic multilayered armor system, display a ballistic performance against the threat of 7.62 mm projectile within the backface signature (indentation < 44 mm) required by the standard.
- This ballistic performance comparable to that of the same thickness Kevlar™ laminate as MAS second layer, was for the first time interpreted as been related to a combination of the following impact energy mechanisms: (i) capture of fragments; (ii) fibrils separation; (iii) fiber pullout; (iv) composite delamination; (v) fiber breaking; and (vi) matrix rupture.
- The better adherence of GOCF to the epoxy matrix reduces, comparatively, the amount of absorbed energy by mechanisms (ii), (iii), (iv), and (vi). This results in slightly higher ballistic backface signature but a better integrity for the 30 vol% GOCF composites, which is a necessary condition for armor vest using MAS. The plain CF ballistic performance is similar to other natural fibers.
- It is also ruled out the need of a ductile metal sheet, usually applied as MAS third layer, since the 10 mm thick GOCF composite is enough for the required standard performance.

Author Contributions: All authors contributed equally to this manuscript.

Funding: This research received no external funding.

Acknowledgments: The authors thank the support to this investigation by the Brazilian agencies: CNPq, CAPES and FAPERJ; and UFPA for supplying the mallow fibers.

Conflicts of Interest: The authors declare no conflict of interest.

References

1. Benzait, Z.; Trabzon, L. A review of recent research on materials used in polymer–matrix composites for body armor application. *J. Compos. Mater.* **2018**, *52*, 3241–3263. [CrossRef]
2. Naveen, J.; Jawaid, M.; Zainudin, E.S.; Sultan, M.T.; Yahaya, R. Evaluation of ballistic performance of hybrid Kevlar®/Cocos nucifera sheath reinforced epoxy composites. *J. Text. Inst.* **2019**, *110*, 1179–1189. [CrossRef]
3. Monteiro, S.; Pereira, A.; Ferreira, C.; Pereira Júnior, É.; Weber, R.; Assis, F. Performance of plain woven jute fabric-reinforced polyester matrix composite in multilayered ballistic system. *Polymers* **2018**, *10*, 230. [CrossRef] [PubMed]
4. Assis, F.S.; Pereira, A.C.; da Costa Garcia Filho, F.; Lima, É.P., Jr.; Monteiro, S.N.; Weber, R.P. Performance of jute non-woven mat reinforced polyester matrix composite in multilayered armor. *J. Mater. Res. Technol.* **2018**, *7*, 535–540. [CrossRef]
5. Braga, F.D.O.; Bolzan, L.T.; Ramos, F.J.H.T.V.; Monteiro, S.N.; Lima, É.P., Jr.; Silva, L.C.D. Ballistic efficiency of multilayered armor systems with sisal fiber polyester composites. *Mater. Res.* **2017**, *20*, 767–774. [CrossRef]
6. Silva, A.O.; de Castro Monsores, K.G.; Oliveira, S.D.S.A.; Weber, R.P.; Monteiro, S.N. Ballistic behavior of a hybrid composite reinforced with curaua and aramid fabric subjected to ultraviolet radiation. *J. Mater. Res. Technol.* **2018**, *7*, 584–591. [CrossRef]
7. Braga, F.O.; Bolzan, L.T.; da Luz, F.S.; Lopes, P.H.L.M.; Lima, É.P., Jr.; Monteiro, S.N. High energy ballistic and fracture comparison between multilayered armor systems using non-woven curaua fabric composites and aramid laminates. *J. Mater. Res. Technol.* **2017**, *6*, 417–422. [CrossRef]
8. Nascimento, L.F.C.; Louro, L.H.L.; Monteiro, S.N.; Lima, É.P.; da Luz, F.S. Mallow fiber-reinforced epoxy composites in multilayered armor for personal ballistic protection. *JOM* **2017**, *69*, 2052–2056. [CrossRef]
9. Nascimento, L.F.C.; Holanda, L.I.F.; Louro, L.H.L.; Monteiro, S.N.; Gomes, A.V.; Lima, É.P. Natural mallow fiber-reinforced epoxy composite for ballistic armor against class III-A ammunition. *Metall. Mater. Trans. A* **2017**, *48*, 4425–4431. [CrossRef]
10. Braga, F.O.; Bolzan, L.T.; Lima, É.P., Jr.; Monteiro, S.N. Performance of natural curaua fiber-reinforced polyester composites under 7.62 mm bullet impact as a stand-alone ballistic armor. *J. Mater. Res. Technol.* **2017**, *6*, 323–328. [CrossRef]
11. Monteiro, S.N.; de Oliveira Braga, F.; Pereira Lima, E.; Henrique Leme Louro, L.; Wieslaw Drelich, J. Promising curaua fiber-reinforced polyester composite for high-impact ballistic multilayered armor. *Polym. Eng. Sci.* **2017**, *57*, 947–954. [CrossRef]

12. Monteiro, S.N.; Milanezi, T.L.; Louro, L.H.L.; Lima, É.P., Jr.; Braga, F.O.; Gomes, A.V.; Drelich, J.W. Novel ballistic ramie fabric composite competing Kevlar™ fabric in multilayered armor. *Mater. Des.* **2016**, *96*, 263–269. [CrossRef]
13. Monteiro, S.N.; Candido, V.S.; Braga, F.O.; Bolzan, L.T.; Weber, R.P.; Drelich, J.W. Sugarcane bagasse waste in composites for multilayered armor. *Eur. Polym. J.* **2016**, *78*, 173–185. [CrossRef]
14. Rohen, L.A.; Margem, F.M.; Monteiro, S.N.; Vieira, C.M.F.; Madeira de Araújo, B.; Lima, E.S. Ballistic efficiency of an individual epoxy composite reinforced with sisal fibers in multilayered armor. *Mater. Res.* **2015**, *18*, 55–62. [CrossRef]
15. Luz, F.S.D.; Junior, L.; Pereira, E.; Louro, L.H.L.; Monteiro, S.N. Ballistic test of multilayered armor with intermediate epoxy composite reinforced with jute fabric. *Mater. Res.* **2015**, *18*, 170–177. [CrossRef]
16. Cruz, R.B.D.; Junior, L.; Pereira, E.; Monteiro, S.N.; Louro, L.H.L. Giant bamboo fiber reinforced epoxy composite in multilayered ballistic armor. *Mater. Res.* **2015**, *18*, 70–75. [CrossRef]
17. Monteiro, S.N.; Louro, L.H.L.; Trindade, W.; Elias, C.N.; Ferreira, C.L.; Sousa Lima, E.; Da Silva, L.C. Natural curaua fiber-reinforced composites in multilayered ballistic armor. *Metall. Mater. Trans. A* **2015**, *46*, 4567–4577. [CrossRef]
18. Sanjay, M.R.; Madhu, P.; Jawaid, M.; Senthamaraikannan, P.; Senthil, S.; Pradeep, S. Characterization and properties of natural fiber polymer composites: A comprehensive review. *J. Clean. Prod.* **2018**, *172*, 566–581. [CrossRef]
19. Pickering, K.L.; Efendy, M.A.; Le, T.M. A review of recent developments in natural fibre composites and their mechanical performance. *Compos. Part A Appl. Sci. Manuf.* **2016**, *83*, 98–112. [CrossRef]
20. Thakur, V.K.; Thakur, M.K.; Gupta, R.K. raw natural fiber–based polymer composites. *Int. J. Polym. Anal. Charact.* **2014**, *19*, 256–271. [CrossRef]
21. Monteiro, S.N.; Lopes, F.P.D.; Barbosa, A.P.; Bevitori, A.B.; Da Silva, I.L.A.; Da Costa, L.L. Natural lignocellulosic fibers as engineering materials—An overview. *Metall. Mater. Trans. A* **2011**, *42*, 2963. [CrossRef]
22. Sharma, P.R.; Sharma, S.K.; Antoine, R.; Hsiao, B.S. Efficient removal of arsenic using zinc oxide nanocrystal-decorated regenerated microfibrillated cellulose scaffolds. *ACS Sustain. Chem. Eng.* **2019**, *7*, 6140–6151. [CrossRef]
23. Klemm, D.; Cranston, E.D.; Fischer, D.; Gama, M.; Kedzior, S.A.; Kralisch, D.; Petzold-Welcke, K. Nanocellulose as a natural source for groundbreaking applications in materials science: Today's state. *Mater. Today* **2018**, *21*, 720–748. [CrossRef]
24. Sharma, P.R.; Chattopadhyay, A.; Sharma, S.K.; Geng, L.; Amiralian, N.; Martin, D.; Hsiao, B.S. Nanocellulose from spinifex as an effective adsorbent to remove cadmium (II) from water. *ACS Sustain. Chem. Eng.* **2018**, *6*, 3279–3290. [CrossRef]
25. Du, X.; Zhang, Z.; Liu, W.; Deng, Y. Nanocellulose-based conductive materials and their emerging applications in energy devices—A review. *Nano Energy* **2017**, *35*, 299–320. [CrossRef]
26. Golmohammadi, H.; Morales-Narvaez, E.; Naghdi, T.; Merkoci, A. Nanocellulose in sensing and biosensing. *Chem. Mater.* **2017**, *29*, 5426–5446. [CrossRef]
27. Sabo, R.; Yermakov, A.; Law, C.T.; Elhajjar, R. Nanocellulose-enabled electronics, energy harvesting devices, smart materials and sensors: A review. *J. Renew. Mater.* **2016**, *4*, 297–312. [CrossRef]
28. Klemm, D.; Heublein, B.; Fink, H.P.; Bohn, A. Cellulose: Fascinating biopolymer and sustainable raw material. *Angew. Chem. Int. Ed.* **2005**, *44*, 3358–3393. [CrossRef]
29. Monteiro, S.N.; Lopes, F.P.D.; Ferreira, A.S.; Nascimento, D.C.O. Natural-fiber polymer-matrix composites: Cheaper, tougher, and environmentally friendly. *JOM* **2009**, *61*, 17–22. [CrossRef]
30. Monteiro, S.N.; Aquino, R.C.M.; Lopes, F.P.D. Performance of curaua fibers in pullout tests. *J. Mater. Sci.* **2008**, *43*, 489–493. [CrossRef]
31. Tomczak, F.; Satyanarayana, K.G.; Sydenstricker, T.H.D. Studies on lignocellulosic fibers of Brazil: Part III–Morphology and properties of Brazilian curauá fibers. *Compos. Part A Appl. Sci. Manuf.* **2007**, *38*, 2227–2236. [CrossRef]
32. Zah, R.; Hischier, R.; Leão, A.L.; Braun, I. Curauá fibers in the automobile industry–a sustainability assessment. *J. Clean. Prod.* **2007**, *15*, 1032–1040. [CrossRef]
33. Caraschi, J.C.; Leāto, A.L. Characterization of curaua fiber. *Mol. Cryst. Liq. Cryst. Sci. Technol. Sect. A. Mol. Cryst. Liq. Cryst.* **2000**, *353*, 149–152. [CrossRef]

34. Leao, A.L.; Caraschi, J.C.; Tan, I.H. Curaua fiber—A tropical natural fibers from Amazon potential and applications in composites. *Nat. Polym. Agrofibers Bases Compos.* **2000**, *2000*, 257–272.
35. Callister, W.D., Jr.; Rethwisch, D.G. *Mater. Science and Engineering: An Introduction*, 10th ed.; John Wiley & Sons: New York, NY, USA, 2018; pp. 665–715. ISBN 9781119405498.
36. Alkbir, M.F.M.; Sapuan, S.M.; Nuraini, A.A.; Ishak, M.R. Fibre properties and crashworthiness parameters of natural fibre-reinforced composite structure: A literature review. *Compos. Struct.* **2016**, *148*, 59–73. [CrossRef]
37. Güven, O.; Monteiro, S.N.; Moura, E.A.; Drelich, J.W. Re-emerging field of lignocellulosic fiber–polymer composites and ionizing radiation technology in their formulation. *Polym. Rev.* **2016**, *56*, 702–736. [CrossRef]
38. Mohammed, L.; Ansari, M.N.; Pua, G.; Jawaid, M.; Islam, M.S. A review on natural fiber reinforced polymer composite and its applications. *Int. J. Polym. Sci.* **2015**, *2015*, 1–15. [CrossRef]
39. Faruk, O.; Bledzki, A.K.; Fink, H.P.; Sain, M. Progress report on natural fiber reinforced composites. *Macromol. Mater. Eng.* **2014**, *299*, 9–26. [CrossRef]
40. Shah, D.U. Developing plant fibre composites for structural applications by optimising composite parameters: A critical review. *J. Mater. Sci.* **2013**, *48*, 6083–6107. [CrossRef]
41. Dittenber, D.B.; GangaRao, H.V. Critical review of recent publications on use of natural composites in infrastructure. *Compos. Part A Appl. Sci. Manuf.* **2012**, *43*, 1419–1429. [CrossRef]
42. National Criminal Justice Reference Service. US Department of Justice, & National Institute of Justice. NIJ 0101.04. Ballistic Resistance of Body Armor. 2000. Available online: https://www.ncjrs.gov (accessed on 13 August 2019).
43. Kalia, S.; Kaith, B.S.; Kaur, I. Pretreatments of natural fibers and their application as reinforcing material in polymer composites—A review. *Polym. Eng. Sci.* **2009**, *49*, 1253–1272. [CrossRef]
44. Li, X.; Tabil, L.G.; Panigrahi, S. Chemical treatments of natural fiber for use in natural fiber-reinforced composites: A review. *J. Polym. Environ.* **2007**, *15*, 25–33. [CrossRef]
45. Geim, A.K.; Novoselov, K.S. The rise of graphene. In *Nanoscience and Technology: A Collection of Reviews from Nature Journals*; World Scientific: Singapore; London, UK, 2010; pp. 11–19. [CrossRef]
46. Lee, C.; Wei, X.; Kysar, J.W.; Hone, J. Measurement of the elastic properties and intrinsic strength of monolayer graphene. *Science* **2008**, *321*, 385–388. [CrossRef] [PubMed]
47. Rourke, J.P.; Pandey, P.A.; Moore, J.J.; Bates, M.; Kinloch, I.A.; Young, R.J.; Wilson, N.R. The real graphene oxide revealed: Stripping the oxidative debris from the graphene-like sheets. *Angew. Chem. Int. Ed.* **2011**, *50*, 3173–3177. [CrossRef] [PubMed]
48. Qi, X.; Pu, K.Y.; Li, H.; Zhou, X.; Wu, S.; Fan, Q.L.; Zhang, H. Amphiphilic graphene composites. *Angew. Chem. Int. Ed.* **2010**, *49*, 9426–9429. [CrossRef] [PubMed]
49. Chen, J.; Huang, Z.; Lv, W.; Wang, C. Graphene oxide decorated sisal fiber/MAPP modified PP composites: Toward high-performance biocomposites. *Polym. Compos.* **2018**, *39*, E113–E121. [CrossRef]
50. Sarker, F.; Karim, N.; Afroj, S.; Koncherry, V.; Novoselov, K.S.; Potluri, P. High-Performance Graphene-Based Natural Fiber Composites. *ACS Appl. Mater. Interfaces* **2018**, *10*, 34502–34512. [CrossRef]
51. Kelly, A., Tyson, W.R. *High Strength Materials*; John Wiley & Sons Inc.: Nova York, NY, USA, 1965; pp. 578–583.
52. American Society for Testing and Materials. *D3822-01: Standard Test Methods for Tensile Properties of Single Textile Fibre*; American Society for Testing and Materials: West Conshohocken, PA, USA, 2001. [CrossRef]
53. Loryuenyong, V.; Totepvimarn, K.; Eimburanapravat, P.; Boonchompoo, W.; Buasri, A. Preparation and characterization of reduced graphene oxide sheets via water-based exfoliation and reduction methods. *Adv. Mater. Sci. Eng.* **2013**, *2013*, 1–5. [CrossRef]
54. Fan, J.; Shi, Z.; Zhang, L.; Wang, J.; Yin, J. Aramid nanofiber-functionalized graphene nanosheets for polymer reinforcement. *Nanoscale* **2012**, *4*, 7046. [CrossRef]
55. Zhang, T.; Zhang, D.; Shen, M. A low-cost method for preliminary separation of reduced graphene oxide nanosheets. *Mater. Lett.* **2009**, *63*, 2051–2054. [CrossRef]
56. Hoareau, W.; Trindade, W.G.; Siegmund, B.; Castellan, A.; Frollini, E. Sugar cane bagasse and curaua lignins oxidized by chlorine dioxide and reacted with furfuryl alcohol: Characterization and stability. *Polym. Degrad. Stab.* **2004**, *86*, 567–576. [CrossRef]
57. Spinace, M.A.; Lambert, C.S.; Fermoselli, K.K.; De Paoli, M.A. Characterization of lignocellulosic curaua fibres. *Carbohydr. Polym.* **2008**, *77*, 47–53. [CrossRef]
58. Corrêa, A.C.; de Morais Teixeira, E.; Pessan, L.A.; Mattoso, L.H.C. Cellulose nanofibers from curaua fibers. *Cellulose* **2009**, *17*, 1183–1192. [CrossRef]

59. Sharma, P.R.; Joshi, R.; Sharma, S.K.; Hsiao, B.S. A simple approach to prepare carboxycellulose nanofibers from untreated biomass. *Biomacromolecules* **2017**, *18*, 2333–2342. [CrossRef] [PubMed]
60. Monteiro, S.N.; Lima, É.P.; Louro, L.H.L.; Da Silva, L.C.; Drelich, J.W. Unlocking function of aramid fibers in multilayered ballistic armor. *Metall. Mater. Trans. A* **2015**, *46*, 37–40. [CrossRef]
61. Lee, Y.S.; Wetzel, E.D.; Wagner, N.J. The ballistic impact characteristics of Kevlar® woven fabrics impregnated with a colloidal shear thickening fluid. *J. Mater. Sci.* **2003**, *38*, 2825–2833. [CrossRef]

© 2019 by the authors. Licensee MDPI, Basel, Switzerland. This article is an open access article distributed under the terms and conditions of the Creative Commons Attribution (CC BY) license (http://creativecommons.org/licenses/by/4.0/).

Article

The Relationships between the Working Fluids, Process Characteristics and Products from the Modified Coaxial Electrospinning of Zein

Menglong Wang [1], Tao Hai [1], Zhangbin Feng [1], Deng-Guang Yu [1,*], Yaoyao Yang [1] and SW Annie Bligh [2,*]

[1] School of Materials Science & Engineering, University of Shanghai for Science & Technology, Shanghai 200093, China
[2] Caritas Institute of Higher Education, 2 Chui Ling Lane, Tseung Kwan O, New Territories, Hong Kong 999077, China
* Correspondence: ydg017@usst.edu.cn (D.G.-Y.); abligh@cihe.edu.hk (S.W.A.B.)

Received: 3 April 2019; Accepted: 11 July 2019; Published: 1 August 2019

Abstract: The accurate prediction and manipulation of nanoscale product sizes is a major challenge in material processing. In this investigation, two process characteristics were explored during the modified coaxial electrospinning of zein, with the aim of understanding how this impacts the products formed. The characteristics studied were the spreading angle at the unstable region (θ) and the length of the straight fluid jet (L). An electrospinnable zein core solution was prepared and processed with a sheath comprising ethanolic solutions of LiCl. The width of the zein nanoribbons formed (W) was found to be more closely correlated with the spreading angle and straight fluid jet length than with the experimental parameters (the electrolyte concentrations and conductivity of the shell fluids). Linear equations $W = 546.44L - 666.04$ and $W = 2255.3\theta - 22.7$ could be developed with correlation coefficients of $R_{wl}^2 = 0.9845$ and $R_{w\theta}^2 = 0.9924$, respectively. These highly linear relationships reveal that the process characteristics can be very useful tools for both predicting the quality of the electrospun products, and manipulating their sizes for functional applications. This arises because any changes in the experimental parameters would have an influence on both the process characteristics and the solid products' properties.

Keywords: coaxial electrospinning; length of straight fluid jet; spreading angle; nanoribbons; linear relationship

1. Introduction

For polymer processing engineering, a key requirement is to be able to accurately interrelate the experimental conditions and the properties of the final products [1–3]. This is particularly challenging when the final products are nanoparticles or nanofibers [4–9]. Both of them can be generated by electrohydrodynamic atomization (EHDA) using electrostatic energy [10–16], and while there are numerous reports of such fabrication processes, it remains the case that it is extremely difficult to predict the outcome of a given experiment.

Figure 1 presents a schematic depicting the experimental parameters that exert a significant influence on the final polymer nanofibers generated in the simplest electrospinning experiment, which involves a single working fluid. In electrospinning, the working fluid and electrostatic energy are brought together at the nozzle of the spinneret [17–21]. After exiting the spinneret, the working fluid is forced to bend and whip, and during this process it is solidified into nanofibers at an extremely rapid speed [22–27]. Based on this working procedure, the key experimental parameters can be divided into three categories (see Figure 1). Correspondingly, the resultant nanofiber diameter (D) is a function of the working fluid properties (w), the operational conditions (o), and the environmental parameters (e): i.e., $D = f(w,o,e)$.

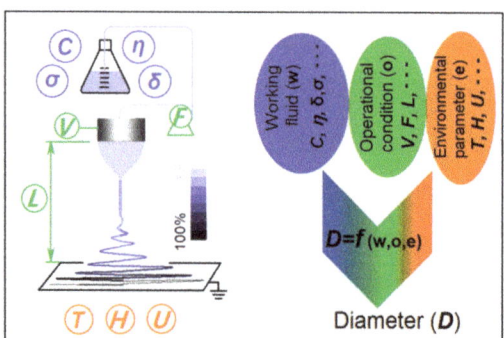

Figure 1. A diagram showing the single-fluid electrospinning process and the experimental parameters exerting influence on the diameters of the polymer nanofibers generated.

Over the past two decades, electrospinning has developed very rapidly, with potential applications of polymer nanofibers having been proposed in a myriad of scientific fields [28–31]. In addition, the simple single-fluid electrospinning process has advanced to two-fluid coaxial and side-by-side processes, and tri-fluid coaxial and combined coaxial/side-by-side processes. It has proven to be possible to perform the electrospinning process even when one or more of the working fluids cannot on its own be processed: For instance, in modified coaxial electrospinning, a spinnable core solution is partnered with an unspinnable sheath fluid. These novel processes permit the production of nanofibers with increasingly complicated nanostructures [32–37]. As a result, in the literature, there are numerous publications that explore the influence of a single parameter on the nanofibers or nanostructures prepared by electrospinning, elucidating relationships for manipulating the products (mainly in terms of diameter, but also for other properties such as morphology, surface smoothness and the details of the nanostructure) [38–43]. However, there are a number of experimental parameters that can simultaneously exert an influence on the final products [44–47]. For example, the properties of working fluid include polymer concentration (C), viscosity (η), surface tension (δ), and conductivity (σ); the operational conditions include the applied voltage (V), the fluid flow rate (F), and the fiber collection distance (L); and the environmental parameters comprise of temperature (T), humidity (H) and the possible vacuum (U) (Figure 1).

Thus, although a lot of effort has been expended to predict and manipulate the diameters of electrospun nanofibers, the results are typically far from satisfactory [48,49]. During electrospinning, almost all the experimental parameters can influence the working process, and furthermore, they are not independent variables, and can also influence each other. For example, the flow rate of the working fluid and the applied voltage need to be matched, or droplets of working fluid may fall directly onto the fiber collector. Thus, although many mathematical models have been put forward for a particular working fluid [48], they often fail to be applicable to other situations. Although the experimental parameters have drawn intensive attention, it is strange that the detailed steps in the process of electrospinning have received very limited attention. These include the formation of the Taylor cone, the ejection of a straight fluid jet, and also the bending and whipping region [50,51]. These individual steps are influenced by all the experimental parameters, and thus can be directly controlled by researchers. It is hypothesized that the nature of each of these stages of the spinning process should have a distinct relationship with the final nanofiber properties, particularly their sizes.

Here, for the first time, we design a method to elucidate the interrelated relationships between the working fluids, the electrospinning process characteristics, and the nanoribbons formed during the modified coaxial electrospinning of zein. Zein is one of the best understood plant proteins. Extracted from maize, it has been widely used as a coating for candy, nuts, fruit, and pharmaceuticals. Zein can be processed into resins and other bioplastic polymers, which can be extruded or rolled into a variety of plastic products [52]. Zein has good processability using both these traditional technologies and advanced technologies such as electrospinning and electrospraying [53–55]. Here, it was selected as a model system for a detailed exploration of the individual stages in the electrospinning process. A series of modified coaxial electrospinning processes were carried out and several types of zein nanoribbons were fabricated. The working processes were digitally recorded and quantitatively described in terms of length of the straight fluid jet and the spreading angle of the unstable region, and these were interrelated with both the initial conductivity and the final zein nanoribbon widths.

2. Materials and Methods

2.1. Materials

Zein (98% purity) was purchased from Shanghai Hewu Biotechnology Co., Ltd. (Shanghai, China). Anhydrous ethanol and lithium chloride were obtained from the Husheng Reagent Co., Ltd. (Shanghai, China). Water was double distilled just before use.

2.2. Modified Coaxial Electrospraying

The core fluid consisted of 28 g zein in 100 mL of a 75%/25% (*v*/*v*) ethanol/water mixture, which showed a yellow color. Four LiCl solutions in ethanol (at 0, 5, 10, and 20 mg/mL) were utilized as the sheath fluids, and the resultant nanoribbons were labeled as Z1, Z2, Z3, and Z4, respectively. The conductivities of the sheath fluids were assessed using a DDS-11 digital conductivity meter (Shanghai Rex Co-perfect Instrument Co., Ltd., Shanghai, China).

A homemade system was employed to conduct all the electrospinning processes. This consisted of two syringe pumps (KDS100 and KDS200, Cole–Parmer, Vernon Hills, IL, USA), a power supply (ZGF200, 60 kV/2 mA, Wuhan Huatian Corp., Wuhan, China), a homemade concentric spinneret, and an aluminum-coated flat piece of cardboard as the collector. The ambient temperature and humidity were 21 ± 4 °C and 53 ± 6%, respectively. All the working processes were recorded using a digital camera (PowerShot A490; Canon, Tokyo, Japan). Following optimization, the collection distance and voltage were fixed at 20 cm and 17 kV, respectively.

2.3. Morphology of the Prepared Nanoparticles

The surface morphologies of the electrospun products were observed by scanning electron microscopy (SEM; Quanta FEG450, FEI Corporation, Hillsboro, OR, USA) at 20 kV acceleration voltage. Before observation, the samples were sputter-coated with gold under vacuum. The images were analyzed using the ImageJ software (National Institutes of Health, Bethesda, MD, USA), with measurements taken at over 100 different places to determine the average ribbon diameter.

3. Results and Discussion

3.1. Implementation of Modified Coaxial Electrospinning

Traditionally, coaxial electrospinning is carried out using an electrospinnable sheath fluid to encapsulate either a core liquid which may be spinnable or unspinnable [18,24]. Some years ago, Yu and co-workers expanded this concept to develop the modified coaxial process, with an unspinnable liquid as the sheath fluid (Figure 2).

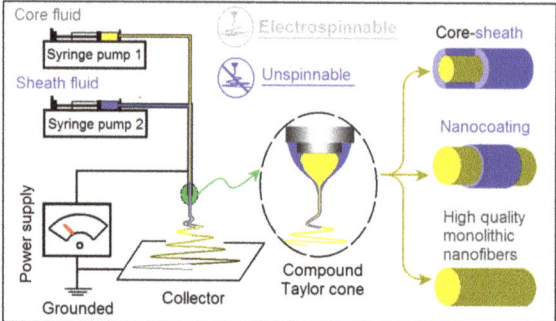

Figure 2. The modified coaxial electrospinning process, which permits a range of novel structures to be obtained through the unspinnable sheath fluid.

The homemade concentric spinneret and the electrospinning apparatus used in this work are shown in Figure 3. The spinneret (Figure 3a) consists of a narrow metal capillary (inner diameter 0.3 mm, wall thickness 0.1 mm) nested into an outer capillary (inner diameter of 1.2 mm, wall thickness 0.2 mm). Two syringe pumps were employed to drive the core and shell liquids to the spinneret (Figure 3b). The yellow zein solution was directly guided to the inner needle of the spinneret through a plastic syringe, whereas the sheath LiCl solution was pumped to the spinneret through elastic silicon tubing. An alligator clip connects the spinneret to the power supply and carries electrostatic energy to the working fluid (Figure 3c).

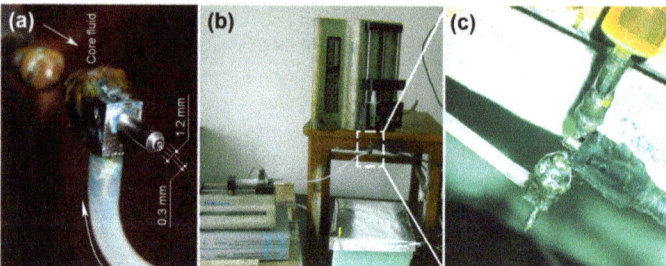

Figure 3. The apparatus used for modified coaxial electrospinning: (**a**) The home-made concentration spinneret; (**b**) the arrangement of apparatus; and (**c**) the connection of the power supply and working fluids with the spinneret.

3.2. The Working Processes and the Resultant Zein Nanoribbons

The electrospinning process consists of three successive stages: The formation of a Taylor cone, the straight fluid jet emitted from the Taylor cone, and the unstable region, which is composed of numerous bending and whipping loops. The formation of the Taylor cone is a balance between the electrical force exerted on the droplets exiting the spinneret, and the surface tension of the working fluids. When the conductivity of the working fluid increases, the electrical forces should increase correspondingly. Thus, an increase in the LiCl concentration in the sheath fluid is expected to result in a stronger electrical force being applied to the working fluids. Under the same applied voltage and spinneret-to-collector distance, this force will greatly change the behaviors of the working fluids. Digital photographs of these are given in Figure 4. As the LiCl concentration increased from 0 to 5, and from 10 to 20 mg/mL, the length of the straight fluid jet decreased from 3.3 ± 0.4, to 2.9 ± 0.3, and from 2.4 ± 0.3 to 2.2 ± 0.2 mm, respectively. Meanwhile, the spreading angles of the unstable region increased from 51 ± 4° to 59 ± 5°, and from 68 ± 4° to 77 ± 6°.

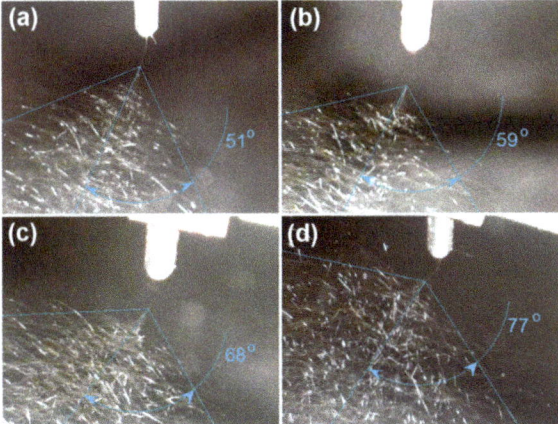

Figure 4. The changes of spreading angle and the length of straight fluid jet with the increase of LiCl in the sheath solution (mg/mL): (**a**) 0; (**b**) 5; (**c**) 10; (**d**) 20.

SEM images of the resultant nanoribbons and their diameter distributions are shown in Figure 5. All the ribbons have a linear morphology. No beads or spindles are found in the ribbons, suggesting the core zein solution has good electrospinnability. Nanoribbons Z1, Z2, Z3, and Z4 have an estimated width of 1.12 ± 0.14, 0.91 ± 0.12, 0.58 ± 0.09, and 0.52 ± 0.07 µm, respectively.

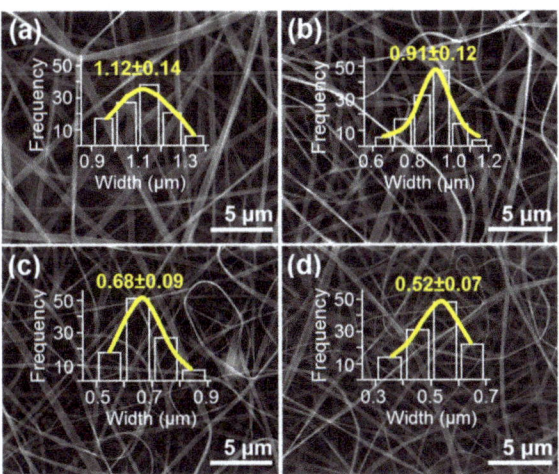

Figure 5. SEM images of resultant zein nanoribbons, with their width distributions. (**a**) Z1; (**b**) Z2; (**c**) Z3; (**d**) Z4.

3.3. The Influence of Conductivity on the Behavior of the Working Fluids

Although a single-step and straightforward process for creating nanoribbons, the electrospinning process is in fact very complicated. This complexity is reflected in two ways. First, the process involves the overlap of multiple disciplines such as hydrodynamic science, polymer science and rheology, and electric dynamics. Second, a small change in the working fluid properties can greatly influence the process and its products.

As the concentration of LiCl increased, the conductivity of the sheath solution also rose (Figure 6a). This increase in conductivity will make the solution subject to stronger electrical forces, which in turn alter the behavior of both the sheath and core working fluids. The length of the straight fluid jets gradually decreased with conductivity in a linear fashion, as shown in Figure 6b: $L = 3.38 - 5.25 \times 10^4\ \sigma$, with a correlation coefficient of $R_{L\sigma}^2 = 0.9761$. Similarly, the spreading angle of the unstable region gradually increased with conductivity (Figure 6c). A linear equation $\theta = 48.775 + 0.011\ \sigma$ can be fitted to the data, giving a correlation coefficient of $R_{\theta\sigma}^2 = 0.9296$. The clearly linear nature of the plots in Figure 6b, c and the high R^2 values obtained show there are clear causal relationships here.

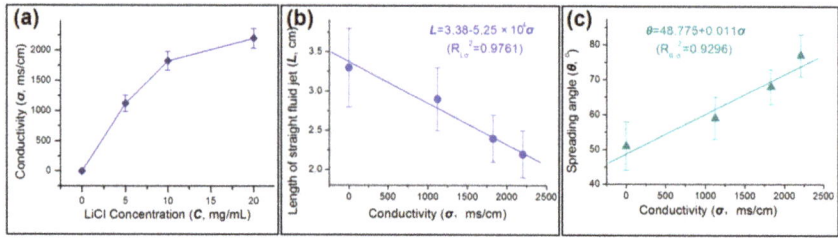

Figure 6. The influence of the sheath fluid conductivity on the behavior of the working fluids: (**a**) The relationship between LiCl concentration and solution conductivity; (**b**) the decrease in the length of the straight fluid jet with an increase of conductivity; (**c**) the increase of spreading angle with rising conductivity.

In literature, numerous publications have investigated the electrospinnability of a certain working fluid, which is mainly determined by the type of filament-forming polymer, its concentrations in the working fluid, and the applied voltage. After the past two decades' effort, near 200 polymers can be processed into fibers using electrospinning. However, few efforts have been focused on the behaviors of working fluids within their electrospinnable windows. Knowledge about the adaptability of working fluids under the high electrical field should be useful for manipulating the fluid processing process in a more intentional manner.

3.4. The Effect of Sheath Working Fluid Properties on the Width of Zein Nanoribbons

A wide variety of experimental parameters have been investigated in terms of their effect on the properties of electrospun fibers, and the solution conductivity of working fluid is recognized as being of major importance [56]. In this study, the electrolyte LiCl was added only into the sheath working fluid, because charges are always concentrated on the surface of the Taylor cone. The width of the zein ribbons produced is clearly correlated with the LiCl concentration in the sheath fluid, with a good fit to the data obtained with the linear equation $W = 1033.2C - 26.9$ ($R_1^2 = 0.9297$; Figure 7a). A similar linear equation is observed when plotting ribbon width as a function of sheath solution conductivity ($W = 11177.86 - 0.29\sigma$; $R_2^2 = 0.9639$). These linear equations suggested that the LiCl concentration and conductivity of sheath working fluid directly influenced the width of the zein nanoribbons fabricated. These equations can hence be exploited to predict the size of the products from this electrospinning process, and provide useful information for optimizing the working processes.

It is a common strategy to optimize the experimental conditions through simultaneous investigations on several levels of an experimental parameter, just as here with the LiCl concentration. However, only a small part of the related publications has taken a further step to disclose the inherent relationship between the vital properties of the working fluid with the final product's quality. Here, the conductivity of sheath LiCl solution showed a better linear relationship with the width of zein ribbons than the LiCl concentration. Thus, among many other solution properties such as surface tension, viscosity, and rheological properties, conductivity is the most important property of LiCl solution that exerted influences on both the working processes, and also the resultant ribbons' quality.

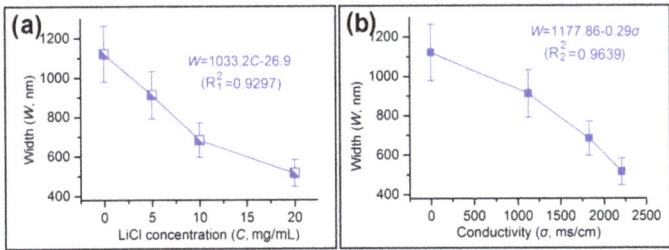

Figure 7. Correlations between the width of electrospun zein nanoribbons with: (**a**) The LiCl concentration; and (**b**) the conductivity of the sheath fluid.

3.5. Correlations between the Width of Electrospun Zein Nanoribbons and the Detailed Steps of Electrospinning

The detailed observations of the electrospinning process discussed in Section 3.2 have a very close relationship with the size of the zein ribbons produced (Figure 8). A linear equation relates the width of the ribbons to the length of the straight fluid jet ($W = 546.44L - 666.04$; $R_{wl}^2 = 0.9845$). Similarly, a linear equation $W = 2255.3\theta - 22.7$ connects the width of the ribbons with the spreading angle ($R_{w\theta}^2 = 0.9924$). These relationships show that these parameters can be very useful tools for predicting the properties of the ribbons fabricated.

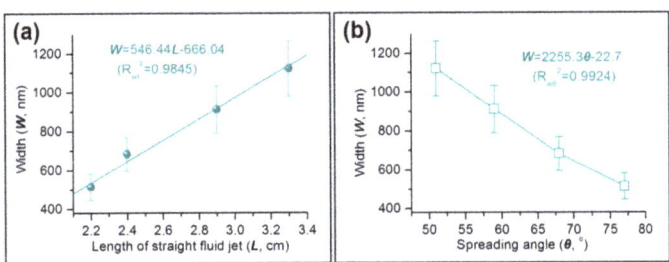

Figure 8. The correlations between the width of electrospun zein nanoribbons and: (**a**) The length of the straight fluid jet; and (**b**) the spreading angle of the unstable zone.

Right from the rebirth of electrospinning, a wide variety of experimental parameters have been studied to disclose their potential roles during the electrospinning processes. These parameters can all be manipulated by the researchers directly and changed within a certain range, which are concluded in Figure 1. However, these parameters often result in interrelated influences. For example, an increase of LiCl concentration resulted in a larger conductivity, but also changed the working fluid's surface tension, viscosity, and exerted on the effect of applied voltage. Thus, although many publications have reported the relationships between a certain experimental parameter and the final nanoproducts' size. It is difficult to disclose their relationship in an accurate manner. In contrast, the process characteristics, similarly as the final product to be influenced systematically from all the experimental parameters, have the essential advantages over the processing parameters in predicting the final nanoproducts' size, and in providing useful information for accurate and robust manipulation of the processing process.

3.6. The Role of Process Characteristics

A schematic diagram of the modified coaxial electrospinning process is presented in Figure 9. Initially, the sheath LiCl solution surrounds the core zein solution to form a compound Taylor cone. The two fluids come through the straight fluid jet and enter the unstable region together. During the early stages of the unstable region, the sheath solution will be evaporated, and then later, the core zein solution will be gradually dried during the drawing processes. A series of different forces will be exerted on the working fluids, such as the force between the two electrodes (F_1) and gravity (G, which can often be neglected). Within the bending and looping fluid jets, repulsive forces will include those between different loops (F_2) and those within the different parts of a single loop (F_3). It is the F_3 forces that draw and narrow the working fluids. The spreading angle will be a parameter that reflects the combined actions of F_1, F_2, and F_3. An increase in sheath solution conductivity will increase all three forces. An increase in F_1 would act to decrease the spreading angle. However, an increase in F_2 would make the fluid travel time increase during the drawing process, and thus provide a trend of enlarging the spreading angle. In addition, an increase in F_3 would make the loops larger, and correspondingly increase the spreading angle. Thus, the combined effects of F_2 and F_3 appear to have a more marked influence on the electrospinning process than F_1, and as a result, the greater the conductivity of the sheath fluid, the larger the spreading angle observed. Similarly, another process characteristic, i.e., the length of straight fluid jet, has received the influences of LiCl concentration directly and comprehensively.

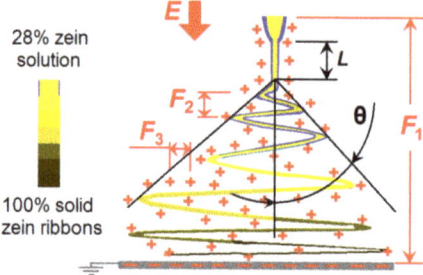

Figure 9. A diagram showing the formation mechanism of electrospun nanoribbons through the modified coaxial electrospinning.

In the biomedical applications of electrospun nanofibers, whether for tissue engineering or advanced drug delivery systems, the accurate manipulation of nanofiber diameter is very important for the fibers' functional performances [57–59]. This work reveals that the process characteristics (the length of straight fluid jet and the spreading angle of unstable region) have a close linear relationship with the final nanoribbon width, and can provide useful information for manipulating the working processes, and developing products with the desired physical properties.

4. Conclusions and Perspectives

Using an electrospinnable zein solution as the core fluid and LiCl solutions as the sheath working liquids, a series of modified coaxial electrospinning processes were performed, and a number of zein nanoribbons successfully prepared. The nanoribbon width (W) was found to be directly correlated with the concentration of LiCl (C) and the conductivity of the sheath fluid (σ), with linear relationships of the form $W = 1033.2C - 26.9$ ($R_1^2 = 0.9297$), and $W = 11177.86 - 0.29\sigma$ ($R_2^2 = 0.9639$) determined. Further, the width of the zein nanoribbons (W) were found to have still closer linear relationships with the spreading angle in the unstable region (θ), and the length of the straight fluid jet (L) ($W = 546.44L - 666.04$ and $W = 2255.3\theta - 22.7$; $R_{wl}^2 = 0.9845$ and $R_{w\theta}^2 = 0.9924$, respectively).

Today, electrospun nanofibers are rapidly approaching commercial applications in several fields such as drug delivery, food packaging, water treatment, and air filtration [60,61], and the production of electrospun nanofibers on a large scale is now possible [62]. Two important issues will require attention for accelerating nanofiber-based commodities to the market. One is the accurate and robust control of the processing process during electrospinning. The second is the prediction and maintenance of the nanofiber quality. For resolving these two issues, the characteristics of the working process itself offer a powerful source of information, and have advantages over the processing parameters (i.e., those that can be manipulated directly by the operator). This is because these working process characteristics manifest the simultaneous influence of all the processing parameters, as do the fibers produced. Thus, it is anticipated that they can act as a useful tool for stabilizing the working process, for systematic manipulation of the processing parameters, and for accurately predicting the resultant nanofiber size. Similar observations have been noted for electrospraying, an alternative electrohydrodynamic atomization process [63].

Author Contributions: Conceptualization, D.G.Y.; data curation, M.W.; formal analysis, Y.Y. and T.H.; funding acquisition, D.G.Y.; investigation, M.W., T.H. and Z.F.; methodology, D.G.Y.; project administration, D.G.Y. and S.W.A.B.; writing—original draft, M.W. and D.G.Y.; writing—review and editing, D.G.Y. and S.W.A.B.

Funding: The National Natural Science Foundation of China (No.51373101 and 51803121) and USST college student innovation projects (SH10252194 & 10252324/330) are appreciated.

Conflicts of Interest: The authors declare no conflict of interest.

References

1. Dotivala, A.C.; Puthuveetil, K.P.; Tang, C. Shear force fiber spinning: Process parameter and polymer solution property considerations. *Polymers* **2019**, *11*, 294. [CrossRef] [PubMed]
2. Vass, P.; Démuth, B.; Hirsch, E.; Nagy, B.; Andersen, S.K.; Vigh, T.; Verreck, G.; Csontos, I.; Nagy, Z.K.; Marosi, G. Drying technology strategies for colontargeted oral delivery of biopharmaceuticals. *J. Control. Release* **2019**, *296*, 162–178. [CrossRef] [PubMed]
3. Wu, H.; Zhao, S.; Ding, W.; Han, L. Studies of interfacial interaction between polymer components on helical nanofiber formation via coelectrospinning. *Polymers* **2018**, *10*, 119. [CrossRef] [PubMed]
4. Vicente, A.C.B.; Medeiros, G.B.; do Carmo Vieira, D.; Garcia, F.P.; Nakamura, C.V.; Muniz, E.C.; Corradini, E. Influence of process variables on the yield and diameter of zeinpoly (n-isopropylacrylamide) fiber blends obtained by electrospinning. *J. Mol. Liquids* **2019**. [CrossRef]
5. Acik, G.; Cansoy, C.E.; Kamaci, M. Effect of flow rate on wetting and optical properties of electrospun poly (vinyl acetate) microfibers. *Colloid Polym. Sci.* **2019**, *297*, 77–83. [CrossRef]
6. Liu, M.; Zhang, Y.; Sun, S.; Khan, A.R.; Ji, J.; Yang, M.; Zhai, G. Recent advances in electrospun for drug delivery purpose. *J. Drug Target.* **2019**, *27*, 270–282. [CrossRef] [PubMed]
7. Chakrabarty, A.; Teramoto, Y. Recent advances in nanocellulose composites with polymers: A guide for choosing partners and how to incorporate them. *Polymers* **2018**, *10*, 517. [CrossRef]
8. Haider, A.; Haider, S.; Kang, I.K. A comprehensive review summarizing the effect of electrospinning parameters and potential applications of nanofibers in biomedical and biotechnology. *Arab. J. Chem.* **2018**, *11*, 1165–1188. [CrossRef]
9. Okutan, N.; Terzi, P.; Altay, F. Affecting parameters on electrospinning process and characterization of electrospun gelatin nanofibers. *Food Hydrocolloids* **2014**, *39*, 19–26. [CrossRef]
10. Chen, B.Y.; Lung, Y.C.; Kuo, C.C.; Liang, F.C.; Tsai, T.L.; Jiang, D.H.; Satoh, T.; Jeng, R.J. Novel multifunctional luminescent electrospun fluorescent nanofiber chemosensorfilters and their versatile sensing of pH, temperature, and metal ions. *Polymers* **2018**, *10*, 1259. [CrossRef]
11. Jiang, D.H.; Tsai, P.C.; Kuo, C.C.; Jhuang, F.C.; Guo, H.C.; Chen, S.P.; Tung, S.H. Facile preparation of Cu/Ag Core/Shell electrospun nanofibers as highly stable and flexible transparent conductive electrodes for optoelectronic devices. *ACS Appl. Mat. Interfaces* **2019**, *11*, 10118–10127. [CrossRef] [PubMed]
12. Yu, D.G.; Zheng, X.L.; Yang, Y.; Li, X.Y.; Williams, G.R.; Zhao, M. Immediate release of helicid from nanoparticles produced by modified coaxial electrospraying. *Appl. Surf. Sci.* **2019**, *473*, 148–155. [CrossRef]
13. Li, X.Y.; Zheng, Z.B.; Yu, D.G.; Liu, X.K.; Qu, Y.L.; Li, H.L. Electrosprayed sperical ethylcellulose nanoparticles for an improved sustainedrelease profile of anticancer drug. *Cellulose* **2017**, *24*, 5551–5564. [CrossRef]
14. Wang, K.; Wen, H.F.; Yu, D.G.; Yang, Y.; Zhang, D.F. Electrosprayed hydrophilic nanocomposites coated with shellac for colonspecific delayed drug delivery. *Mat. Design* **2018**, *143*, 248–255. [CrossRef]
15. Liu, Z.P.; Zhang, L.L.; Yang, Y.Y.; Wu, D.; Jiang, G.; Yu, D.G. Preparing composite nanoparticles for immediate drug release by modifying electrohydrodynamic interfaces during electrospraying. *Powder Technol.* **2018**, *327*, 179–187. [CrossRef]

16. Yew, C.; Azari, P.; Choi, J.; Muhamad, F.; PingguanMurphy, B. Electrospun polycaprolactone nanofibers as a reaction membrane for lateral flow assay. *Polymers* **2018**, *10*, 1387. [CrossRef]
17. Kijeńska, E.; Swieszkowski, W. General requirements of electrospun materials for tissue engineering: Setups and strategy for successful electrospinning in laboratory and industry. *Electrospun Mat. Tissue Eng. Biomed. Appl.* **2017**. [CrossRef]
18. Zhou, H.; Shi, Z.; Wan, X.; Fang, H.; Yu, D.G.; Chen, X.; Liu, P. The relationships between the process parameters and the polymeric nanofibers fabricated using a modified coaxial electrospinning. *Nanomaterials* **2019**, *9*, 843. [CrossRef]
19. Chlanda, A.; Kijeńska, E.; Rinoldi, C.; Tarnowski, M.; Wierzchoń, T.; Swieszkowski, W. Structure and physicomechanical properties of low temperature plasma treated electrospun nanofibrous scaffolds examined with atomic force microscopy. *Micron* **2018**, *107*, 79–84. [CrossRef]
20. Wu, Y.H.; Li, H.P.; Shi, X.X.; Wan, J.; Liu, Y.F.; Yu, D.G. Effective utilization of the electrostatic repulsion for improved alignment of electrospun nanofibers. *J. Nanomat.* **2016**, *2016*, 2067383. [CrossRef]
21. Yao, C.H.; Yang, S.P.; Chen, Y.S.; Chen, K.Y. Electrospun poly (γ–glutamic acid)/β–tricalcium phosphate composite fibrous mats for bone regeneration. *Polymers* **2019**, *11*, 227. [CrossRef] [PubMed]
22. Liu, Y.; Liang, X.; Wang, S.; Qin, W.; Zhang, Q. Electrospun antimicrobial polylactic acid/tea polyphenol nanofibers for foodpackaging applications. *Polymers* **2018**, *10*, 561. [CrossRef] [PubMed]
23. Caimi, S.; Wu, H.; Morbidelli, M. PVdF-HFP and ionic-liquid-based, freestanding thin separator for lithium-ion batteries. *ACS Appl. Energy Mat.* **2018**, *1*, 5224–5232. [CrossRef]
24. Yu, D.G.; Li, J.J.; Williams, G.R.; Zhao, M. Electrospun amorphous solid dispersions of poorly watersoluble drugs: A review. *J. Control. Release* **2018**, *292*, 91–110. [CrossRef] [PubMed]
25. Jin, M.; Yu, D.G.; Wang, X.; Geraldes, C.F.G.C.; Williams, G.R.; Annie Bligh, S.W. Electrospun contrast agent-loaded fibers for colon-targeted MRI. *Adv. Healthc. Mat.* **2016**, *5*, 977–985. [CrossRef] [PubMed]
26. Lee, H.; Inoue, Y.; Kim, M.; Ren, X.; Kim, I. Effective formation of welldefined polymeric microfibers and nanofibers with exceptional uniformity by simple mechanical needle spinning. *Polymers* **2018**, *10*, 980. [CrossRef]
27. Domokos, A.; Balogh, A.; Dénes, D.; Nyerges, G.; Ződi, L.; Farkas, B.; Marosi, G.; Nagy, Z.K. Continuous manufacturing of orally dissolving webs containing a poorly soluble drug via electrospinning. *Euro. J. Pharma. Sci.* **2019**, *130*, 91–99. [CrossRef] [PubMed]
28. Saghazadeh, S.; Rinoldi, C.; Schot, M.; Kashaf, S.S.; Sharifi, F.; Jalilian, E.; Nuutila, K.; Giatsidis, G.; Mostafalu, P.; Derakhshandeh, H. Drug delivery systems and materials for wound healing applications. *Adv. Drug Deliv. Rev.* **2018**, *127*, 138–166. [CrossRef]
29. Yang, W.; Zhang, M.; Li, X.; Jiang, J.; Sousa, A.M.; Zhao, Q.; Pontious, S.; Liu, L. Incorporation of tannic acid in foodgrade guar gum fibrous mats by electrospinning technique. *Polymers* **2019**, *11*, 141. [CrossRef]
30. Zhang, Y.; Zhang, Y.; Zhu, Z.; Jiao, X.; Shang, Y.; Wen, Y. Encapsulation of thymol in biodegradable nanofiber via coaxial eletrospinning and applications in fruit preservation. *J. Agric. Food Chem.* **2019**, *67*, 1736–1741. [CrossRef]
31. Lv, S.; Zhao, X.; Shi, L.; Zhang, G.; Wang, S.; Kang, W.; Zhuang, X. Preparation and properties of SCPLA/PMMA transparent nanofiber air filter. *Polymers* **2018**, *10*, 996. [CrossRef] [PubMed]
32. Liu, X.; Yang, Y.; Yu, D.G.; Zhu, M.J.; Zhao, M.; Williams, G.R. Tunable zeroorder drug delivery systems created by modified triaxial electrospinning. *Chem. Eng. J.* **2019**, *356*, 886–894. [CrossRef]
33. Yang, Y.; Li, W.; Yu, D.G.; Wang, G.; Williams, G.R.; Zhang, Z. Tunable drug release from nanofibers coated with blank cellulose acetate layers fabricated using triaxial electrospinning. *Carbohydr. Polym.* **2019**, *203*, 228–237. [CrossRef] [PubMed]
34. Guarino, V.; Caputo, T.; Calcagnile, P.; Altobelli, R.; Demitri, C.; Ambrosio, L. Core/shell cellulose-based microspheres for oral administration of ketoprofen lysinate. *J. Biomed. Mat. Res. Part B* **2018**, *106*, 2636–2644. [CrossRef] [PubMed]
35. Costantini, M.; Colosi, C.; Święszkowski, W.; Barbetta, A. Coaxial wetspinning in 3d bioprinting: State of the art and future perspective of microfluidic integration. *Biofabrication* **2018**, *11*, 012001. [CrossRef] [PubMed]
36. Caimi, S.; Timmerer, E.; Banfi, M.; Storti, G.; Morbidelli, M. Core-shell morphology of redispersible powders in polymer-cement waterproof mortars. *Polymers* **2018**, *10*, 1122. [CrossRef] [PubMed]

37. Yu, D.G.; Yang, C.; Jin, M.; Williams, G.R.; Zou, H.; Wang, X.; Bligh, S.A. Medicated Janus fibers fabricated using a tefloncoated sidebyside spinneret. *Colloids Surf. B* **2016**, *138*, 110–116. [CrossRef]
38. Hou, J.; Wang, Y.; Xue, H.; Dou, Y. Biomimetic growth of hydroxyapatite on electrospun CA/PVP core–shell nanofiber membranes. *Polymers* **2018**, *10*, 1032. [CrossRef]
39. Liao, Y.; Loh, C.H.; Tian, M.; Wang, R.; Fane, A.G. Progress in electrospun polymeric nanofibrous membranes for water treatment: Fabrication, modification and applications. *Progr. Polym. Sci.* **2018**, *77*, 69–94. [CrossRef]
40. Martins, V.D.; Cerqueira, M.A.; Fuciños, P.; GarridoMaestu, A.; Curto, J.M.; Pastrana, L.M. Active bilayer cellulosebased films: Development and characterization. *Cellulose* **2018**, *25*, 6361–6375. [CrossRef]
41. Naeem, M.; Lv, P.; Zhou, H.; Naveed, T.; Wei, Q. A novel in situ selfassembling fabrication method for bacterial celluloseelectrospun nanofiber hybrid structures. *Polymers* **2018**, *10*, 712. [CrossRef]
42. Yoon, J.; Yang, H.S.; Lee, B.S.; Yu, W.R. Recent progress in coaxial electrospinning: New parameters, various structures, and wide applications. *Adv. Mat.* **2018**, *30*, 1704765. [CrossRef]
43. Li, J.; Xu, S.; Hassan, M.; Shao, J.; Ren, L.F.; He, Y. Effective modeling and optimization of pvdf–ptfe electrospinning parameters and membrane distillation process by response surface methodology. *J. Appl. Polym. Sci.* **2019**, *136*, 47125. [CrossRef]
44. Liu, Y.Q.; He, C.H.; Li, X.X.; He, J.H. Fabrication of beltlike fibers by electrospinning. *Polymers* **2018**, *10*, 1087. [CrossRef]
45. Abudula, T.; Saeed, U.; Salah, N.; Memic, A.; AlTuraif, H. Study of electrospinning parameters and collection methods on size distribution and orientation of PLA/PBS hybrid fiber using digital image processing. *J. Nanosci. Nanotechnol.* **2018**, *18*, 8240–8251. [CrossRef]
46. Xia, H.; Chen, T.; Hu, C.; Xie, K. Recent advances of the polymer micro/nanofiber fluorescence wave guide. *Polymers* **2018**, *10*, 1086. [CrossRef]
47. Szabó, E.; Démuth, B.; Nagy, B.; Molnár, K.; Farkas, A.; Szabó, B.; Balogh, A.; Hirsch, E.; Marosi, G.; Nagy, Z. Scaleup preparation of drugloaded electrospun polymer fibres and investigation of their continuous processing to tablet form. *Express Polym. Lett.* **2018**, *12*, 436–451. [CrossRef]
48. Rutledge, G.C.; Fridrikh, S.V. Formation of fibers by electrospinning. *Adv. Drug Deliv. Rev.* **2007**, *59*, 1384–1391. [CrossRef]
49. Yeo, L.Y.; Friend, J.R. Electrospinning carbon nanotube polymer composite nanofibers. *J. Exp. Nanosci.* **2006**, *1*, 177–209. [CrossRef]
50. Wang, Q.; Yu, D.G.; Zhang, L.L.; Liu, X.K.; Deng, Y.C.; Zhao, M. Electrospun hypromellosebased hydrophilic composites for rapid dissolution of poorly watersoluble drug. *Carbohydr. Polym.* **2017**, *174*, 617–625. [CrossRef]
51. Hai, T.; Wan, X.; Yu, D.G.; Wang, K.; Yang, Y.; Liu, Z.P. Electrospun lipidcoated medicated nanocomposites for an improved drug sustainedrelease profile. *Mat. Design* **2019**, *162*, 70–79. [CrossRef]
52. Liu, Z.P.; Zhang, Y.Y.; Yu, D.G.; Wu, D.; Li, H.L. Fabrication of sustained-release zein nanoparticles via modified coaxial electrospraying. *Chem. Eng. J.* **2018**, *334*, 807–816. [CrossRef]
53. Dehcheshmeh, M.A.; Fathi, M. Production of coreshell nanofibers from zein and tragacanth for encapsulation of saffron extract. *Int. J. Bio. Macromol.* **2019**, *122*, 272–279. [CrossRef]
54. Ghalei, S.; Asadi, H.; Ghalei, B. Zein nanoparticle-embedded electrospun PVA nanofibers as wound dressing for topical delivery of anti-inflammatory diclofenac. *J. Appl. Polym. Sci.* **2018**, *135*, 46643. [CrossRef]
55. Deng, L.; Zhang, X.; Li, Y.; Que, F.; Kang, X.; Liu, Y.; Feng, F.; Zhang, H. Characterization of gelatin/zein nanofibers by hybrid electrospinning. *Food Hydrocolloids* **2018**, *75*, 72–80. [CrossRef]
56. Wu, Y.H.; Yang, C.; Li, X.Y.; Zhu, J.Y.; Yu, D.G. Medicated nanofibers fabricated using NaCl solutions as shell fluids in a modified coaxial electrospinning. *J. Nanomat.* **2016**, *2016*, 8970213. [CrossRef]
57. Kouhi, M.; Fathi, M.; Reddy, V.J.; Ramakrishna, S. Bredigite reinforced electrospun nanofibers for bone tissue engineering. *Mat. Today Proc.* **2019**, *7*, 449–454. [CrossRef]
58. Ngadiman, N.; Yusof, N.; Idris, A.; Fallahiarezoudar, E.; Kurniawan, D. Novel processing technique to produce three dimensional polyvinyl alcohol/maghemite nanofiber scaffold suitable for hard tissues. *Polymers* **2018**, *10*, 353. [CrossRef]
59. Sill, T.J.; von Recum, H.A. Electrospinning: Applications in drug delivery and tissue engineering. *Biomaterials* **2008**, *29*, 1989–2006. [CrossRef]

60. Gao, S.; Tang, G.; Hua, D.; Xiong, R.; Han, J.; Jiang, S.; Zhang, Q.L.; Huang, C. Stimuli-responsive bio-based polymeric systems and their applications. *J. Mat. Chem. B* **2019**, *7*, 709–729. [CrossRef]
61. Lv, D.; Zhu, M.; Jiang, Z.; Jiang, S.; Zhang, Q.; Xiong, R.; Huang, C. Green electrospun nanofibers and their application in air filtration. *Macromol. Mat. Eng.* **2018**, *303*, 1800336. [CrossRef]
62. Duan, G.; Greiner, A. Air-blowing-assisted coaxial electrospinning toward high productivity of core/sheath and hollow fibers. *Macromol. Mat. Eng.* **2019**, *304*, 1800669. [CrossRef]
63. Huang, W.; Hou, Y.; Lu, X.; Gong, Z.; Yang, Y.; Lu, X.J.; Liu, X.L.; Yu, D.G. The process–property–performance relationship of medicated nanoparticles prepared by modified coaxial electrospraying. *Pharmaceutics* **2019**, *11*, 226. [CrossRef]

© 2019 by the authors. Licensee MDPI, Basel, Switzerland. This article is an open access article distributed under the terms and conditions of the Creative Commons Attribution (CC BY) license (http://creativecommons.org/licenses/by/4.0/).

Article

Effects of Fiber Surface Grafting with Nano-Clay on the Hydrothermal Ageing Behaviors of Flax Fiber/Epoxy Composite Plates

Anni Wang [1,2,3], Guijun Xian [1,2,3,*] and Hui Li [1,2,3]

1. Key Lab of Structures Dynamic Behavior and Control of the Ministry of Education, Harbin Institute of Technology, Harbin 150090, China
2. Key Lab of Smart Prevention and Mitigation of Civil Engineering Disasters of the Ministry of Industry and Information Technology, Harbin Institute of Technology, Harbin 150090, China
3. School of Civil Engineering, Harbin Institute of Technology, Harbin 150090, China
* Correspondence: gjxian@hit.edu.cn; Tel./Fax: +86-(451)-8628-3120

Received: 14 June 2019; Accepted: 11 July 2019; Published: 31 July 2019

Abstract: Flax fiber has high sensitivity to moisture, and moisture uptake leads to the decrease of mechanical properties and distortion in shape. This paper attempts to graft flax fabric with nano-clay, with assistance from a silane-coupling agent, in order to improve hygrothermal resistance. The nano-clay grafted flax fabric reinforced epoxy (FFRP) composite produced through vacuum assisted resin infusion (VARI) process were subjected to 80% RH chamber for 12 weeks at 20, 40 and 70 °C, respectively. Moisture uptake, dimensional stability, and tensile properties was studied as a function of humidity exposure. Through SEM and FTIR, the effects of hygrothermal exposure was elucidated. In comparison to control FFRP plates, nano-clay grafting decreases saturation moisture uptake and the coefficient of diffusion of FFRP by 38.4% and 13.2%, respectively. After exposure for six weeks, the retention rate of the tensile modulus of the nano-clay grafted flax fiber based FFRP increased by 33.8% compared with that of the control ones. Nano-clay grafting also reduces the linear moisture expansion coefficient of FFRPs by 8.4% in a radial direction and 10.9% in a weft direction.

Keywords: flax fiber; nano-clay; water uptake; hygrothermal properties

---

1. Introduction

Flax fiber is a natural fiber that is biodegradable, renewable and environment-friendly compared to traditional carbon fiber and glass fiber. Flax fiber possesses relatively higher tensile strength compare to other natural fibers, which are considered as a high performance natural fiber. Although its water resistance properties are not very good compared to thermoplastic polymer, epoxy resin has been widely used as a resin matrix for polymeric composites. Epoxy resin has good wettability with flax fiber, which provides good interface properties of flax fiber reinforced epoxy polymer (FFRP) composites. FFRPs are widely used in decorative materials, automobiles and other fields due to their high specific modulus. However, due to the high hydrophilicity of flax fibers, their poor durability limits the development of FFRPs.

The special chemical composition and structure of flax fiber lead to high hydrophilicity. Flax fiber is composed of plant cells, whose main structure are cell walls [1,2]. Like most plant cells, flax fiber cell walls consist mainly of cellulose, hemicellulose and pectin [3–5]. Cellulose, hemicellulose and lignin are made up of macromolecular chains of glucose and contain a large number of hydroxyl groups, which can adsorb water molecules [6]. Cellulose, called micro fibrils, is wrapped by hemicelluloses and lignin, and glued together or linked by hydrogen bonds [7]. Crystalline cellulose cannot store water, but water molecules can store them inside the amorphous hemicellulose and lignin [8,9]. In addition, flax fiber cells contain cell cavities that can store water. Thus, compared to traditional fiber such as carbon fiber, flax fiber shows high water absorbability.

Owing to the water absorption of fibers and the storage of water molecules at the interface between fiber and polymer, FFRPs also exhibit high water absorption [10,11]. Researchers have done a lot of research on the water absorption process of nature fiber reinforced polymer composites (NFRPs). The water absorption process of NFRPs is consistent to Fick law at lower temperature [12–14]. Both saturated water absorption and rate of water absorption of NFRP composites samples increase as the fiber volume fraction increase [14,15]. At the same time, the diffusion rate of water molecules in NFRP is related to temperature. The higher the temperature, the faster the diffusion rate. The researchers also studied the deterioration of mechanical properties of NFRP in a hygrothermal environment. On the other hand, the absorbed moisture results in more detrimental effects on the mechanical properties of NFRPs since the water not only interacts with fiber and polymer matrices, physically, i.e., plasticization, and/or chemically, i.e., hydrolysis, as in the unfilled system, but it also attacks the fiber–matrix interface [10]. Thus, the decrease of mechanical properties of NFRP caused by water molecules entering composite materials. Hongguang Wang et al. put the ramie fiber reinforced composites at 20 °C and 40 °C under 100% RH, and found that only after 1 day, both of the flexural strength and modulus were reduced dramatically and the deterioration rate of strength and modulus slowed down with the extension of immersion time [16].

Surface treatment of fiber is a good way to improve the properties of FRPs by improving the interface properties [17–19]. In order to promote the application of natural fiber, its hygrothermal resistance properties need to be improved by fiber treatment. At present, the main methods to improve the hygrothermal properties are as follows: removing the active hydroxyl groups on the surface of the fibers by chemical reaction, reducing the adsorption sites of water molecules; coating hydrophobic coatings on the surface of the fibers, hindering the diffusion of water molecules in the fibers. H. Alamri et al. used n-SiC fill cellulose fiber reinforced epoxy eco-nanocomposites and found that saturated water absorption of the composites decreased with the increase of n-SiC content [20]. Anna Dilfi K.F. et al. studied the durability of jute fiber reinforced epoxy composites treated by alkali and silane coupling agents and found that the deterioration of mechanical properties of jute fiber reinforced composites after chemical treatment are less than that of untreated ones [21]. Gao Ma et al. found that both alkali and silane treatments of jute fiber reduced water absorption and enhanced the tensile strength of the resulting jute fabric/epoxy composites [22].

The hydrophobicity and lamellar structure of nano-clay makes it possible to improve the durability of flax fibers after grafting onto the surface of flax fibers. Nano-clay is a kind of special nano-material with a large specific surface area, which is composed of two tetrahedral silicon atoms and eight sides of aluminum or magnesium hydroxide [23]. Nano-clay exhibits a hydrophobic lamellar structure, so polymer-clay nanocomposites have received much attention due to significant increase in mechanical properties, and a moisture barrier [24]. Polymer nanocomposites contain relatively small amounts (typically less than 5 wt. %) of nanometer-sized filler particles, which, if properly dispersed, have been found to cause significant reductions in both gas and water vapor permeability [25]. Neetu Malik et al. mixed biodegradable polymer polycaptalactone (PCL) and organic modified montmorillonite clay (OMMT) and found that with an increase in weight percentage of OMMT within the bio polymer films, the moisture absorption value of bio-nanocomposite films reduced rapidly from 34.4% to 22.3% [26].

In this paper, the method of improving the durability of flax fiber reinforced composites by nano-modification of flax fiber is studied. The effects of nano-clay grafting on the water absorption process as well as deterioration of mechanical properties of the related FRPs in hygrothermal environment were investigated. The mechanism of surface grafted nano-clay on durability of FFRPs was studied.

2. Experimental

2.1. Materials

The organic nano-clay, belonging to a high purity montmorillonite organic ammonium derivative, was purchased from Lingshou Huarun Mineral Factory (Shijiazhuang City, China). The nano-clay used in the present study is an organically treated montmorillonite (OMMT) with ammonium. Bi-directional flax fiber fabrics was purchased from Harbin Flax Textile Co., Ltd., (Harbin, Heilongjiang Province, China). The density of fabric is 1.5 g/cm^3 and nominal thickness is 0.16 mm. The silane coupling agent used in the current work is 3-Triethoxysilylpropylamine (APTES, KH550), purchased from Chengong Silicon Company (Nanjing, China) with purity of 98%. The epoxy resin used is room temperature impregnating adhesive (TS), purchased from Shandong Dagong Composite Material Co., Ltd (Linyi, China).

2.2. Surface Grafting of Flax Fabric

According to the author's previous work [27], the preparation process of nano-clay grafted flax fiber/epoxy resin composite is shown in Figure 1. The details can be found in Ref. [28]. Nano-clay were dispersed into a solvent (ethanol: distilled water = 4:1 by weight) with an ultrasonic bath. The organic nano-clay content in the dispersion medium is 1.3 wt. %. After 1 h of ultrasonic treatment at room temperature, flax fabric and 1wt % KH550 (to the dispersion medium) was added to the solution and sonicated for 15 min at room temperature. Finally, the fiber was washed in distilled water for 5 min. The composites were prepared by vacuum-assisted resin infusion process.

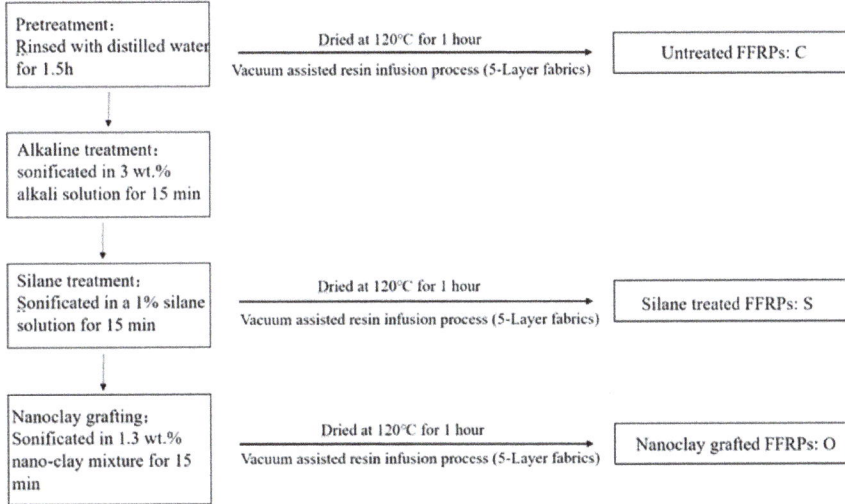

Figure 1. Methods of nano-clay grafted onto flax fiber and preparation of flax fiber reinforced polymer composites (FFRPs).

Figure 2 shows the SEM pictures of untreated fiber and nano-clay grafted flax fiber. When compared to the untreated one, lamellar nano-clay can be seen on the surface of the flax fiber. Due to its hydrophobicity, the presence of nano-clay enhances the barrier properties of the materials by creating tortuous pathways for water molecules to diffuse into flax fiber, which leads to a reduction in absorbed water and the coefficient of diffusion. Lamellar and hydrophobic nano-clays inhibit the diffusion of water molecules more obviously than other nano materials.

Figure 2. SEM photos of flax fiber: (**a**) untreated fiber (**b**) nano-clay grafted fiber.

2.3. Hydrothermal Environment Conditions

In the experiment, three different humid and hot environments were prepared using a saturated salt solution (Table 1). At 20 °C and 40 °C, the saturated potassium bromide solution can have ambient humidity of 80 RH%. At 70 °C, the humidity is 80% with saturated potassium chloride solution. As shown in Figure 3, the humidity chamber is prepared with a saturated salt solution in glass boxes at different temperatures. The humidity and temperature in the sealed glass box are monitored. When the temperature and humidity are stable, FFRP samples are placed in the sealed box. The condensed water droplets are prevented from falling on to the surface of the sample during the experiment. Before placing into the hygrothermal environment, the FFRP samples were dried in an oven at 70 °C for 48 h.

Table 1. Preparation of a hydrothermal environment with a saturated salt solution.

Temperatures	20 °C	40 °C	70 °C
Saturated salt solvent for 80% RH	Potassium bromide (KBr)	Potassium bromide (KBr)	Potassium chloride (KCl)

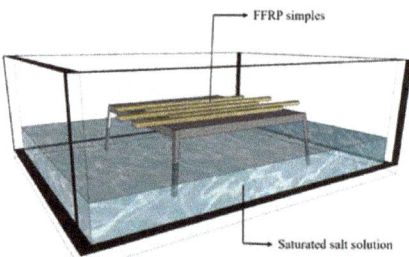

Figure 3. Schematic diagram showing preparation of a hydrothermal environment with a saturated salt solution.

2.4. Characterization

2.4.1. Moisture Uptake

According to ASTM D5229, the sample for moisture uptake is 76 mm × 25 mm, more than 5 g. Each group contains eight samples. Moisture uptake was detected by periodically recording the mass of the sample. Samples taken out of the humidity chamber were weighed using an electronic balance with accuracy of 0.01 mg. The presented data are an average for eight coupons. The immersion periods were set as 4 h, 8 h, 12 h, 1 day, 2 days, 4 days, 1 week, 2 weeks, 3 weeks, 4 weeks, 5 weeks, 8 weeks and 12 weeks.

2.4.2. Fourier Transform Infrared Test

Flax fiber fabrics were cut into powder—about 2 mg—to make fiber specimens. Fourier transform infrared (FTIR) spectra of the fiber specimens (control and treated flax fiber yarn) were recorded on a spectrometer (Spectrum 100, Perkin Elmer Instruments, Boston, MA, USA) at a range of 400–4000 cm^{-1}.

2.4.3. FBG Monitoring

Fiber Bragg Grating (FBG) Demodulator is manufactured by Shanghai Qipeng Engineering Materials Technology Co. (Shanghai, China) The FBG sensor for temperature monitoring was encapsulated in a steel capillary. With protection of the steel capillary, the FBG sensor was in a strain-free condition and affected only by the temperatures. To measure the internal stress during the aging, the FBG sensors were embedded in the interlayers of the FRP wet layups. The FBG sensors were carefully located in the fiber direction or perpendicular to the fibers and placed in the middle of the plate along the depth of the layer and near the central part in the planar direction. The fabricated resin samples were cured with the same curing procedure as the FRP samples. In this work, FFRPs were embedded with two FBG sensors—one along the fiber direction and one in the perpendicular direction. The fellow equations describe the strain measurement with the FBG sensors:

$$\Delta \lambda = \Delta \lambda_1 + \Delta \lambda_2 \tag{1}$$

$$\Delta \lambda_1 = \alpha_\varepsilon \Delta \varepsilon \tag{2}$$

$$\Delta \lambda_2 = \alpha_T \Delta T \tag{3}$$

where $\Delta \lambda$ is FBG wavelength change; $\Delta \lambda_1$ is strain-induced FBG wavelength change; $\Delta \lambda_2$ is temperature-induced FBG wavelength change; α_ε is strain sensitivity coefficient, which is 1.2 pm/µε for the FBG used in this experiment; α_T is temperature sensitivity coefficient, which is 10 pm/°C for FBG used in this experiment; $\Delta \varepsilon$ is strain value; ΔT is temperature change.

The wavelength change obtained by the FBG can be converted into strain:

$$\Delta \varepsilon = \frac{(\Delta \lambda - \Delta \lambda_2)}{\alpha_\varepsilon} \tag{4}$$

2.4.4. Mechanical Property Test

Tensile properties of the FFRP plate samples were tested according to ASTM D3039. The dimensions of the specimens were 250 mm × 15 mm. The crosshead speed is set as 2 mm/min. The samples were removed from the hygrothermal environment chamber at regular time intervals, i.e., 2, 4, and 6 weeks. Five samples were repeated for one condition.

2.4.5. Scanning Electron Microscope (SEM) Test

For the SEM test, all the specimens were sputter coated with gold for 15 min (Gatan Model 682 Precision etching coating system) before SEM analysis to improve their electrical conductivity. The control and treated FFRP samples were observed through a scanning electron microscope with accelerated voltages of 20–30 V.

3. Results and Discussion

3.1. Moisture Absorption of FFRPs

The water absorption percentage of FFRPs in hydrothermal environments can be calculated by the following equation:

$$\Delta M(t) = \frac{m_t - m_0}{m_0} \times 100 \tag{5}$$

where ΔM is moisture uptake, m_0 and m_t are the mass of specimen before and after exposure time of t.

Figure 4 shows the percentage of weight gain of untreated (C), silane coupling agent treated (S) and 1.3 wt. % nano-clay grafted flax fiber reinforced epoxy composites (O) at 70 °C under 80% RH from 0 to 3 months. Fick's diffusion model describes the dynamic equilibrium of the water absorption process when the material reaches a certain degree of water content. In the initial stage, the material's water absorption is proportional to the square root of exposure time. With the passage of time, the moisture absorption rate decreases dramatically, and finally reaches a dynamic balance. As shown in Figure 4, the moisture uptake process for all samples is linear in the beginning, then levels off, indicating a Fick diffusion process. R^2 represents the deviation between the experimental data and fitting results with the Fick's model.

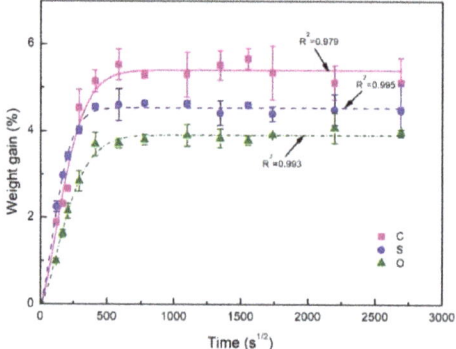

Figure 4. Water absorption process of FFRPs at 70 °C under 80% RH. (C: untreated FFRPs; S: silane coupling agent treated FFRPs; O: nano-clay grafted FFRPs).

According to Fick's law, the diffusion of water in materials is controlled by the concentration gradient of diffused substances. To obtain the water uptake and diffusion parameters, the curve fitting method was adopted with two-stage water uptake models:

$$M_t = M_m \left\{ 1 - exp\left[-7.3\left(\frac{Dt}{h^2}\right)^{0.75} \right] \right\} \qquad (6)$$

where M_m is the maximum weight gain, M_t is the weight gain at time t and h is the half of thickness of the composite.

Using Equation (2), the water diffusion parameters of the FFRPs are obtained as shown in Table 2. The nano-clay grafted flax leads to a remarkable decrease in saturated moisture uptake by 38.4% and 15.4% compared with the control and silane-treated ones. Similarly, nano-clay grafting also results in the reduction of the diffusion coefficient by 13.2% and 56.6% than that of the control and silane-treated ones. It is worth noting that silane-treated FFRPs show higher coefficient of diffusion and lower water uptake compared with the untreated ones. The lower water uptake of silane treated FFRPs is attributed to the reduced hydroxyl groups of flax fibers, which were reacted with silane. Note that the silane reaction is not complete, and some silane coupling agents did not form a network. Those relatively low molecules existing between the fiber and resin matrix accelerate water absorption. As a result, the FFRPs with silane-treated FFRPs show a higher coefficient of diffusion.

The water absorption process of nano-clay grafted FFRPs is also affected by temperature. Figure 5 shows the water absorption process of nano-clay grafted FFRPs at 20, 40 and 70 °C under 80% RH, which all followed a Fickian diffusion process. After aging for three months, the samples at 40 and 70 °C were saturated with water and had the same saturated water absorption, while the samples at 20 °C under 80% RH reach saturated water absorption and had lower saturated water absorption. As shown in Table 3, with the increase of temperature, the diffusion rate of water molecules increases. As the temperature increases, the higher the activation energy of water molecules is, and the faster is water saturation of FFRPs. Ana Espert et al. got the same results by immersion of wood fibers/polypropylene composites in water at three different temperatures [14].

Table 2. Maximum water uptake and diffusion coefficient (D) of FFRPs at 70 °C under 80% RH.

Samples	M_∞ (%)	$D \times 10^{-6}$ (mm^2/s)
Untreated FFRPs	5.41	3.03
Silane treated FFRPs	4.53	6.04
Nano-clay grafted FFRPs	3.91	2.62

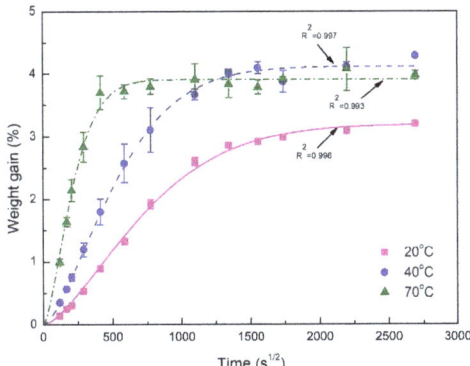

Figure 5. Water absorption process of nano-clay grafted FFRPs at 20 °C, 40 °C and 70 °C under 80% RH.

Table 3. Maximum water uptake and diffusion coefficient (D) of nano-clay grafted FFRPs at 20, 40 and 70 °C under 80% RH.

Hydrothermal Environments	M_∞ (%)	$D \times 10^{-6}$ (mm^2/s)
20 °C, 80% RH	3.20	0.22
40 °C, 80% RH	4.11	0.43
70 °C, 80% RH	3.91	2.62

3.2. FTIR Observation

As shown in Figure 6, FTIR test results display the fundamental OH stretching vibration of untreated fiber and silane treated fiber [20]. In the FTIR results of untreated flax fibers, the peak at 3295 (cm^{-1}) indicated hydroxyl groups adsorbed by hydrogen bonds and the peak at 3359 (cm^{-1}) indicated free hydroxyl or amino groups. Because the untreated flax fibers contain no amino group, this peak completely represents the vibration of hydroxyl group [28,29]. FTIR results of silane coupling agent treated fiber showed that the peak at 3320 (cm^{-1}) represented hydroxyl groups adsorbed by hydrogen bond, and the peak at 3419 (cm^{-1}) represents free hydroxyl or amino groups [28,29]. Because silane coupling agent treated flax fibers contain a large number of amino groups, and the absorption peaks become wider, indicating that there are overlapping groups, free hydroxyl and amino groups are represented here. Because the hydroxyl absorption peaks of the two fibers are similar, it can be considered that the hydroxyl content of silane treated flax fibers is lower than that of untreated flax fibers. The saturated water absorption of silane-treated FFRP decreases due to the decrease of hydroxyl content and the decrease of water adsorption sites. This phenomenon also can be explained by the chemical reaction between flax fibers and nano-clay. Based on previous research by the author, the interface properties of flax fiber and resin was improved using this chemical treatment. Reduction of water absorption could be attributed to the complete adhesion and wettability between the flax fibers and the polymer matrix, which may have less gaps and flaws at the interface [30]. The increase of interfacial bonding reduces the amount of water molecules stored at the interface of FFRP [10]. This chemical reaction was presented in a previous research work by the author [27].

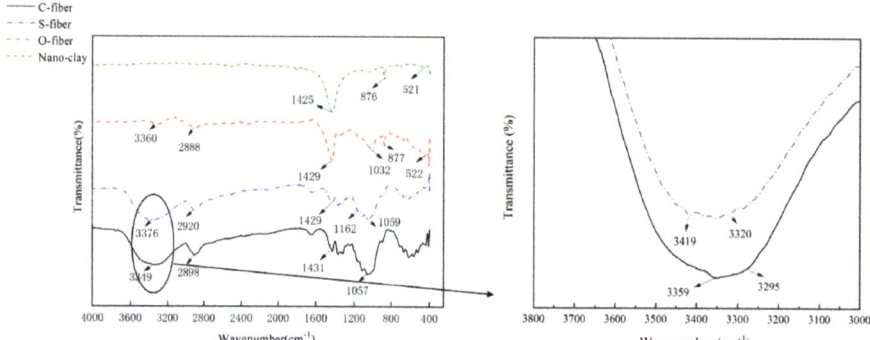

Figure 6. FTIR test results of flax fiber (C-fiber: untreated fiber; S-fiber: silane treated fiber; O-fiber: nano-clay grafted fiber) [24].

Figure 7a, b show the mechanism of improving the hydrothermal ageing behaviors of nano-clay grafted FFRP. Water molecules enter FFRPs along the fiber direction and are perpendicular to the fiber direction. As shown in Figure 7a, water molecules exist in untreated FFRPs in two forms: (1) free water stored in the cell compartment of fiber, fiber and interfacial space, and micro cracks; (2) bound water adsorbed on the fiber surface and cell wall. As shown in Figure 7b, after the nano-clay grafted, the interfacial adhesion properties of FFRP are improved. Therefore, water molecules in the interfacial and cracks are reduced. The nano-clay is grafted onto the surface of the fiber, making the surface of the fiber hydrophobic, increasing the diffusion path of water molecules. In Figure 7c, water molecules enter the interface between the fibers and epoxy, and form the first layer of water molecules by hydrogen bonding on the surface of unmodified flax fibers. Then the water molecules enter the first layer of water molecules in the form of the second layer of water molecules. As shown in Figure 7d, chemical grafting occupies the hydroxyl groups on the surface of the fiber, reducing the adsorption of water molecules on the cell wall of the flax fiber. However, with the increase of aging time and temperature, water molecules will gradually destroy the chemical bonds between the silane and flax fibers and adsorb on the surface of the fibers.

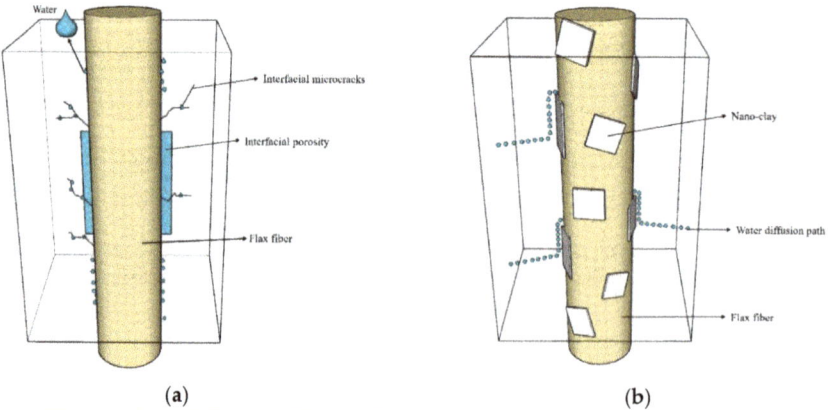

Figure 7. *Cont.*

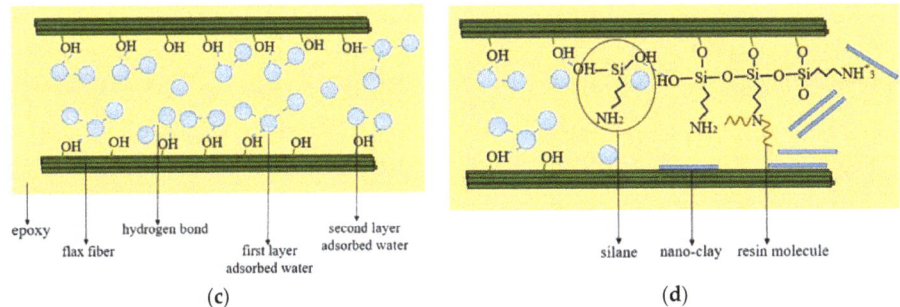

Figure 7. Diagram of water molecule existence and diffusion path in FFRPs (**a**) untreated FFRPs; (**b**) nano-clay grafted FFRPs; model of water molecule interaction on flax fiber-epoxy interface (**c**) untreated flax fibers (**d**) nano-clay grafted flax fibers.

3.3. FBG Monitoring

Figure 8 shows the radial and latitudinal strain values of FFRP calculated by the above formulas, which vary with the time it is placed in the 70 °C under 80% RH environment. Because FBG is sensitive to strain and temperature changes, there are fluctuations in the test results. In order to facilitate the analysis of the results, FBG is used to process the collected data, and the results are simplified to 100 data points without changing the trend. In Figure 8, O/W and C/W represent the weft strain (with less fibers) change process of nano-clay grafted and untreated FFRP; O/R and C/R represents radial strain test results of FBG in nano-clay grafted FFRP and untreated FFRP. The maximum strain in the radial and weft directions of the nano-clay grafted FFRPs are smaller than that of the untreated FFRPs.

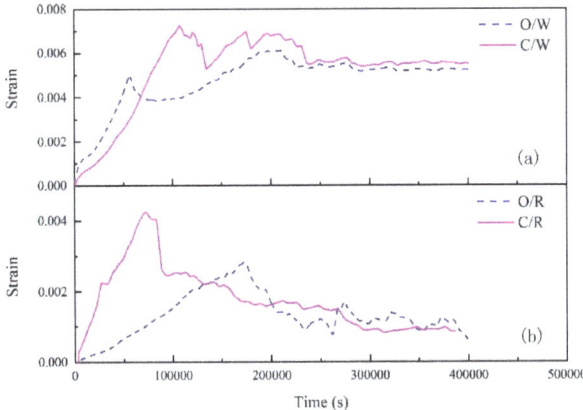

Figure 8. Effect of water absorption on the internal strain of untreated FFRP and nano-clay grafted FFRPs at 70 °C under 80% RH (**a**) the weft strain (**b**) the radial strain.

As the FFRP is placed in a hygrothermal environment for extended periods of time, water molecules gradually enter the fibers, causing them to expand. With the increase of water absorption, the water molecules diffuse into the specimen through the resin, fibers and the interfaces between them. The water ingress plasticizes the resin matrix, and even the fibers, leading to relaxation of the internal strain [31]. Consequently, the internal tension strain gradually reduces. When FFRP reaches saturated water absorption, the FFRP internal strain tends to balance after a certain fluctuation. As shown in Figure 8, the nano-clay-modified FFRP causes a decrease in the swelling amount of the fiber due to a decrease in saturated water absorption. Nano-clay grafted FFRPs show better dimensional stability.

According to the definition of the linear moisture expansion coefficient, when the laminate absorbs moisture, it produces line strain in the main direction of the material, as per the following formula [32,33]:

$$\beta_x = \frac{\varepsilon_x}{C} \tag{7}$$

$$\beta_y = \frac{\varepsilon_y}{C} \tag{8}$$

where β_x is linear moisture expansion coefficient in the x direction; ε_x is strain in the x direction; β_y is linear moisture expansion coefficient in the y direction; ε_y is strain in the y direction; C is water absorption concentration.

As shown in Table 4, the radial and latitudinal linear moisture expansion coefficient of untreated FFRPs and nano-clay grafted FFRPs are calculated by the maximum strain and saturated water absorption, respectively. The linear moisture expansion coefficient of nano-clay-grafted FFRP in two main directions is smaller than that of untreated FFRP. This is because the presence of the silane coupling agent film and the nano-clay constrains expansion of the fiber.

Table 4. Linear moisture expansion coefficient of FFRPs.

	Maximum Strain (Radial)	Radial Expansion Coefficient	Maximum Strain (Weft)	Weft Expansion Coefficient
Untreated FFRPs	0.0042	0.0777	0.007	0.1296
Nano-clay grafted FFRPs	0.0028	0.0717	0.005	0.1282

3.4. Tensile Properties

Figure 9a,b show the degradation of tensile properties of FFRPs. The degradation of tensile strength is smaller and the degradation of the tensile modulus is larger. The tensile strength of FFRP is affected by the flax fiber strength and the interfacial bond strength between the fiber and resin. The mechanical properties of flax fibers are influenced by the composition, structure and number of defects in a fiber. Under stress, tensile failure occurs by intercellular and/or intracellular modes [34]. Cellular stress is mainly determined by cellulose content and the angle between cellulose microfibers and the axis. When water molecules enter the fibers, moisture in fiber influences the degree of crystalinity and the crystalline orientation of fibers whereby it results in higher amounts and better orientation of crystalline cellulose in fibers. The absorption of water in the pores and amorphous regions of the fibers serves to reduce interfibrillar cohesion and to relieve internal fiber stresses [35]. Cellulose microfibers are embedded in hemicellulose, wax, etc. Hydrogen bonds play a key role in their combination. Water ingress deteriorates the hydrogen bonds, leading to higher elongation and strength, but lower modulus. Besides, increase in tensile strength of flax fiber is due to the availability of free water molecules, providing a plasticizing effect, which is advantageous to the strength of cellulose fibers [15]. However, the plasticization effect of water weakens the fiber/matrix bonding, resulting in interfacial failure [36]. Therefore, when water molecule acts on the composite, the tensile strength of the composite decreases slightly.

As shown in Figure 9a, after a six-week immersion in 70 °C under 80% RH environment, the tensile strength of untreated FFRPs (C) reduced by 13.5%, and that of silane treated FFRPs (S) and nano-clay grafted FFRPs (O) decreased by 15.8% and 15.6% respectively. The tensile strength retention rates of C, S and O were 88.0%, 84.1%, and 84.3%. As mentioned above, the tensile properties of FFRP depend on the tensile strength of the fibers and the strength of fibers depend on the content and angle of cellulose. Thus, surface modification has little effect on the degradation of tensile strength of FFRP in a short time. On the other hand, the interface is the medium of stress transfer between fibers, and also affects the tensile strength of FFRP to some extent. Cellulosic fibers can absorb water from the environment and can swell. This causes shear stress at the interface, which favors ultimate debonding of the fibers, which in turn causes a reduction in tensile strength [36]. The silane coupling agent forms a thin layer of macromolecule at the interface between the fiber and the resin, but the coupling agent is highly sensitive to water molecules [22]. When water molecules enter the early stage of the composite, under the action of high temperature, the Si–O–C bond between fiber and silane is not stable towards hydrolysis [37]. As a result, some of the coupling agent molecules that do not form macromolecules are easily hydrolyzed [38]. After partial hydrolysis of silane coupling agent molecules, the interfacial properties of modified flax fiber composites decreased before those of unmodified composites. Therefore, in the early stage of aging, the tensile strength of the untreated FFRPs decreased less than that of the silane treated FFRPs. For the same reason, the nanoclay grafted FFRP has also been treated with silane coupling agent, so the short-term degradation of tensile properties shows the same regularity as the silane treated FFRP. However, many papers show that the tensile strength of NFRP modified by the silane coupling agent is less than that of the unmodified ones after prolonged soaking time in a hygrothermal environment [22,39,40]. It can be predicted that the tensile properties of the nano-claymodified flax fibers are less degraded under the long-term action of water molecules.

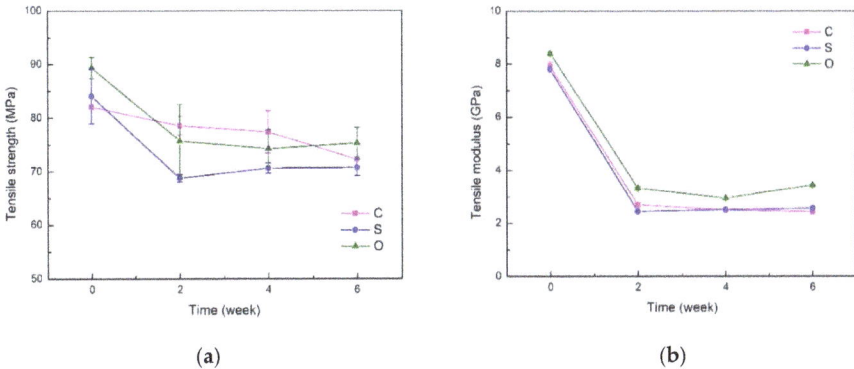

Figure 9. Effect of water absorption on the tensile properties (a) tensile strength (b) tensile modulus for FFRPs at 70 °C under 80% RH (C: untreated FFRPs; S: silane treated FFRPs; O: nano-clay grafted FFRPs).

Figure 9b shows the degradation of tensile modulus at 70 °C under an 80% RH environment. The tensile modulus degrades more than the strength. The tensile modulus of untreated FFRPs (C) reduced by 69.2%, and that of silane treated FFRPs (S) and nano-clay grafted FFRPs (O) decreased by 67% and 59.4% respectively. The tensile modulus retention rates of C, S and O were 30.7%, 32.9%, and 41.1%. After the water molecules enter flax fibers, because water molecules can exist in the amorphous structure [8,9], it makes the amorphous structure soften, resulting in a decrease in the modulus of the fiber. Another reason for the decrease in the modulus of the wet sample can be explained by the weakening of the cellulose structure of the natural fiber by the water molecules in the cellulose network structure, where water acts as a plasticizer and allows the cellulose molecules to move freely. Therefore, the quality of the cellulose is softened and the dimensions of the fiber can be easily changed by application of force [41]. The decrease of the absorption of water molecules reduces the plasticizing effect of the composite, so the decrease of the modulus of FFRP grafted by nano-clay is reduced.

The stress-strain curve of the FFRP tensile test is shown in Figure 10. In this picture, C, S and O represent various FFRP samples. 0W and 6W represent FFRPs before aging and placed at 70 °C under 80% RH environment for 6W, respectively. Curves extended into the nonlinear region in all cases. The stress-strain curve can be divided into three parts: (1) the first linear part, which is the deformation of each cell wall; (2) the second non-linear part, which is the elastic-plastic deformation of the fibers, is the rearrangement of the amorphous part (mainly made of pectin and hemicellulose) in the thickest cell wall (S2); and (3) the final approximately linear part, which is elastic response of cellulose microfibers to applied tensile strain [1]. The elastic linear area, where the damage is irreversible, reduces as a function of the water ageing [42]. The ultimate strain of untreated and nano-clay grafted FFRPs are increased after being placed in hygrothermal environment for 6 weeks. Water molecules can combine with hydroxyl bonds to act as plasticizers, which makes the material more ductile [15]. In addition, the significant increase in failure strain is due to the decomposition of the cellulose structure after the aging process, resulting in increased ductility of the flax fibers [43]. On the other hand, after the water molecules enter the FFRP, they occupy the pores and defects inside, which increases the ultimate strain of FFRP. At the same time, this increase is attributed to the lubrication of the water molecules, which may slide against each other during loading, resulting in more deformation and elongation [44]. As shown in Figure 10, the ultimate tensile strain of the nano-clay grafted FFRP after being placed at 70 °C under 80% humidity environment for six weeks was smaller than that of untreated FFRP. The reason is obvious. The saturated water absorption of the nano-clay grafted FFRP is lower than untreated FFRP, and the plasticization of FFRP by water molecules is reduced, so the increase in ultimate strain is reduced.

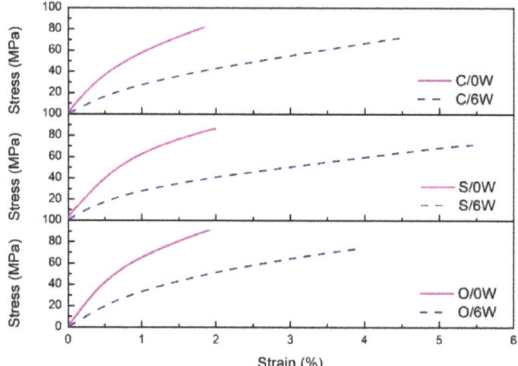

Figure 10. Effect of the water absorption on the stress-strain curves of different FFRPs at 70 °C under 80% RH.

Figure 11a,b show the degradation of tensile strength and modulus of nano-clay treated FFRPs in 20, 40 and 70 °C under 80% RH. Similar to the former, FFRPs also show the same degradation law at different temperatures, that is, tensile strength degradation is less, while tensile modulus degradation is greater. With the increase of temperature, the degradation of tensile properties of FFRPs increase, which is due to the increase of temperature, accelerating the movement of water molecules, increasing the diffusion rate of water, accelerating the aging of FFRP. Figure 11a shows the change in tensile strength of nano-clay-grafted FFRP over a six-week period in three different temperatures under 80% RH environments. After six-week immersion in 20, 40 and 70 °C under 80% RH environment, the tensile strength of nano-clay grafted FFRPs (O) decreased by 3.0%, 10.0%, 15.7%. When the nano-clay grafted FFRP is placed in an environment of 20 °C under 80% RH, the tensile strength increases during the first two weeks. Because of the slower diffusion rate of water molecules in a lower temperature, there are a few of water molecules inside the FFRP, and this part of the water molecules enhances the flax fiber without breaking the interface bonding between flax fiber and epoxy. Therefore, the FFRP tensile strength increased during the first two weeks. Subsequently, due to the prolongation of time, the water molecules gradually entered the FFRP, causing the debonding of the interface and damage inside the fiber. Therefore, after four weeks aging, the strength of the FFRP decreased. In addition, due to the difference in the diffusion rate of water molecules and the deterioration of the FFRP at high temperatures, the tensile strength of FFRP does not degenerate after four weeks at 20 and 40 °C under 80% RH, and the FFRP at 70 °C under 80% RH no longer degenerates in two weeks. It is worth noting that although the saturated water absorption of FFRP is approximately the same under different temperature environments, the effects on the tensile properties of FFRP are different. The higher the temperature, the more severe the aging of the composite material. This is because the high temperature accelerates the movement of the water molecules and also increases the deterioration of the composite material. Figure 11b shows the same results, which are the higher the temperature, the greater the modulus drop. After six-week immersion in 20, 40 and 70 °C under an 80% RH environment, the tensile modulus of nano-clay grafted FFRPs decreased by 36.9%, 47.6%, 59.5%.

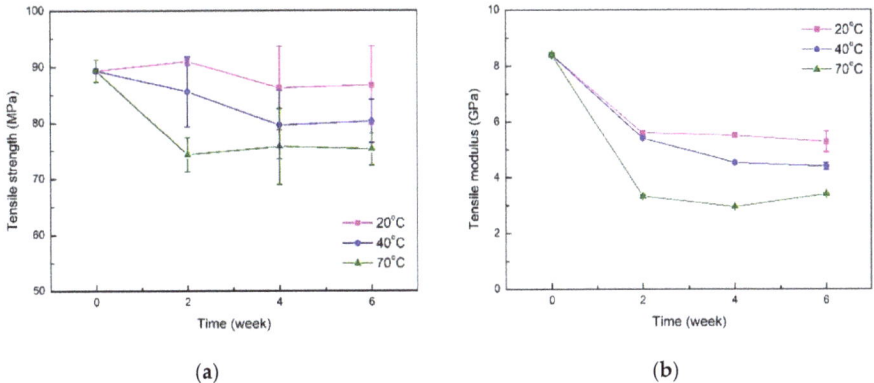

Figure 11. Effect of the water absorption on the tensile properties: (a) tensile strength (b) tensile modulus for nano-clay grafted FFRPs at different temperature under 80% RH.

The stress-strain curve of the nano-clay grafted FFRP tensile test after immersion in different environments is shown in Figure 12. Here, 20 °C/0W, 40 °C/0W and 70 °C/0W represent the stress-strain curve of nano-clay grafted FFRP before immersion; 20 °C/6W, 40 °C/6W and 70 °C/6W represent the stress-strain curves of nano-clay grafted FFRP subjected to exposure at 20, 40 and 70 °C under 80% RH for 6 weeks. The result also shows the same result. When FFRP is placed in the hygrothermal environment with a higher temperature, the greater the ultimate strain of FFRP.

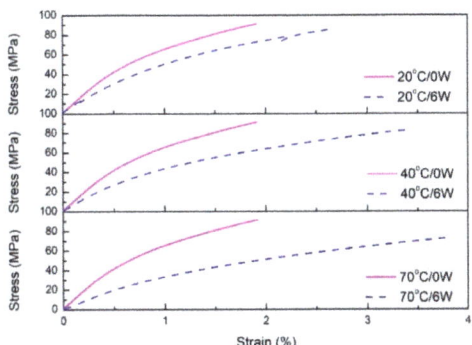

Figure 12. Effect of water absorption on the stress-strain curves of nano-clay grafted FFRPs at different temperatures under 80% RH.

Elongation at break of FFRPs show the same results. Table 5 shows the elongation at break of FFRPs. After immersion in 70 °C under an 80% RH environment for six weeks, the elongation at break of FFRPs increases. Among them, the nano-clay grafted FFRP has the smallest change in elongation at break. On the other hand, as the temperature increases, the greater the increase in elongation at break.

Table 5. Tensile properties of FFRPs.

Hydrothermal Environments	Properties	Untreated FFRPs		Silane Treated FFRPs		Nano-Clay Grafted FFRPs	
		0W	6W	0W	6W	0W	6W
20 °C, 80% RH	Elongation at break	–	–	–	–	0.88%	1.38%
	Strength (MPa)	82.1	–	84.1	–	89.3	86.7
	Modulus (GPa)	7.9	–	7.8	–	8.4	5.3
40 °C, 80% RH	Elongation at break	–	–	–	–	0.88%	1.85%
	Strength (MPa)	82.1	–	84.1	–	89.3	80.3
	Modulus (GPa)	7.9	–	7.8	–	8.4	4.4
70 °C, 80% RH	Elongation at break	0.84%	2.36%	0.86%	3.51%	0.88%	1.96%
	Strength (MPa)	82.1	72.3	84.1	70.7	89.3	75.3
	Modulus (GPa)	7.9	2.4	7.8	2.6	8.4	3.4

3.5. SEM Observation

Figure 13a–d show the SEM test results of tensile fracture of FFRPs after aging. Figure 13a showed tensile fracture of untreated FFRP before aging in which the fibers were pulled out of the resin but the gap between the fibers and the resin was small. The tensile fracture picture (Figure 13b) of untreated FFRP after six weeks of aging showed that the direct gap between fiber and resin was larger, which indicated that the interface between unmodified fiber and resin was destroyed by water molecules. Figure 13c showed tensile fracture of nano-clay grafted FFRPs before aging. The interfacial adhesion between modified fibers and resins increased. Thus, the interfacial debonding of fiber and resin is less, and most of them are broken by fiber fracture. While after six weeks immersion at 70 °C under 80% RH, nano-clay grafted fibers (Figure 13d) are pulled out of the resin, but there is still a small amount of resin attached to the surface of the fiber, and the gap between the fiber and the resin is small. For the nano-clay grafted FFRPs, the entry of water molecule has a certain effect on the bonding between fiber and resin, but the damage is weaker than that of the untreated ones. Because plant fibers have a multi-stack structure, overall swelling results from the local swelling of each component and each cell-wall layer. As each component of S2 layer has a different swelling behavior, differential swelling stresses may induce structural damage of fibers and thus degrade the mechanical properties.

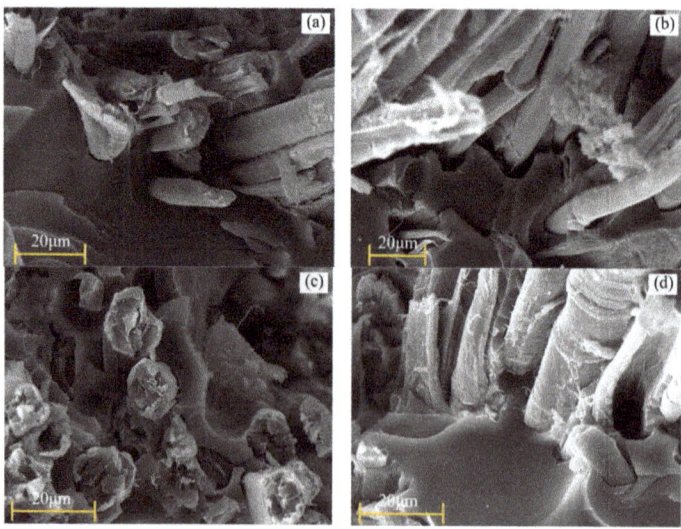

Figure 13. SEM photos of nano-clay grafted FFRPs tensile fracture: (a) untreated FFRP before aging; (b) untreated FFRP immersion in 70 °C under 80% RH for six weeks; (c) nano-clay grafted FFRP before aging (d) nano-clay grafted FFRP immersion in 70 °C under 80% RH for six weeks.

4. Conclusions

In this article, the effects of grafting of nano-clay on the hydrothermal resistance of the flax fiber reinforced epoxy composite plate were investigated. The moisture uptake, dimension change and tensile properties of the composite plates were tested. The following conclusions can be drawn based on the testing results and analysis:

(1) The introduction of nano-clay onto the flax fiber reduced the saturated moisture uptake and the coefficient of diffusion by 38.4% and 13.2% of FFRP compared to the control samples. The introduction of lamellar nano-clay is expected to reduce the hydrophilicity of the fiber surface and increases the diffusion path of water molecules.

(2) Nano-clay grafted FFRPs show better dimensional stability than the untreated ones. The linear moisture expansion coefficient of nano-clay grafted FFRP in radial and weft directions is smaller than that of untreated FFRP.

(3) After exposure for six weeks, the retention rate of the tensile modulus of the nano-clay grafted flax fiber-based FFRP was increased by 33.8% compared with that of the control ones, while the retention rate of tensile strength has a little decrease by nano-clay grafting. On immersion in a hydrothermal environment, the degradation of tensile strength of FFRP is not obvious while the ultimate strain of FFRPs is increased.

Author Contributions: G.X., A.W., and H.L. conceived and designed the experiments; G.X. and A.W. performed the experiment, analyzed the data and wrote the paper; G.X. and H.L. monitored the experimental process.

Funding: This research was funded by Chinese MIIT Special Research Plan on Civil Aircraft with grant No. MJ-2015-H-G-103 and the National Natural Science Foundation of China with Grant No. 51878223.

Conflicts of Interest: The authors declare no conflict of interest.

References

1. Yan, L.; Chouw, N.; Jayaraman, K. Flax fibre and its composites—A review. *Compos. Part. B Eng.* **2014**, *56*, 296–317. [CrossRef]
2. Charlet, K.; Eve, S.; Jernot, J.P.; Gomina, M.; Breard, J. Tensile deformation of a flax fiber. *Procedia Eng.* **2009**, *1*, 233–236. [CrossRef]
3. Baley, C. Analysis of the flax fibres tensile behaviour and analysis of the tensile stiffness increase. *Compos. Part. A Appl. Sci. Manuf.* **2002**, *33*, 939–948. [CrossRef]
4. Charlet, K.; Jernot, J.P.; Eve, S.; Gomina, M.; Bréard, J. Multi-scale morphological characterisation of flax: From the stem to the fibrils. *Carbohydr. Polym.* **2010**, *82*, 54–61. [CrossRef]
5. Bourmaud, A.; Morvan, C.; Bouali, A.; Placet, V.; Perré, P.; Baley, C. Relationships between micro-fibrillar angle, mechanical properties and biochemical composition of flax fibers. *Ind. Crop. Prod.* **2013**, *44*, 343–351. [CrossRef]
6. Dicker, M.P.M.; Duckworth, P.F.; Baker, A.B.; Francois, G.; Hazzard, M.K.; Weaver, P.M. Green composites: A review of material attributes and complementary applications. *Compos. Part. A Appl. Sci. Manuf.* **2014**, *56*, 280–289. [CrossRef]
7. Lefeuvre, A.; Duigou, A.L.; Bourmaud, A.; Kervoelen, A.; Morvan, C.; Baley, C. Analysis of the role of the main constitutive polysaccharides in the flax fibre mechanical behaviour. *Ind. Crop. Prod.* **2015**, *76*, 1039–1048. [CrossRef]
8. Xian, Y.; Chen, F.; Li, H.; Wang, G.; Cheng, H.; Cao, S.J.F. The effect of moisture on the modulus of elasticity of several representative individual cellulosic fibers. *Fiber Polym.* **2015**, *16*, 1595–1599. [CrossRef]
9. Le Duigou, A.; Bourmaud, A.; Baley, C. In-situ evaluation of flax fibre degradation during water ageing. *Ind. Crop. Prod.* **2015**, *70*, 204–210. [CrossRef]
10. Azwa, Z.N.; Yousif, B.F.; Manalo, A.C.; Karunasena, W. A review on the degradability of polymeric composites based on natural fibres. *Mater. Des.* **2013**, *47*, 424–442. [CrossRef]
11. Mokhothu, T.H.; John, M.J. Review on hygroscopic aging of cellulose fibres and their biocomposites. *Carbohydr. Polym.* **2015**, *131*, 337–354. [CrossRef] [PubMed]

12. Alamri, H.; Low, I.M. Mechanical properties and water absorption behaviour of recycled cellulose fibre reinforced epoxy composites. *Polym. Test.* **2012**, *31*, 620–628. [CrossRef]
13. Scida, D.; Assarar, M.; Poilâne, C.; Ayad, R. Influence of hygrothermal ageing on the damage mechanisms of flax-fibre reinforced epoxy composite. *Compos. Part. B Eng.* **2013**, *48*, 51–58. [CrossRef]
14. Espert, A.; Vilaplana, F.; Karlsson, S. Comparison of water absorption in natural cellulosic fibres from wood and one-year crops in polypropylene composites and its influence on their mechanical properties. *Compos. Part. A Appl. Sci. Manuf.* **2004**, *35*, 1267–1276. [CrossRef]
15. Dhakal, H.N.; Zhang, Z.Y.; Richardson, M.O.W. Effect of water absorption on the mechanical properties of hemp fibre reinforced unsaturated polyester composites. *Compos. Sci. Technol.* **2007**, *67*, 1674–1683. [CrossRef]
16. Wang, H.; Xian, G.; Li, H.; Sui, L.J.F. Durability study of a ramie-fiber reinforced phenolic composite subjected to water immersion. *Fiber Polym.* **2014**, *15*, 1029–1034. [CrossRef]
17. Chukov, D.; Nematulloev, S.; Zadorozhnyy, M.; Tcherdyntsev, V.; Stepashkin, A.; Zherebtsov, D. Structure, mechanical and thermal properties of polyphenylene sulfide and polysulfone impregnated carbon fiber composites. *Polymers* **2019**, *11*, 684. [CrossRef]
18. Chukov, D.I.; Stepashkin, A.A.; Gorshenkov, M.V.; Tcherdyntsev, V.V.; Kaloshkin, S.D. Surface modification of carbon fibers and its effect on the fiber–matrix interaction of uhmwpe based composites. *J. Alloy. Compd.* **2014**, *586*, S459–S463. [CrossRef]
19. Stepashkin, A.A.; Chukov, D.I.; Cherdyntsev, V.V.; Kaloshkin, S.D. Surface treatment of carbon fibers-fillers for polymer matrixes. *Inorg. Mater. Appl. Res.* **2014**, *5*, 22–27. [CrossRef]
20. Alamri, H.; Low, I.M. Effect of water absorption on the mechanical properties of n-sic filled recycled cellulose fibre reinforced epoxy eco-nanocomposites. *Polym. Test.* **2012**, *31*, 810–818. [CrossRef]
21. Dilfi KF, A.; Balan, A.; Hong, B.; Xian, G.; Thomas, S. Effect of surface modification of jute fiber on the mechanical properties and durability of jute fiber-reinforced epoxy composites. *Polym. Compos.* **2018**, *39*, E2519–E2528. [CrossRef]
22. Ma, G.; Yan, L.; Shen, W.; Zhu, D.; Huang, L.; Kasal, B. Effects of water, alkali solution and temperature ageing on water absorption, morphology and mechanical properties of natural frp composites: Plant-based jute vs. Mineral-based basalt. *Compos. Part. B Eng.* **2018**, *153*, 398–412. [CrossRef]
23. Mohan, T.P.; Kanny, K. Mechanical and thermal properties of nanoclay-treated banana fibers. *J. Nat. Fibers* **2017**, *14*, 718–726. [CrossRef]
24. Tzounis, L.; Debnath, S.; Rooj, S.; Fischer, D.; Mäder, E.; Das, A.; Stamm, M.; Heinrich, G. High performance natural rubber composites with a hierarchical reinforcement structure of carbon nanotube modified natural fibers. *Mater. Des.* **2014**, *58*, 1–11. [CrossRef]
25. Tan, B.; Thomas, N.L. A review of the water barrier properties of polymer/clay and polymer/graphene nanocomposites. *J. Membr. Sci.* **2016**, *514*, 595–612. [CrossRef]
26. Malik, N.; Shrivastava, S.; Ghosh, S.B. Moisture absorption behaviour of biopolymer polycapralactone (pcl)/organo modified montmorillonite clay (ommt) biocomposite films. In *IOP Conference Series: Materials Science and Engineering*; IOP Publishing: Bristol, UK, 2018; Volume 346.
27. Wang, A.; Xia, D.; Xian, G.; Li, H. Effect of nanoclay grafting onto flax fibers on the interfacial shear strength and mechanical properties of flax/epoxy composites. *Polym. Compos.* **2019**. [CrossRef]
28. Shih, Y.-F.; Jeng, R.-J. Ipns based on unsaturated polyester/epoxy: Iv. Investigation on hydrogen bonding, compatability and interaction behavior. *Polym. Int.* **2004**, *53*, 1892–1898. [CrossRef]
29. Wu, H.-D.; Chu, P.P.; Ma, C.-C.M.; Chang, F.-C. Effects of molecular structure of modifiers on the thermodynamics of phenolic blends: An entropic factor complementing pcam. *Macromolecules* **1999**, *32*, 3097–3105. [CrossRef]
30. Cui, Y.-H.; Lee, S.; Tao, J. Effects of alkaline and silane treatments on the water-resistance properties of wood-fiber-reinforced recycled plastic composites. *J. Vinyl Addit. Techn.* **2008**, *14*, 211–220. [CrossRef]
31. Gonon, P.; Sylvestre, A.; Teysseyre, J.; Prior, C. Combined effects of humidity and thermal stress on the dielectric properties of epoxy-silica composites. *Mater. Sci. Eng. B* **2001**, *83*, 158–164. [CrossRef]
32. Ramezani-Dana, H.; Casari, P.; Perronnet, A.; Fréour, S.; Jacquemin, F.; Lupi, C. Hygroscopic strain measurement by fibre bragg gratings sensors in organic matrix composites—Application to monitoring of a composite structure. *Compos. Part. B Eng.* **2014**, *58*, 76–82. [CrossRef]

33. Le Duigou, A.; Merotte, J.; Bourmaud, A.; Davies, P.; Belhouli, K.; Baley, C. Hygroscopic expansion: A key point to describe natural fibre/polymer matrix interface bond strength. *Compos. Sci. Technol.* **2017**, *151*, 228–233. [CrossRef]
34. Reddy, N.; Yang, Y. Biofibers from agricultural byproducts for industrial applications. *Trends Biotechnol.* **2005**, *23*, 22–27. [CrossRef] [PubMed]
35. Hearle, J.W.S. The fine structure of fibers and crystalline polymers. Iii. Interpretation of the mechanical properties of fibers. *J. Appl. Polym. Sci.* **1963**, *7*, 1207–1223. [CrossRef]
36. Joseph, P.V.; Rabello, M.S.; Mattoso, L.H.C.; Joseph, K.; Thomas, S. Environmental effects on the degradation behaviour of sisal fibre reinforced polypropylene composites. *Compos. Sci. Technol.* **2002**, *62*, 1357–1372. [CrossRef]
37. Xie, Y.; Hill, C.A.S.; Xiao, Z.; Militz, H.; Mai, C. Silane coupling agents used for natural fiber/polymer composites: A review. *Compos. Part. A Appl. Sci. Manuf.* **2010**, *41*, 806–819. [CrossRef]
38. Pavlidou, S.; Papaspyrides, C.D. The effect of hygrothermal history on water sorption and interlaminar shear strength of glass/polyester composites with different interfacial strength. *Compos. Part. A Appl. Sci. Manuf.* **2003**, *34*, 1117–1124. [CrossRef]
39. Apolinario, G.; Ienny, P.; Corn, S.; Léger, R.; Bergeret, A.; Haudin, J.-M. *Effects of Water Ageing on the Mechanical Properties of Flax and Glass Fibre Composites: Degradation and Reversibility*; Springer Netherlands: Dordrecht, The Netherlands, 2016; pp. 183–196.
40. Molaba, T.P.; Chapple, S.; John, M.J. Aging studies on flame retardant treated lignocellulosic fibers. *J. Appl. Polym. Sci.* **2016**, *133*. [CrossRef]
41. Dhakal, H.; Zhang, Z.; Bennett, N.; Lopez-Arraiza, A.; Vallejo, F. Effects of water immersion ageing on the mechanical properties of flax and jute fibre biocomposites evaluated by nanoindentation and flexural testing. *J. Compos. Mater.* **2014**, *48*, 1399–1406. [CrossRef]
42. Le Duigou, A.; Bourmaud, A.; Davies, P.; Baley, C. Long term immersion in natural seawater of flax/pla biocomposite. *Ocean Eng.* **2014**, *90*, 140–148. [CrossRef]
43. Methacanon, P.; Weerawatsophon, U.; Sumransin, N.; Prahsarn, C.; Bergado, D.T. Properties and potential application of the selected natural fibers as limited life geotextiles. *Carbohydr. Polym.* **2010**, *82*, 1090–1096. [CrossRef]
44. Chilali, A.; Zouari, W.; Assarar, M.; Kebir, H.; Ayad, R. Effect of water ageing on the load-unload cyclic behaviour of flax fibre-reinforced thermoplastic and thermosetting composites. *Compos. Struct.* **2018**, *183*, 309–319. [CrossRef]

© 2019 by the authors. Licensee MDPI, Basel, Switzerland. This article is an open access article distributed under the terms and conditions of the Creative Commons Attribution (CC BY) license (http://creativecommons.org/licenses/by/4.0/).

Article

Flax, Basalt, E-Glass FRP and Their Hybrid FRP Strengthened Wood Beams: An Experimental Study

Bo Wang [1], Erik Valentine Bachtiar [2], Libo Yan [1,2,*], Bohumil Kasal [1,2] and Vincenzo Fiore [3]

1. Department of Organic and Wood-Based Construction Materials, Technische Universität Braunschweig, Hopfengarten 20, 38102 Braunschweig, Germany
2. Center for Light and Environmentally-friendly Structures, Fraunhofer Wilhelm-Klauditz-Institut, Bienroder Weg 54E, 38108 Braunschweig, Germany
3. Department of Engineering, University of Palermo, Viale delle Scienze, Edificio 6, 90128 Palermo, Italy
* Correspondence: l.yan@tu-braunschweig.de or libo.yan@wki.fraunhofer.de; Tel.: +49-531-22077-25

Received: 27 June 2019; Accepted: 25 July 2019; Published: 29 July 2019

Abstract: In this study, the structural behavior of small-scale wood beams externally strengthened with various fiber strengthened polymer (FRP) composites (i.e., flax FRP (FFRP), basalt FRP (BFRP), E-glass FRP ("E" stands for electrical resistance, GFRP) and their hybrid FRP composites (HFRP) with different fiber configurations) were investigated. FRP strengthened wood specimens were tested under bending and the effects of different fiber materials, thicknesses and the layer arrangements of the FRP on the flexural behavior of strengthened wood beams were discussed. The beams strengthened with flax FRP showed a higher flexural loading capacity in comparison to the beams with basalt FRP. Flax FRP provided a comparable enhancement in the maximum load with beams strengthened with glass FRP at the same number of FRP layers. In addition, all the hybrid FRPs (i.e., a combination of flax, basalt and E-glass FRP) in this study exhibited no significant enhancement in load carrying capacity but larger maximum deflection than the single type of FRP composite. It was also found that the failure modes of FRP strengthened beams changed from tensile failure to FRP debonding as their maximum bending load increased.

Keywords: flax FRP; basalt FRP; glass FRP; wood beam; bending; hybrid FRP

1. Introduction

With an increasing concern on the energy conservation and environment protection, wood as a natural and sustainable construction material has returned to the spotlight after a long time flagging [1]. Compared with other conventional construction and building materials, wood has several shortcomings, e.g., relatively low tensile stiffness and strength compared to steel and low compression stiffness and strength compared to concrete. Wood is also susceptible to biological degradations, such as from fungi, bacteria and insects [2], which weaken its mechanical properties. To overcome the inferior mechanical properties of wood elements, fiber reinforced polymer (FRP) composite [3–5] can be one of the solutions. FRP has been widely utilized in the past two decades for rehabilitation and reinforcing of existing structures. FRP materials such as glass or carbon FRP have high strength-to-weight ratio, corrosion-resistance and provide design flexibility [6–8].

The commonly utilized FRP composites as reinforcement for wood beams are carbon FRP (CFRP), E-glass FRP (GFRP) and aramid FRP (AFRP) [3–5,9–12]. However, the production processes of these fibers are energy-intensive and the initial costs are still high. Recently, mineral-based natural FRP, such as basalt FRP (BFRP), has been introduced. BFRP has low material cost, high fire resistance, good thermal, electrical and sound insulating properties [13–15]. Furthermore, basalt fiber also has high tensile properties (e.g., tensile strength of 1850–4800 MPa) [14]. However, similar to glass fiber, the production of basalt fiber also requires a large amount of energy because of the high melting point of basalt rocks (1300 °C–1700 °C) [13].

As an alternative to glass, carbon and basalt fiber materials, the ecological and economical plant-based FRPs (e.g., flax or jute FRP) have been introduced in civil engineering. Various investigations on plant-based fibers (e.g., flax) have shown that as a single fiber, they have comparable specific mechanical properties (e.g., specific tensile strength and stiffness) compared to those of man-made E-glass fiber [6]. However, this is somewhat misleading since the length of natural fibers are limited, while carbon or glass fiber can be manufactured to have an endless length. The natural fibers are used in the forms of yarns, which will generally have lower mechanical properties compared to the ones of individual fibers.

Nevertheless, several investigations using the natural fibers in FRP as a reinforcement in civil engineering application have been carried out. Huang et al. [16] investigated flax FRP (FFRP) strengthened reinforced concrete (RC) beams. Their results revealed that the FFRP increased the ultimate load and maximum strain as well as the ductility of RC beams significantly. It also showed a better interfacial compatibility with the RC beams compared to GFRP and CFRP strengthened RC beams. Yan et al. [17] investigated the flexural properties of plain concrete beams externally strengthened with FFRP. It has been shown that the bending load capacity of plain concrete beams increased by 100%, 230% and 327% and their fracture energy were increase by 3500%, 4200% and 8160% with two-, four- and six-layer FFRP reinforcement [17]. In addition, FFRP has been used as external confining materials of natural aggregate concrete [18], recycled aggregate concrete [19] and fiber reinforced concrete [20,21].

In literature, a large number of studies have investigated FRP as an external reinforcement of wood structures, but only very few have considered plant-based FRPs. For example, Speranzini et al. [22] investigated solid wood beams externally strengthened with carbon, glass, basalt, hemp and flax FRP under a four-point bending test. No significant difference was observed on the loading capacity of the different FRP composites (i.e., the increase of the bending strength were 42.3%, 24.6%, 23.2%, 24.0% and 35.4% for carbon, glass, basalt, hemp and flax FRP, respectively) although there was a large difference in the tensile strength of these FRPs (i.e., 479, 142, 245, 36 and 25 MPa for carbon, glass, basalt, hemp and flax FRPs, respectively). According to the author, flax and hemp fibers may have better adhesion to wood compared to other FRPs. Borri et al. [23] investigated flax and basalt FRP strengthened low-grade (bending strength of 18.4 MPa) and high-grade (bending strength of 41.3 MPa) wood beams. The tensile strengths of FFRP and BFRP in the study was 240 MPa and 1880 MPa, respectively. The results showed an increase of bending strength of 38.6% and 65.8%, and maximum mid-span deflection of 58.2% and 40.2% respectively by two-layer FFRP and BFRP strengthened low-grade wood beams. Moreover, the strength increases were 29.2% and 25.9%, the increases of maximum mid-deflection were 9.1% and 14.5% respectively for two-layer FFRP and BFRP strengthened high-grade wood beams. This study concluded that both BFRP and FFRP provided the beams with higher strength and better ductile behavior. Similar results can be found in another research by Borri et al. [24] for flax and basalt FRP. André et al. [25] applied FFRP and GFRP with similar fabric density (i.e., 230 g/m^2 for flax and 250 g/m^2 for glass) perpendicular to grain on wood beams. It is reported that the maximum bending load of the entire specimen strengthened with GFRP (45.1 kN) was 23% higher than that one strengthened with FFRP (36.0 kN).

Realizing the advantages and disadvantages of using different types of fibers in FRP, hybrid FRP (HFRP) was proposed in the literature. Hybrid FRP, which consists of two or more combinations of strengthened fibers or fabrics, was designed to inherit the advantages and minimize the disadvantages of the combined fibers. Kim et al. [26] investigated HFRP made of carbon and glass fabrics to retrofit RC beams. The results showed that the HFRP contributed to higher ultimate bending strength and ductility of the RC beams compared to the single type of CFRP or GFRP. The maximum load in bending of RC strengthened with GFRP–CFRP (G GFRP attached at the tension surface of the RC beam) specimens was 6.6% and 3.9% higher than the one strengthened with two-layer CFRP (CC) and two-layer GFRP (GG), respectively. Moreover, the maximum mid-span deflection was also 27.4% and 18.5% higher than that of CC and GG specimens.

Compared with man-made fiber/fabric materials in conventional FRP composites (e.g., E-glass and carbon), plant-based fiber/fabric has a lower price and positive ecological impact [27], but it has lower mechanical properties as it has been mentioned before. In order to balance the performance and the cost for proper material design, several studies have investigated the hybridization of a plant-based fabric with a man-made one in FRP composite [28,29]. Gupta et al. [29] have summarized the mechanical properties of this hybrid material reinforcing thermoset polymers. It was concluded that the tensile, flexural and impact strengths of hybrid FRP were higher than those of the single type natural fabric FRP. However, the application of the hybrid FRP with natural fabric for reinforcing wood beams have been scarcely investigated before. Throughout the literature, only very few studies have investigated HFRP strengthened wood beams. Yang et al. [30] strengthened wood beams with hybrid carbon and glass FRP. Compared to the wood beams strengthened by GFRP or CFRP alone, the HFRP provided a larger energy dissipation for wood beams.

In this study, the flexural behavior of flax FRP strengthened wood beams were investigated. The results were compared with man-made E-glass and mineral-based basalt FRPs. Additionally, hybrid flax/glass/basalt FRPs were also investigated and compared with single type of FRPs (i.e., FFRP, BFRP and GFRP). Various different FRP materials (i.e., FFRP, GFRP and BFRP), FRP thickness (i.e., one-, two- and three-layer) and the arrangement of FRP in the HFRP were considered as experimental variables. As complementary initial investigations, tensile and bending test of flat coupon single type fiber FRPs were also carried out. Furthermore, since the interfacial bonding of fiber/epoxy and FRP/wood are also critical points for the flexural behavior of beams, the microstructures of these interfaces from the fractured specimens were examined under light and scanning electron microscopes.

2. Materials and Methods

2.1. Materials

Flax, basalt and E-glass were selected to represent the plant-based, mineral-based and conventional man-made fiber/fabric material for FRP composites, respectively. Among plant-based fibers, flax has comparable specific tensile properties with a lower unit price compared to those of E-glass fiber [6]. In addition, flax has a short growing cycle (harvested within 100 days after sowing the seeds). It also has a large annual production, which is required due to its broad applications, e.g., for household textiles, sails or tents, etc. [6]. For mineral-based fibers, basalt is generally used as a replacement of dangerous asbestos fibers and probably the only mineral-based fiber type that is available on the market [27]. Furthermore, basalt fiber also has tensile properties close to those of carbon fibers (e.g., for tensile strength, basalt fiber: 1850–4800 MPa and carbon fiber: 3000–5000 MPa) [14]. E-glass is one of the most widely used fibers as it is cheaper than carbon or aramid fibers and it has relatively high tensile strength (1800–3500 MPa).

In this study, bidirectional woven flax fabric (FlaxPly BL 550 from Lineo, Valliquerville, France, seven single-strand yarn threads per cm in the fabric weft and warp directions) (Figure 1a), unidirectional E-glass fabric (S15EU910, Saertex GmbH & Co. KG, Saerbech, Germany) (Figure 1b) and randomly distributed basalt mat (HG Europe, Milano, Italy) (Figure 1c) were investigated as FRP fabric materials. Based on the supplier data sheets, the areal density of flax, E-glass and basalt fabrics are 550 g m^{-2}, 600 g m^{-2} and 220 g m^{-2}, respectively. The nominal fiber thicknesses for one layer of flax, basalt and glass fabrics were 1.2 mm, 0.7 mm and 0.9 mm, respectively. However, it has to be mentioned that these nominal fiber thicknesses were only rough approximations as they are highly dependent on the pressure applied during measurement and the weaving structure of the fabrics. The FRPs were manufactured with a two-component epoxy polymer PRIME^{TM} 20LV epoxy resin and Prime 20 Slow hardener by Gurit Company, Zullwil, Switzerland. The tensile strength, tensile modulus and strain at failure of the cured epoxy were 73 MPa, 3.5 GPa and 3.5%, respectively. Although, some other adhesives (such as phenolic [31,32] or melamine [33] based adhesives), which are commonly used as adhesives for wood or other cellulosic materials, can be used as a matrix. Epoxy resin was selected in this study since it has been proven to have higher mechanical properties and chemical resistance than the other adhesives [6,34]. Epoxy is also the most commonly used polymer in FRP composites [7,8,20]. The structural wood beams, which were strengthened by the FRPs, were manufactured from Douglas Fir (*Pseudotsuga menziesii* Mirb.) with a dimension of 600 mm (length) × 40 mm (width) × 35 mm (height). The length direction of the beam was along the fiber direction of the wood (Figure 1d). The average density of the wood beams was 577 ± 33 kg·m^{-3}.

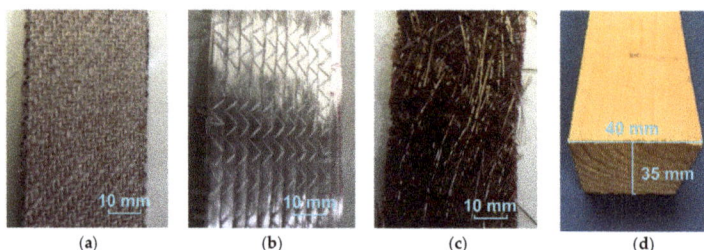

Figure 1. Photos of testing materials: (**a**) flax fabric, (**b**) glass fabric, (**c**) basalt mat and (**d**) wood beam.

2.2. Manufacture of FRP and FRP–Wood Specimens

The FRP manufacture process in this study was conducted through hand wet lay-up process and two kinds of specimens were produced: (1) FRP laminates for tensile and bending test and (2) FRP strengthened wood beams for bending tests. Initially, the epoxy resin and hardener were mixed with a ratio of 1:0.26 by weight for five minutes. The first layer of the fabric was placed on a flat and water-proofed plastic foil surface. It was then saturated with the epoxy mixture by using a brush. To avoid excess epoxy resin on the fabric, the saturation process was conducted slowly and directly stopped as soon as the fabric reached the saturation point. After that, the next layer was laid on the top of the first one and slowly saturated again with the epoxy. This process was repeated until the targeted number of layers was reached. Similar steps were used for the hybrid FRP. The fabrics were laid one by one in the intended order. All the epoxy-impregnated FRP composites were then cured at a room temperature (20 ± 3 °C) for seven days before they were cut to laminates for the flat-coupon tensile and flexural tests. No external pressure was applied on the FRP composites during the curing process. For tensile and bending tests, the FRP was cut into the appropriate size after curing. For the production of FRP strengthened wood specimens, the fabrics were cut firstly into strips with the size of 600 mm × 40 mm and the surface of the beams were coated by epoxy. Then, the strips were applied directly on the wood beams. While the basalt mat was arbitrarily applied on the wood beam due to its random orientation, the main fiber direction of the glass fabric and the warp direction of the flax fabric were always applied along the grain of the wood.

2.3. Test Matrix

A total of 39 small-scale wood beam specimens (three wood beams and 36 FRP strengthened wood beams) were tested under a three-point bending test according to DIN 52186 [35]. Table 1 shows the test matrix of the specimens used in this study. In the specimen name for each specimen type, W indicates wood, while F, B and G denote flax, basalt and glass as the type of the fabric for the FRP composites, respectively. The number of the FRP layers are denoted by 1L, 2L and 3L, i.e., one-, two- and three-layer. For hybrid FRP composite strengthened wood, the combination of F, B and G denotes the sequence of the arrangement of the FRP composite, i.e., 3L-GBF indicates the arrangement of the FRP, which is the outer layer (glass), middle layer (basalt), and the inner layer (flax) attached to the wood beams.

Table 1. Matrix of the specimens.

Specimen	Name [1]	Number of the FRP Laminates	Number of Replications
Wood	W	0	3
FFRP–wood	W_1L-F	1	3
	W_2L-F	2	3
	W_3L-F	3	3
BFRP–wood	W_1L-B	1	3
	W_2L-B	2	3
	W_3L-B	3	3
GFRP–wood	W_1L-G	1	3
	W_2L-G	2	3
	W_3L-G	3	3
Hybrid–wood	W_3L-GBF	3	3
	W_3L-BFG	3	3
	W_3L-BGF	3	3

1. W for wood; L for layers; B, G and F for basalt, glass and flax, respectively.

The mechanical properties of the different FRP composites were determined before the bending test of FRP-wood beams. Flat coupon tensile and bending tests were carried out for the FRP laminates according to ASTM D 3039 [36] and ASTM D 790 [37], respectively. For both tests, FRP composites with three different fabric materials (i.e., flax, glass and basalt) and three different layers (i.e., one-, two- and three-layer) were tested. For each specimen type, 10 specimens were prepared with the size of 250 mm in length × 25 mm in width and 150 mm in length × 25 mm in width for tensile and bending tests, respectively. The final thicknesses of the FRP laminates were determined by averaging the thickness of the laminates at three different locations. These thicknesses are presented as results in Table 2.

Table 2. Testing result of flat coupon tensile test and standard three-point bending test of FRP laminates.

Name [1]	Number of Replications	Nominal Fiber Thickness [2]	Thickness	Elastic Modulus	Strength	Strain at Peak Load
		mm	mm	GPa	MPa	%
Tensile span of extensometer = 140 mm, testing speed = 2.5 mm/min							
1L_B_Te	10	0.7	1.11	6.2 (± 0.8)	49.6 (± 8.2)	0.92 (± 0.22)
2L_B_Te	9	1.4	3.11	6.1 (± 0.7)	61.1 (± 9.4)	1.15 (± 0.15)
3L_B_Te	10	2.1	3.41	6.0 (± 0.4)	56.3 (± 6.1)	1.03 (± 0.12)
1L_F_Te	10	1.2	1.81	4.8 (± 0.3)	41.7 (± 5.5)	1.29 (± 0.31)
2L_F_Te	8	2.4	3.07	5.4 (± 0.2)	48.2 (± 1.7)	1.30 (± 0.07)
3L_F_Te	6	3.6	4.33	5.6 (± 0.1)	76.8 (± 2.1)	1.69 (± 0.12)
1L_G_Te	10	0.9	1.06	19.3 (± 1.5)	377.1 (± 55.7)	2.12 (± 0.68)
2L_G_Te	5	1.7	1.71	23.3 (± 0.7)	493.6 (± 46.0)	2.18 (± 0.29)
3L_G_Te	10	2.6	2.72	22.4 (± 1.0)	449.1 (± 38.8)	2.09 (± 0.47)
Bending span = 100 mm; testing speed = 1%/min, maximum strain before stop = 5%							
1L_B_Be	10	0.7	1.11	5.8 (± 0.5)	79.6 (± 7.2)	2.07 (± 0.24)
2L_B_Be	10	1.4	3.11	6.3 (± 0.5)	156.8 (± 11.5)	2.74 (± 0.15)
3L_B_Be	10	2.1	3.41	5.8 (± 0.5)	139.9 (± 11.8)	2.65 (± 0.18)
1L_F_Be	9	1.2	1.81	3.7 (± 0.7)	60.3 (± 10.0)	2.26 (± 0.36)
2L_F_Be	10	2.4	3.07	5.1 (± 0.2)	94.6 (± 7.1)	3.37 (± 0.25)
3L_F_Be	10	3.6	4.33	4.8 (± 0.2)	90.3 (± 3.0)	3.23 (± 0.23)
1L_G_Be	10	0.9	1.06	8.0 (± 0.6)	90.4 (± 6.8)	1.90 (± 0.20)
2L_G_Be	10	1.7	1.71	18.1 (± 2.6)	331.0 (± 31.3)	2.80 (± 0.15)
3L_G_Be	10	2.6	2.72	16.9 (± 2.1)	525.0 (± 50.9)	4.18 (± 0.36)

1. L for layer; B, F, G for basalt, flax and glass, respectively; Te and Be for tensile and bending, respectively.
2. Approximated nominal fiber thicknesses. The values depend on the pressure applied during the measurement and the different weaving structures of the fabrics.

2.4. Test Instrumentation

Zwick 1474 Test Machine (from ZwickRoell GmbH & Co. KG, Ulm, Germany) with a load cell capacity of 100 kN was used for flat coupon tensile test (Figure 2a), bending test (Figure 2b) for FRP laminates and three-point bending test for FRP strengthened wood beams (Figure 3). The testing machine was equipped with a standard extensometer (with an initial distance of 140 mm) to record the displacement of the sample during the test. The tensile tests were carried out with a displacement-controlled rate of 2.5 mm/min. The bending tests on FRP laminates were performed with a span distance of 100 mm and based on the standard, the testing rate was calculated as:

$$R = ZL^2/6d \qquad (1)$$

where,

- R rate of crosshead motion, mm/min
- Z rate of straining of the outer fibric, 0.01%/min
- L support span, mm
- d thickness of the specimen, mm.

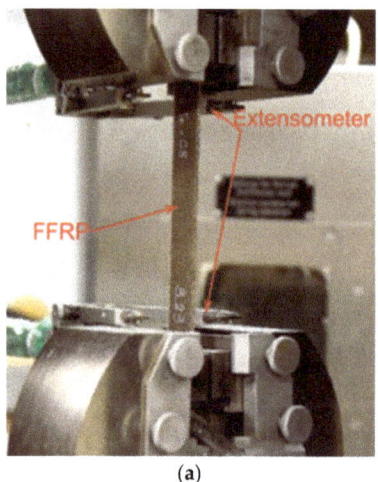 

(a) (b)

Figure 2. (a) Flat coupon tensile test and (b) bending test for fiber reinforced polymer (FRP) laminates.

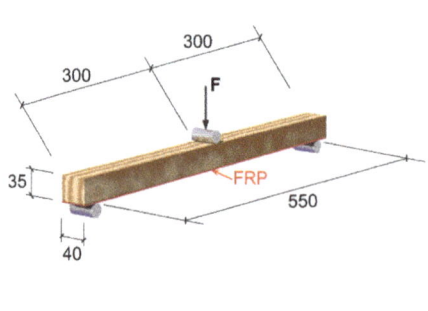 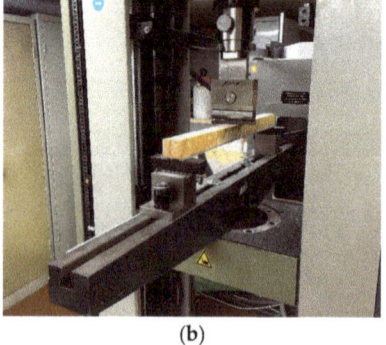

(a) (b)

Figure 3. Test setup of bending test for FRP–wood beams (unit: mm).

All tests for FRP laminates were conducted until failure or the maximum strain of 5% was reached.

The span of FRP strengthened wood beams tested under bending loading was 550 mm. The load was applied at the middle of beams with a loading rate of 12 mm/min until failure. The apparent flexural elastic modulus of the FRP–wood beams was calculated through the following equation, which is adapted from DIN 52186 [35]:

$$E = \frac{L^3}{4bd^3} \cdot \frac{\Delta F}{\Delta D} \quad (2)$$

where

E	flexural elastic modulus, GPa
L	support span, mm
b	width of the tested beam, mm
d	depth of the tested beam, mm
ΔF	difference of force between 20% to 40% of the maximum bending loading, kN
ΔD	difference of mid-span displacement at the corresponding bending loading, mm

After the mechanical tests, the fracture areas of the FRP–wood beams were observed with a light microscope (ZEISS 47 50 57 from Carl Zeiss Jena GmbH, Jena, Germany) and a scanning electron microscope (SEM, JSM-6700F, JEOL LTD, Tokyo, Japan). The specimens for the SEM were vacuum-coated with gold by evaporation process in BAL-TEC SCD 050 sputter coater.

2.5. Data Analysis Method

During the analysis and the interpretation of the data, the results were only compared based on the average value. The readers must be cautioned that these comparisons were only preliminary in character due to the comparing of the average values. No statistical analysis of the data was possible due to the limited number of specimens. Matching of specimens (for a pairwise comparison) is impossible for wood samples due to the variability within the material itself as well as variability between the specimens.

3. Results and Discussion

3.1. Tensile and Bending Tests for FRP Laminates

The results of the tested FRP laminates under tensile and bending loadings are presented in Table 2. For each specimen type, eight to ten specimens were successfully tested, except for 3L_F_Te and 2L_G_Te, where six and five specimens were successfully tested, respectively. The averaged value and the standard deviation of these successfully tested specimens are presented in the table. Furthermore, Figures S1 and S2 show the tensile and flexural stress–strain curves of the specimens during the tests, respectively. In these table and figures, indices Te and Be refer to tensile and bending tests, respectively.

Under bending loading, the maximum strengths of BFRP (79.6–156.8 MPa) were in general higher than FFRP (60.3–94.6 MPa) at any number of investigated fabric layers. Under tensile loading, however, FFRP (41.7–76.8MPa) had a comparatively similar strength than those of BFRP (49.6–61.1 MPa). Based on previous studies, the tensile strength of BFRP can be reached at around 1000 MPa (e.g., 707 MPa by Reyes-Araiza et al. [38] and 1282 MPa by Quagliarini et al. [39]). The low strength of BFRP obtained in this study was suspected due to the thin nominal fabric thickness, which led to a low areal density, and the random distribution of the basalt fibers in the mat. When compared with GFRP, FFRP presented significantly lower tensile and bending properties, and lower strain at peak load. This was expected since flax yarn consists of multiple bundles of short fibers, while glass yarn may have continuous fibers. Flax fibers may also contain natural defects [6], which cannot be avoided. Similar results were reported by Zhang et al. [40]. Their results showed that 10-layer FFRP had tensile strength of about 220 MPa and tensile failure strain of 0.85%, which was much lower than 10L-GFRP with tensile strength of about 700 MPa and tensile failure strain of 1.41%.

The number of fabric layers also influenced the mechanical properties of the overall FFRP. A relatively similar tensile strength was observed for one-layer and two-layer FFRP (41.7 and 48.2 MPa, respectively). However, the three-layer FFRP provided distinctly higher tensile strength (76.8 MPa). Under bending, on the other hand, 1L-FFRP (60.3 MPa) had a lower strength compared to the 2L- and 3L-FFRP (94.6 and 90.3 MPa, respectively). The strains at the peak load of the FFRP specimens also followed the same pattern. Under tensile loading, 3L-FFRP was observed to have a higher maximum strain (1.69%) compared to 1L- and 2L-FFRPs (1.29% and 1.30%, respectively). In contrast, the 1L-FFRP specimens had the highest strain at failure under bending load (2.26% compared to 3.37% and 3.23% for 2L- and 3L-FFRP, respectively). Besides the number of layers and the type of loading (tension or bending loadings), the inconsistency of the produced fiber volume fraction of the FRP using hand lay-up method may have contributed to the current finding. Moreover, under bending loading the thickness of the specimen strongly influenced the results. When it was bent in a same span length, a thicker specimen produced more internal shear, thus, it was stiffer and failed faster compared to a thinner specimen.

3.2. Bending Tests for FRP Strengthened Wood Beams

3.2.1. Effect of FRP Thicknesses on the Bending Behavior of FRP Strengthened Wood Beams

Figure S3 shows the representative load–displacement curves of wood beams strengthened with a different number of layers of B-, F- and GFRP. The results together with the calculated improvement of the properties due to the FRP reinforcements (unstrengthened wood beams as the reference) are also presented in Table 3.

The load capacity improvement increased with an increasing number of FRP layers for all wood beams strengthened with a single type of FRP. FFRP strengthened wood beams had maximum bending load capacities of 4.5, 5.5 and 6.2 kN for one, two and three layers, respectively. These corresponded to 60.7%, 96.4% and 121.4% load capacity improvement compared to unstrengthened wood beams, which had an average maximum load capacity of 2.8 kN. However, the improvement of the load bearing capacity was not linearly proportional to the increasing number of FRP layers. Similar to the FFRP, the load capacity improvements of one, two and three layers of GFRP were 71.4%, 117.9% and 132.1%. As the number of FRP layers increased, the gradient of the capacity improvement declined. However, surprisingly it was observed that the gradient of the capacity improvement of BFRP increased, i.e., 14.3%, 50.0% and 107.1% for one, two and three layers of BFRP. The reason could be due to the change of the failure mode, which is often governed by the weakest components in the FRP–wood composite beams. Under bending, the load is transferred to the compression and tension loadings. The compression loading was on the top part of the specimen, which was carried by the wood, while tensile loading was on the bottom part carried by the FRP. The tensile rupture of FRP may have initiated the overall failure of the composite if the FRP laminates were too thin (e.g., the 1L-FRP) or did not have enough strength (e.g., flax and basalt). With the increasing number of layers in the FRP, the tensile capacity of the FRP increased, which may have led to a shifting of the failure mode. The FRP–wood composite may then have failed due to the yielding failure of the wood in the compression zone or the delamination of the FRP–wood interface due to the induced internal shear loading. Further discussions of the different failure modes are given in Section 3.3.

Table 3. Test results and relevant improvements of three-point bending test on wood beams.

FRP Type	Layer	Name	Maximum Load Capacity Fmax kN	Load Capacity Improvement D_F %	Elastic Modulus E GPa	Elastic Modulus Improvement D_E %	Maximum Mid-Span Deflection D Mm	Deflection Improvement D_d %
None	0	W	2.8 (± 0.8)	—	9.0	—	12.7 (± 1.0)	—
Flax	1	W_1L-F	4.5 (± 0.6)	60.7	12.7	40.5	16.4 (± 2.6)	29.1
	2	W_2L-F	5.5 (± 0.3)	96.4	12.6	39.8	18.8 (± 6.0)	48.0
	3	W_3L-F[1]	6.2 (± 0.0)	121.4	12.9	42.4	21.2 (± 3.2)	66.9
Basalt	1	W_1L-B[1]	3.2 (± 0.3)	14.3	10.1	12.0	21.8 (± 8.6)	71.7
	2	W_2L-B	4.2 (± 0.7)	50.0	10.0	10.5	21.5 (± 7.3)	69.3
	3	W_3L-B	5.8 (± 0.3)	107.1	11.0	21.7	20.7 (± 3.1)	63.0
Glass	1	W_1L-G	4.8 (± 0.8)	71.4	9.2	1.8	34.2 (± 3.7)	169.3
	2	W_2L-G	6.1 (± 0.1)	117.9	13.3	46.9	31.1 (± 6.9)	144.9
	3	W_3L-G	6.5 (± 0.4)	132.1	15.1	66.6	26.8 (± 5.1)	111.0
Hybrid	3	W_3L-BFG	5.6 (± 0.5)	100.0	13.9	53.3	28.1 (± 4.0)	121.3
	3	W_3L-BGF	5.8 (± 0.3)	107.1	14.6	61.8	30.8 (± 5.8)	142.5
	3	W_3L-GBF	5.9 (± 0.4)	110.7	11.4	26.6	29.8 (± 1.9)	134.6

[1]: one test from each group were not successfully tested.

Table 3 also presents the elastic modulus of the investigated FRP–wood composites. The elastic modulus may have increase up to 66% as the woods were strengthened with the FRP. The influence of FFRP and BFRP thickness on the elastic modulus of the overall beam was less pronounced compared to the one from GFRP. By using one-, two- and three-layer GFRP under the wood beams, the elastic modulus was increased from 9.0 GPa to 9.2, 13.3, and 15.1 GPa, respectively. Among all the specimens, wood beams strengthened with three layers of GFRP had the highest elastic modulus. Compared with BFRP (10.0–11.0 GPa), FFRP strengthened wood beams had a higher elastic modulus (12.6–12.9 GPa). However, these results do not fully follow the results of the tensile tests and bending tests of FRP laminates showed in Section 3.1. FFRP laminates had the lowest tensile and bending modulus (i.e., 4.8–5.6 GPa in tensile and 3.7–5.1 GPa in bending) compared to BFRP (6.0–6.2 GPa in tensile and 5.8–6.3 GPa in bending) and GFRP (19.3–23.3 GPa in tensile and 8.0–18.1 GPa in bending). The reason of these findings was suspected to be due to the different thickness of FRP beams and the compatibility between the fabric and wood.

Based on the cross-section inertia of the beams and also presented in Equation (2), the height of the beam to the power of three highly influences the elastic modulus. The actual thickness of each specimen was considered in the calculation. However, the different thickness of the FRP led to the different height of the FRP–wood specimens. Thus, the cross-sectional FRP–wood ratios were varied between specimens. This may have led to a different stress distribution during bending loading. Higher thickness of the FRP–wood may also have resulted in stiffer beams due to the more pronounced influence from the internal shear loading of the specimen under bending loading.

In addition to that, as a cellulosic natural material, flax has the same chemical components as wood (i.e., cellulose, hemicellulose and lignin). Therefore, similar bonding behavior is expected between flax/epoxy and epoxy/wood. On the other hand, the bonding behavior of glass/epoxy and basalt/epoxy are different. The similar bonding behavior was suspected to give a positive impact of the overall mechanical properties of the FRP–wood beams. This was also supported by the results from HFRP, the highest stiffness was reached when flax connected directly to the wood (14.6 GPa for W_3L-BGF). This reason, however, is only a theory based on the results obtained in this study. Further investigations have to be conducted to support this theory.

3.2.2. Effect of FRP Materials on the Bending Behavior of FRP Strengthened Wood Beams

Figure 4 shows the representative bending load–displacement curves of the unstrengthened wood beam and all types of three-layer FRP strengthened wood beams. Their maximum load, maximum deflection and flexural elastic modulus are presented in Table 3. All the three-layer FRP reinforcements increased the maximum load of wood beams remarkably. The average load capacity of W_3L-F, W_3L-B and W_3L-G were 6.2, 5.8 and 6.5 kN with increments of 121.4%, 107.1% and 132.1%, respectively in comparison to the average load capacity of unstrengthened wood beams. The hybrid FRPs showed similar enhancement in load capacity. The maximum bending load of W_3L-BFG, W_3L-BGF and W_3L-GBF were 5.6, 5.8 and 5.9 kN, respectively. Among these tested FRPs, the best performance based on the maximum mid-span deflection was observed from HFRP strengthened wood beams. The W_3L-BGF had the highest maximum strain increment by 142.5%, followed by W_3L-GBF (134.6%) and W_3L-BFG (121.3%), which were higher than that of FFRP (66.9%), BFRP (63.0%) and GFRP (111.0%).

When comparing the FFRP to BFRP and GFRP, it was found that FFRP laminates had higher ultimate strain than BFRP under tensile loading. Therefore, FFRP provided a larger enhancement in deflection than BFRP for FRP strengthened wood beams. FFRP had only a slightly lower tensile strength to BFRP (41.7 and 49.6 MPa for one-layer FFRP and BFRP, respectively), which was already enough to carry the tensile loads on the tensile area at the bottom of the wood beams. Moreover, FFRP laminates were also thicker than BFRP. As a result, FFRP provided larger enhancement than BFRP in FRP strengthened wood beams. It should be, however, kept in mind that the basalt fabric mat used in this study had a low areal density with short fibers that were orientated randomly. Furthermore, the bending results of FRP strengthened wood beams also showed that FFRP provided similar maximum strength and maximum deflection enhancement with GFRP for wood beams (especially with higher number of FRP layers), although FFRP laminate had much lower tensile strength than GFRP laminates (i.e., 76.8 MPa for three-layer FFRP vs. 449.1 MPa for three-layer GFRP). This was primarily because, at a high number of FRP layers, the failure of the interface between wood and epoxy would have been more decisive on initiating the whole failure of the FRP–wood beams. Thus, having a stronger FRP material such as glass, may not necessarily increase the overall performance of FRP–wood composite. The interface debonding will always initiate failure of the whole composite systems and the maximum capacity of GFRP cannot be fully utilized.

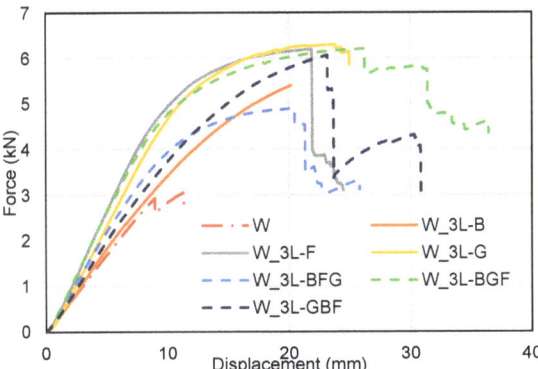

Figure 4. Load mid-span displacement curves of three-layer FRP strengthened wood beams.

3.3. Failure Modes and Microstructure of FRP–Wood Beam System

3.3.1. Failure Modes

The typical failure modes of the reference wood beams and FRP strengthened beams are shown in Figure 5. The reference beams showed a typical tension failure (Figure 5a). The crack was initiated at the mid-span of the tensile zone and then propagated until the complete failure of the beam. For FRP strengthened wood beams, two kinds of failure were observed, i.e., tensile failure and debonding of FRP. The tensile failure in FRP–wood beams (Figure 5b) was initiated at the middle of FRP strips followed by the failure of the tensile zone of wood beams. The debonding of FRP took place at the interface between wood beams and FRP and occurred in either mid-span of the beam (Figure 5c) or at the edge (Figure 5d).

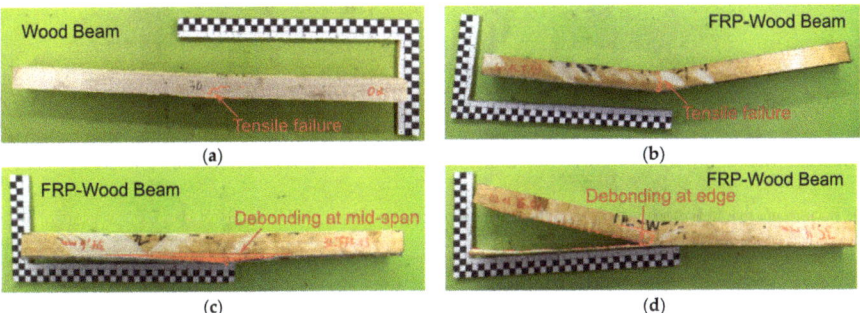

Figure 5. Typical failure modes of FRP–wood beams under three-point bending with different FRP laminates: (**a**) tensile failure for reference wood beam, (**b**) tensile failure (W_1L-F), (**c**) debonding at mid-span (W_3L-F) and (**d**) debonding at edge for FRP strengthened wood beam (W_1L-G).

Table 4 presents the general failure modes for all specimens tested in this study. It can be observed that most FRP strengthened wood beams with low maximum bending loading (e.g., less than 5.8 kN) showed a primarily tensile failure mode. At a higher bending loading, interface debonding was observed. The reason has already been discussed previously that weakest parts (between wood, FRP and the FRP–wood interface) of the composite will decide the failure mode. Thus, by changing of the number of layers the failure mode may be also changed. Exception can be found in W_1L-G with maximum bending load of 4.8 kN, where debonding failure was observed and W_3L-B with maximum bending load of 5.8 kN, where tensile failure was observed. W_3L-GBF had also a slightly higher maximum bending load of 5.9 kN, but tensile failure was observed This may have been due to the relatively low manufacturing quality and repeatability through the hand wet lay-up process. Further investigations on the relation between the failure mode and the tensile strength of FRP strengthened beams should be carried out in the future

Table 4. General failure modes for control and FRP strengthened wood beams.

FRP Type		Number of FRP Layers		
	0	1	2	3
without FRP	Tensile failure	—	—	—
FFRP	—	Tensile failure	Tensile failure	Debonding at mid-span
BFRP	—	Tensile failure	Tensile failure	Tensile failure
GFRP	—	Debonding at edge	Debonding at edge	Debonding at edge
HFRP W_3L-BFG	—	—	—	Debonding at edge
W_3L-BGF	—	—	—	Debonding at edge
W_3L-GBF	—	—	—	Tensile failure

3.3.2. Microstructure

Figure 6 shows the light microstructures of FFRP, GFRP, and BFRP as well as a hybrid FRP strengthened wood beams (W_3L-GBF). The interface between epoxy/wood, flax yarn, glass and basalt fiber structure can be clearly observed under the light microscope. As can be seen, no gaps were found in the interfaces between epoxy and wood. Such interface eased the transfer of the bending load from wood to the FRP fabric. However, several air bubbles were observed in the FRP. These air bubbles might be regarded as defects which may result in stress concentration at the FRP/wood interface. This should be further identified in a future study. The presence of the air bubbles may explain why W_1L-G had a lower bending load of 5.4 kN with debonding failure compared to W_3L-B and W_3L-GBF.

The scanning electron microscope analysis (SEM) was used for the observation of the interface of fabric/epoxy (or fiber/epoxy). Figure 7 shows the example of fracture surface from basalt FRP after tensile failure in the mid-span. In Figure 7a, no obvious gap between the fiber and the matrix was observed, which indicated a good interfacial bond between the fiber and the matrix. The close-up image of the fiber/epoxy interface in Figure 7b shows that only a small amount of epoxy remained on the basalt fiber after the tensile failure of BFRP strengthened wood beams. This indicates that the fiber was pulled out from epoxy matrix during the test. The reasons can be the smooth surface of the basalt fiber or the low wetting behavior between epoxy and basalt fiber. Similar pull-out failure can be also found in FFRP and GFRP. Therefore, methods to increase the surface roughness of fiber (e.g., with alkali solution for flax [7]) or to improve the wetting behavior between fiber and polymer are possibilities that could improve the interface bond between fibers and polymer in FRP composites.

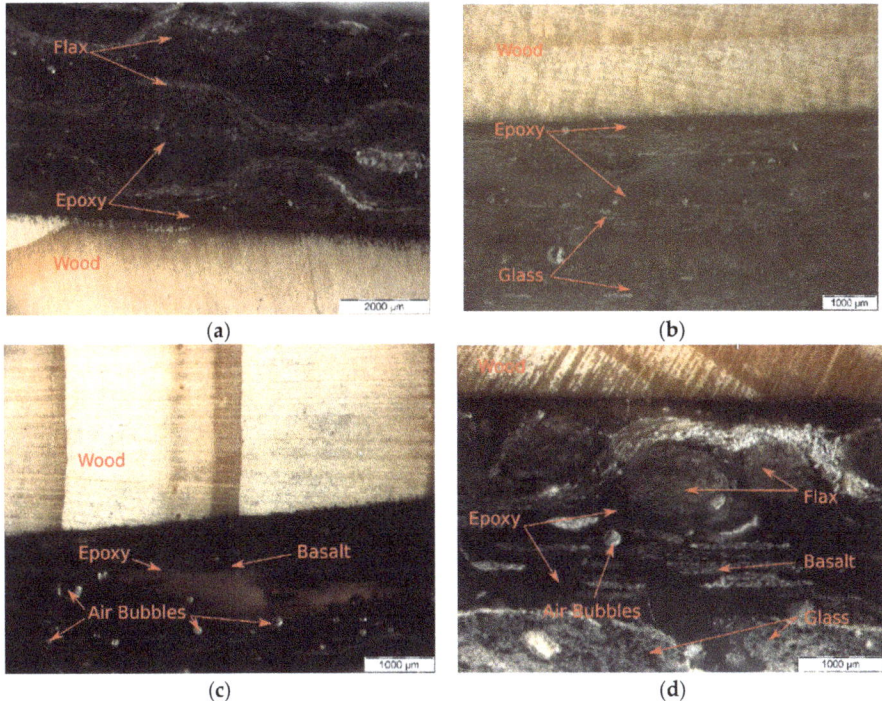

Figure 6. Light microstructure of (**a**) flax (F)FRP–wood, (**b**) E-glass (G)FRP–wood, (**c**) basalt (B)FRP–wood and (**d**) hybrid FRP–wood.

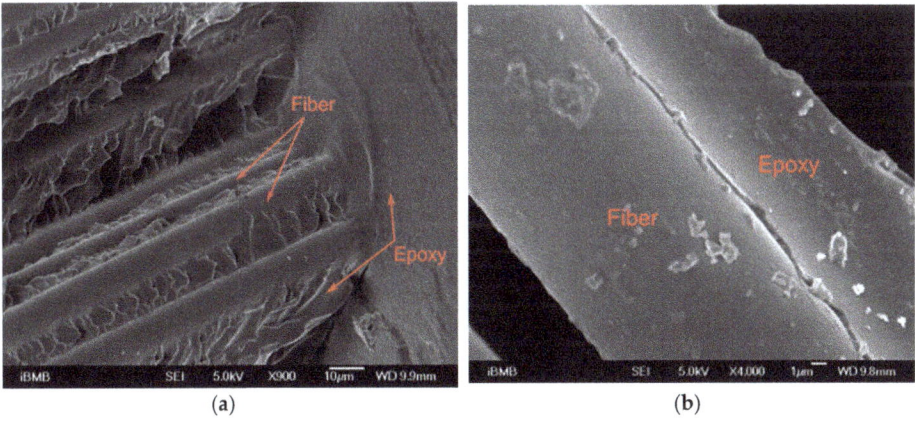

Figure 7. The fracture surface from SEM of (**a**) the interface between epoxy and basalt fiber, (**b**) a close-up of the interface.

4. Conclusions

This study presented the structural behavior of wood beams externally strengthened with various FRP composites. The effects of fabric materials, FRP thicknesses and the sequence of arrangement of the FRP laminas on the flexural behavior of FRP strengthened wood beams were investigated through three-point bending tests. It was shown that the load bearing capacity of the beam under bending was increased as the number of FRP layers increased. The beam strengthened with HFRP had an average higher maximum deflection before failure, yet relatively similar maximum bending loading and elastic modulus compared to the ones strengthened with single type FRPs. It was also observed that the failure modes of FRP strengthened wood beams changed from tensile failure to FRP debonding as the number of layers and maximum bending load increased. This was an indication that the interface between epoxy and wood became more decisive as the FRP became stronger. Under the light microscope, air bubbles were observed in the FRP, which may create inhomogeneity and stress concentration in the cross section of the FRP and could have led to the premature failure of the FRP and the whole beam structure. Under scanning electron microscope, fiber pull-out failure was observed at fracture area of the FRP. The failure was suspected mainly due to the combined smooth surface and the low wetting behavior of the fiber. Improvement can be made by increasing the surface roughness and by improving the wetting properties of the fiber.

Supplementary Materials: The following are available online at http://www.mdpi.com/2073-4360/11/8/1255/s1.

Author Contributions: L.Y. and B.K. managed the project and provide critical comments on the experiments and paper writing. B.W. and E.V.B. drafted manuscript. L.Y., B.W. and V.F. designed the experimental works. All the authors discussed the results and edited the manuscript.

Funding: The research was financially supported by Fachagentur Nachwachsende Rohstoffe e. V. (FNR, Agency for Renewable Resources) founded by Bundesministerium für Ernährung und Landwirtschaft (BMEL, The Federal Ministry of Food and Agriculture of Germany), under the Grant Award: 22011617 and by Bundesministerium für Bildung und Forschung (BMBF, Federal Ministry of Education and Research of Germany) (Grant No.: 01DS18023).

Acknowledgments: The authors acknowledge support by the German Research Foundation and the Open Access Publication Funds of the Technische Universität Braunschweig. The first author also acknowledges the PhD scholarship awarded by the Chinese Scholarship Council (CSC).

Conflicts of Interest: The authors declare no conflict of interest.

References

1. Walberg, D. *Massiv-und Holzbau bei Wohngebäuden: Vergleich von Massiven Bauweisen mit Holzfertigbauten aus Kostenseitiger, Bautechnischer und Nachhaltiger Sicht*; Arbeitsgemeinschaft für Zeitgemäßes Bauen: Kaunas, Lithuania, 2015.
2. Rowell, R.M.; Barbour, R.J. *Archaelogical Wood Properties, Chemistry, and Preservation*; ACS: Washington, DC, USA, 1990.
3. Johns, K.C.; Lacroix, S. Composite reinforcement of timber in bending. *Can. J. Civ. Eng.* **2000**, *27*, 899–906. [CrossRef]
4. Plevris, N.; Triantafillou, T.C. FRP-Reinforced Wood as Structural Material. *J. Mater. Civ. Eng.* **1992**, *4*, 300–317. [CrossRef]
5. Triantafillou, T.C.; Deskovic, N. Prestressed FRP Sheets as External Reinforcement of Wood Members. *J. Struct. Eng.* **1992**, *118*, 1270–1284. [CrossRef]
6. Yan, L.B.; Chouw, N.; Jayaraman, K. Flax fibre and its composites—A review. *Compos. Part B Eng.* **2014**, *56*, 296–317. [CrossRef]
7. Yan, L.B.; Chouw, N. Compressive and flexural behaviour and theoretical analysis of flax fibre reinforced polymer tube encased coir fibre reinforced concrete composite. *Mater. Des. (1980–2015)* **2013**, *52*, 801–811. [CrossRef]
8. Yan, L.B.; Chouw, N. Natural FRP tube confined fibre reinforced concrete under pure axial compression. *Thin-Walled Struct.* **2014**, *82*, 159–169. [CrossRef]
9. Lopez-Anido, R.; Michael, A.P.; Sandford, T.C. Experimental characterization of FRP composite-wood pile structural response by bending tests. *Mar. Struct.* **2003**, *16*, 257–274. [CrossRef]
10. Borri, A.; Corradi, M.; Grazini, A. A method for flexural reinforcement of old wood beams with CFRP materials. *Compos. Part B Eng.* **2005**, *36*, 143–153. [CrossRef]
11. Borri, A.; Corradi, M.; Grazini, A. FRP reinforcement of wood elements under bending loads. In Proceedings of the 10th International Conference on Structural Faults and Repair, London, UK, 1–3 July 2003.

12. Tingley, D.A. The Stress-Strain Relationships in Wood and Fiber-Reinforced Plastic Laminae of Reinforced Glued-Laminated Wood Beams. Ph.D. Thesis, Oregon State University, Corvallis, OR, USA, 1996.
13. Van de Velde, K.; Kiekens, P.; van Langenhove, L. Basalt fibres as reinforcement for composites. In Proceedings of the 10th International Conference on Composites/nano Engineering, New Orleans, LA, USA, 1 January 2003.
14. Zoghi, M. *The International Handbook of FRP Composites in Civil Engineering*; CRC Press: Florida, FL, USA, 2013.
15. Fiore, V.; Scalici, T.; Di Bella, G.; Valenza, A. A review on basalt fibre and its composites. *Compos. Part B Eng.* **2015**, *74*, 74–94. [CrossRef]
16. Huang, L.; Yan, B.; Yan, L.B.; Xu, Q.; Tan, H.Z.; Kasal, B. Reinforced concrete beams strengthened with externally bonded natural flax FRP plates. *Compos. Part B Eng.* **2016**, *91*, 569–578. [CrossRef]
17. Yan, L.B.; Su, S.; Chouw, N. Microstructure, flexural properties and durability of coir fibre reinforced concrete beams externally strengthened with flax FRP composites. *Compos. Part B Eng.* **2015**, *80*, 343–354. [CrossRef]
18. Yan, L.B. Plain concrete cylinders and beams externally strengthened with natural flax fabric reinforced epoxy composites. *Mater. Struct.* **2016**, *49*, 2083–2095. [CrossRef]
19. Yan, B.; Huang, L.; Yan, L.B.; Gao, C.; Kasal, B. Behavior of flax FRP tube encased recycled aggregate concrete with clay brick aggregate. *Constr. Build. Mater.* **2017**, *136*, 265–276. [CrossRef]
20. Yan, L.B.; Chouw, N. Experimental study of flax FRP tube encased coir fibre reinforced concrete composite column. *Constr. Build. Mater.* **2013**, *40*, 1118–1127. [CrossRef]
21. Yan, L.B.; Chouw, N.; Jayaraman, K. Effect of column parameters on flax FRP confined coir fibre reinforced concrete. *Constr. Build. Mater.* **2014**, *55*, 299–312. [CrossRef]
22. Speranzini, E.; Tralascia, S. Engineered lumber: LVL and solid wood reinforced with natural fibres. In Proceedings of the WCTE, Trentino, Italy, 20–24 June 2010; pp. 1685–1690.
23. Borri, A.; Corradi, M.; Speranzini, E. Reinforcement of wood with natural fibers. *Compos. Part B Eng.* **2013**, *53*, 1–8. [CrossRef]
24. Borri, A.; Corradi, M.; Speranzini, E. Bending Tests on Natural Fiber Reinforced Fir Wooden Elements. *Adv. Mater. Res.* **2013**, *778*, 537–544. [CrossRef]
25. André, A.; Johnsson, H. Flax Fiber-Reinforced Glued-Laminated Timber in Tension Perpendicular to the Grain: Experimental Study and Probabilistic Analysis. *J. Mater. Civ. Eng.* **2010**, *22*, 827–835. [CrossRef]
26. Kim, H.S.; Shin, Y.S. Flexural behavior of reinforced concrete (RC) beams retrofitted with hybrid fiber reinforced polymers (FRPs) under sustaining loads. *Compos. Struct.* **2011**, *93*, 802–811. [CrossRef]
27. Yan, L.B.; Wang, B.; Kasal, B. Can Plant-Based Natural Flax Replace Basalt and E-Glass for Fiber-Reinforced Polymer Tubular Energy Absorbers? A Comparative Study on Quasi-Static Axial Crushing. *Front. Mater.* **2017**, *4*, 635. [CrossRef]
28. Huang, L.; Gao, C.; Yan, L.; Kasal, B. Reliability assessment of confinement models of carbon fiber reinforced polymer-confined concrete. *J. Reinf. Plast. Compos.* **2016**, *12*, 996–1026. [CrossRef]
29. Gupta, M.K.; Srivastava, R.K. Mechanical Properties of Hybrid Fibers-Reinforced Polymer Composite: A Review. *Polym. Plast. Technol. Eng.* **2016**, *55*, 626–642. [CrossRef]
30. Yang, Y.L.; Liu, J.W.; Xiong, G.J. Flexural behavior of wood beams strengthened with HFRP. *Constr. Build. Mater.* **2013**, *43*, 118–124. [CrossRef]
31. Fiorelli, J.; Dias, A.A. Fiberglass-reinforced glulam beams: Mechanical properties and theoretical model. *Mater. Res.* **2006**, *9*, 263–269. [CrossRef]
32. Fiorelli, J.; Dias, A.A. Glulam beams reinforced with FRP externally-bonded: Theoretical and experimental evaluation. *Mater. Struct.* **2011**, *44*, 1431–1440. [CrossRef]
33. Marinella, F.; Giovanni, M.; Maurizio, P. Flexural behaviour of glulam timber beams reinforced with FRP cords. *Constr. Build. Mater.* **2015**, *95*, 54–64.
34. Barkoula, N.M.; Garkhail, S.K.; Peijs, T. Biodegradable composites based on flax/polyhydroxybutyrate and its copolymer with hydroxyvalerate. *Ind. Crop. Prod.* **2010**, *31*, 34–42. [CrossRef]
35. *Testing of Wood, Bending Test: DIN 52186*; Deutsches Institue für Normung: Berlin, Germany, 1978.
36. *Standard Test Method for Tensile Properties of Polymer Matrix Composite Materials: ASTM D 3039*; ASTM International: West Conshohocken, PA, USA, 2000.
37. *Standard Test Methods for Flexural Properties of Unreinforced and Reinforced Plastics and Electrical Insulating Materials: ASTM. D 790*; ASTM International: West Conshohocken, PA, USA, 2007.

38. Reyes-Araiza, J.L.; Manzano-Ramírez, A.; Rubio-Avalos, J.C.; González-Sosa, E.; Pérez-Robles, J.F.; Arroyo-Contreras, M.; Signoret, C.; Castillo-Castañeda, E.; Vorobiev, Y.V. Comparative study on tensile behavior of inorganic fibers embedded in unsaturated polyester bisphenol "A"-styrene copolymer. *Inorg. Mater.* **2008**, *44*, 549–554. [CrossRef]
39. Quagliarini, E.; Monni, F.; Lenci, S.; Bondioli, F. Tensile characterization of basalt fiber rods and ropes: A first contribution. *Constr. Build. Mater.* **2012**, *34*, 372–380. [CrossRef]
40. Zhang, Y.L.; Li, Y.; Ma, H.; Yu, T. Tensile and interfacial properties of unidirectional flax/glass fiber reinforced hybrid composites. *Compos. Sci. Technol.* **2013**, *88*, 172–177. [CrossRef]

© 2019 by the authors. Licensee MDPI, Basel, Switzerland. This article is an open access article distributed under the terms and conditions of the Creative Commons Attribution (CC BY) license (http://creativecommons.org/licenses/by/4.0/).

MDPI
St. Alban-Anlage 66
4052 Basel
Switzerland
Tel. +41 61 683 77 34
Fax +41 61 302 89 18
www.mdpi.com

Polymers Editorial Office
E-mail: polymers@mdpi.com
www.mdpi.com/journal/polymers

www.ingramcontent.com/pod-product-compliance
Lightning Source LLC
LaVergne TN
LVHW070443100526
838202LV00014B/1651